W9-BKB-022

NAZI
GERMANY
AND THE
JEWS

VOLUME I

The Years of Persecution, 1933–1939

NAZI GERMANY
AND THE
JEWS

VOLUME I

The Years of Persecution, 1933–1939

SAUL FRIEDLÄNDER

HarperCollins*Publishers*

HarperCollins books may be purchased for educational, business, or sales promotional use. For information please write: Special Markets Department, HarperCollins Publishers, Inc., 10 East 53rd Street, New York, NY 10022.

FIRST EDITION

Designed by Joseph Rutt

Library of Congress Cataloging-in-Publication Data

Friedländer, Saul
    Nazi Germany and the Jews / Saul Friedländer. — 1st ed.
      p.   cm.
    Includes bibliographical references and index.
    Contents: v. 1 The years of persecution, 1933–1939.
    ISBN 0-06-019042-6
    1. Jews—Germany—History—1933–1945. 2. Jews—Persecutions—Germany. 3. Holocaust, Jewish (1939–1945)—Germany. 4. Germany—Ethnic relations. I. Title.
    DS135.G3315F75   1997
    940.53'18—dc20   96-21915

97 98 99 00 01 ❖/RRD 9 8 7 6 5 4 3 2 1

To Omer, Elam, and Tom

I would not wish to be a Jew in Germany.

HERMANN GÖRING, NOVEMBER 12, 1938

Contents

Acknowledgments

In my work on this book I have been assisted in many different ways. The Maxwell Cummings Family of Montreal and the 1939 Club of Los Angeles have endowed chairs, at Tel Aviv University and at UCLA, that facilitated the implementation of this project. Short stays at the Humanities Research Institute at UC Irvine (1992) and at the Getty Center for the History of Art and the Humanities in Los Angeles (1996) provided me with the most invaluable of all privileges: free time. Throughout the years, I have greatly benefited from the vast resources and the generous help offered by the Wiener Library at Tel Aviv University, the University Research Library at UCLA, the Leo Baeck Institute Archives in New York, and the library and archives of the Institut für Zeitgeschichte in Munich.

Friends and colleagues have been kind enough to read parts or the totality of the manuscript, and some have followed it throughout its various stages. From all of them I received much good advice. At UCLA I wish to thank Joyce Appleby, Carlo Ginzburg, and Hans Rogger; at Tel Aviv University, my friends, colleagues, and coeditors of *History & Memory*, particularly Gulie Ne'eman Arad, for her remarkable judgment and constant assistance in this project, as well as Dan Diner and Philippa Shimrat. I also wish to express my gratitude to Omer Bartov (Rutgers), Philippe Burrin (Geneva), Sidra and Yaron Ezrahi (Jerusalem), and Norbert Frei (Munich). Moreover, I am very much indebted to my research assistants: Orna Kenan, Christopher Kenway, and Gavriel Rosenfeld. Needless to say, the usual formula holds: Any mistakes in this book are my own.

The late Amos Funkenstein unfortunately could not read the entire manuscript, but I shared with him my many thoughts and doubts until nearly the end. He gave me much encouragement, and it is infinitely more than a usual debt of gratitude that I owe the closest of my friends, whom I miss more than I can say.

Both Aaron Asher and Susan H. Llewellyn contributed to the editing of this book, which is the first I wrote entirely in English. Aaron, my friend and former publisher, brought his intellectual insights and linguistic skills to bear on a manuscript studded with gallicisms; Sue applied her own stylistic sensibility to a deep understanding of the text. My editor at HarperCollins, Hugh Van Dusen, was a highly experienced and attentive guide whose expert eye followed every phase of this process. The assistant editor, Katherine Ekrem, demonstrated an impressive efficiency, always in the kindest way. And, from the first book I published in the United States, *Pius XII and the Third Reich* (1964), I have been represented by Georges and Anne Borchardt, who became friends.

For thirty-seven years now, Hagith has given me the warmth and the support that are vital to everything I do. This support has never been more decisive than during the long time spent in the preparation of this book. Years ago I dedicated a book to our children, Eli, David, and Michal; this book is dedicated to our grandchildren.

Introduction

Most historians of my generation, born on the eve of the Nazi era, recognize either explicitly or implicitly that plowing through the events of those years entails not only excavating and interpreting a collective past like any other, but also recovering and confronting decisive elements of our own lives. This recognition does not generate any agreement among us about how to define the Nazi regime, how to interpret its internal dynamics, how to render adequately both its utter criminality and its utter ordinariness, or, for that matter, where and how to place it within a wider historical context.[1] Yet, despite our controversies, many of us share, I think, a sense of personal involvement in the depiction of this past, which gives a particular urgency to our inquiries.

For the next generation of historians—and by now also for the one after that—as for most of humanity, Hitler's Reich, World War II, and the fate of the Jews of Europe do not represent any shared memory. And yet, paradoxically, the centrality of these events in present-day historical consciousness seems much greater than it was some decades ago. The ongoing debates tend to unfold with unremitting bitterness as facts are questioned and evidence denied, as interpretations and commemorative endeavors confront one another, and as statements about historical responsibility periodically come to the fore in the public arena. It could be that in our century of genocide and mass criminality, apart from its specific historical context, the extermination of the Jews of Europe is perceived by many as the ultimate standard of evil, against which all degrees of evil may be measured. In these debates, the historian's role is central. For my generation, to partake at one and the same time in the memory and the present perceptions of this past may create an unsettling dissonance; it may, however, also nurture insights that would otherwise be inaccessible.

Establishing a historical acccount of the Holocaust in which the policies of the perpetrators, the attitudes of surrounding society, and the world of the victims could be addressed within an integrated framework remains a major challenge. Some of the best-known historical renditions of these events have focused mainly on the Nazi machinery of persecution and death, paying but scant attention to the wider society, to the wider European and world scene or to the changing fate of the victims themselves; others, less frequently, have concentrated more distinctly on the history of the victims and offered only a limited analysis of Nazi policies and the surrounding scene.[2] The present study will attempt to convey an account in which Nazi policies are indeed the central element, but in which the surrounding world and the victims' attitudes, reactions, and fate are no less an integral part of this unfolding history.

In many works the implicit assumptions regarding the victims' generalized hopelessness and passivity, or their inability to change the course of events leading to their extermination, have turned them into a static and abstract element of the historical background. It is too often forgotten that Nazi attitudes and policies cannot be fully assessed without knowledge of the lives and indeed of the feelings of the Jewish men, women, and children themselves. Here, therefore, at each stage in the description of the evolving Nazi policies and the attitudes of German and European societies as they impinge on the evolution of those policies, the fate, the attitudes, and sometimes the initiatives of the victims are given major importance. Indeed, their voices are essential if we are to attain an understanding of this past.[3] For it is their voices that reveal what was known and what *could* be known; theirs were the only voices that conveyed both the clarity of insight and the total blindness of human beings confronted with an entirely new and utterly horrifying reality. The constant presence of the victims in this book, while historically essential in itself, is also meant to put the Nazis' actions into full perspective.

It is easy enough to recognize the factors that shaped the overall historical context in which the Nazi mass murder took place. They determined the methods and scope of the "Final Solution"; they also contributed to the general climate of the times, which facilitated the way to the exterminations. Suffice it here to mention the ideological radicalization—with fervent nationalism and rabid anti-Marxism (later anti-Bolshevism) as its main propelling drives—that surfaced during the last decades of the nineteenth century and reached its climax after World War I (and the Russian Revolution); the new dimension of massive industrial killing introduced by that war; the growing technological and bureaucratic control exerted by modern societies; and the other major features of modernity itself, which were a dominant aspect of Nazism.[4] Yet, as essential as these conditions were in preparing the ground for the Holocaust—and as such they are an integral part of this history—they nonetheless do not alone constitute the necessary cluster of elements that shaped the course of events leading from persecution to extermination.

With regard to that process, I have emphasized Hitler's personal role and the function of his ideology in the genesis and implementation of the Nazi regime's anti-Jewish measures. In no way, however, should this be seen as a return to earlier reductive interpretations, with their sole emphasis on the role (and responsibility) of the supreme leader But, over time, the contrary interpretations have, it seems to me, gone too far. Nazism was not essentially driven by the chaotic clash of competing bureaucratic and party fiefdoms, nor was the planning of its anti-Jewish policies mainly left to the cost-benefit calculations of technocrats.[5] In all its major decisions the regime depended on Hitler. Especially with regard to the Jews, Hitler was driven by ideological obsessions that were anything but the calculated devices of a demagogue; that is, he carried a very specific brand of racial anti-Semitism to its most extreme and radical limits. I call that distinctive aspect of his worldview "redemptive anti-Semitism"; it is different, albeit derived, from other strands of anti-Jewish hatred that were common throughout Christian Europe, and different also from the ordinary brands of German and European racial anti-Semitism. It was this redemptive dimension, this synthesis of a murderous rage and an "idealistic" goal, shared by the Nazi leader and the hard core of the party, that led to Hitler's ultimate decision to exterminate the Jews.[6]

But Hitler's policies were not shaped by ideology alone, and the interpretation presented here traces the interaction between the Führer and the system within which he acted. The Nazi leader did not take his decisions independently of the party and state organizations. His initiatives, mainly during the early phase of the regime, were molded not only by his world view but also by the impact of internal pressures, the weight of bureaucratic constraints, at times the influence of German opinion at large and even the reactions of foreign governments and foreign opinion.[7]

To what extent did the party and the populace partake in Hitler's ideological obsession? "Redemptive" anti-Semitism was common fare among the party elite. Recent studies have also shown that such extreme anti-Semitism was not unusual in the agencies that were to become central to the implementation of the anti-Jewish policies, such as Reinhard Heydrich's Security Service of the SS (Sicherheitsdienst, or SD).[8] As for the so-called party radicals, they were often motivated by the kind of social and economic resentment that found its expression in extreme anti-Jewish initiatives. In other words, within the party and, as we shall see, sometimes outside it, there were centers of uncompromising anti-Semitism powerful enough to transmit and propagate the impact of Hitler's own drive. Yet, among the traditional elites and within the wider reaches of the population, anti-Jewish attitudes were more in the realm of tacit acquiescence or varying degrees of compliance.

Despite most of the German population's full awareness, well before the war, of the increasingly harsh measures being taken against the Jews, there were but minor areas of dissent (and these were almost entirely for economic and specifically religious-ideological reasons). It seems, however, that the majority of Germans, although undoubtedly influenced by various forms of traditional anti-Semitism and easily accepting the segregation of the Jews, shied away from widespread violence against them, urging neither their expulsion from the Reich nor their physical annihilation. After the attack on the Soviet Union, when total extermination had been decided upon, the hundreds of thousands of "ordinary Germans" (as distinct from the highly motivated SS units, among others) who actively participated in the killings acted no differently from the equally numerous and "ordinary" Austrians, Rumanians, Ukrainians, Balts, and other Europeans who became the most willing operatives of the murder machinery functioning in their midst. Nonetheless, whether they were conscious of it or not, the German and Austrian killers had been indoctrinated by the regime's relentless anti-Jewish propaganda, which penetrated every crevice of society and whose slogans they at least partially internalized, mainly in the context of the war in the East.[9]

By underscoring that Hitler and his ideology had a decisive impact on the course of the regime, I do not mean in any way to imply that Auschwitz was a preordained result of Hitler's accession to power. The anti-Jewish policies of the thirties must be understood in their context, and even Hitler's murderous rage and his scanning of the political horizon for the most extreme options do not suggest the existence of any plans for total extermination in the years prior to the German invasion of the Soviet Union. But at the same time, no historian can forget the end of the road. Thus emphasis is also placed here on those elements that we know from hindsight to have played a role in the evolution toward the fateful outcome. The history of Nazi Germany should not be written only from the perspective of the wartime years and their atrocities, but the heavy shadow cast by what happened during that time so darkens the prewar years that a historian cannot pretend that the later events do not influence the weighing of the evidence and the evaluation of the overall course of that history.[10] The crimes committed by the Nazi regime were neither a mere outcome of some haphazard, involuntary, imperceptible, and chaotic onrush of unrelated events nor a predetermined enactment of a demonic script; they were the result of converging factors, of the interaction between intentions and contingencies, between discernible causes and chance. General ideological objectives and tactical policy decisions enhanced one another and always remained open to more radical moves as circumstances changed.

At the most basic level, in this two-volume account the narration follows the chronological sequence of the events: their prewar evolution in this volume, their monstrous wartime culmination in the next. That overall time frame highlights continuities and indicates the context of major changes; it also makes it possible to shift the narration within a stable chronological span. Such shifts result from the changes in perspective my approach demands, but they also stem from another choice: to juxtapose entirely different levels of reality—for example, high-level anti-Jewish policy debates and decisions next to routine scenes of persecution—with the aim of creating a sense of estrangement counteracting our tendency to "domesticate" that particular past and blunt its impact by means of seamless explanations and standardized renditions. That sense of estrangement seems to me to reflect the perception of the hapless victims of the regime, at least during the thirties, of a reality both absurd and ominous, of a world altogether grotesque and chilling under the veneer of an even more chilling normality.

From the moment the victims were engulfed in the process leading to the "Final Solution," their collective life—after a short period of enhanced cohesion—started to disintegrate. Soon this collective history merged with the history of the administrative and murderous measures of their extermination, and with its abstract statistical expression. The only concrete history that can be retrieved remains that carried by personal stories. From the stage of collective disintegration to that of deportation and death, this history, in order to be written at all, has to be represented as the integrated narration of individual fates.

Although I mention my generation of historians and the insights potentially available to us because of our particular position in time, I cannot ignore the argument that personal emotional involvement in these events precludes a rational approach to the writing of history. The "mythic memory" of the victims has been set against the "rational" understanding of others. I certainly do not wish to reopen old debates, but merely to suggest that German and Jewish historians, as well as those of any other background cannot avoid a measure of "transference" vis-à-vis this past.[11] Such involvement of necessity impinges upon the writing of history. But the historian's necessary measure of detachment is not thereby precluded, provided there is sufficient self-awareness. It may indeed be harder to keep one's balance in the other direction; whereas a constantly self-critical gaze might diminish the effects of subjectivity, it could also lead to other, no lesser risks, those of undue restraint and paralyzing caution.

Nazi persecutions and exterminations were perpetrated by ordinary people who lived and acted within a modern society not unlike our own, a society that had produced them as well as the methods and instruments for the implementation of their actions; the goals of these actions, however, were formulated by a regime, an ideology, and a political culture that were anything but commonplace. It is the relationship between the uncommon and the ordinary, the fusion of the widely shared murderous potentialities of the world that is also ours and the peculiar frenzy of the Nazi apocalyptic drive against the mortal enemy, the Jew, that give both universal significance and historical distinctiveness to the "Final Solution of the Jewish Question."

PART I

A Beginning and
an End

CHAPTER I

Into the Third Reich

I

The exodus from Germany of Jewish and left-wing artists and intellectuals began during the early months of 1933, almost immediately after Adolf Hitler's accession to power on January 30. The philosopher and literary critic Walter Benjamin left Berlin for Paris on March 18. Two days later he wrote to his colleague and friend, Gershom Scholem, who lived in Palestine: "I can at least be certain that I did not act on impulse. . . . Nobody among those who are close to me judges the matter differently."[1] The novelist Lion Feuchtwanger, who had reached the safety of Switzerland, confided in his fellow writer Arnold Zweig: "It was too late for me to save anything. . . . All that was there is lost."[2]

The conductors Otto Klemperer and Bruno Walter were compelled to flee. Walter was forbidden access to his Leipzig orchestra, and, as he was about to conduct a special concert of the Berlin Philharmonic, he was informed that, according to rumors circulated by the Propaganda Ministry, the hall of the Philharmonic would be burned down if he did not withdraw. Walter left the country.[3] Hans Hinkel, the new president of the Prussian Theater Commission and also responsible for the "de-Judaization" of cultural life in Prussia, explained in the April 6 *Frankfurter Zeitung* that Klemperer and Walter had disappeared from the musical scene because there was no way to protect them against the "mood" of a German public long provoked by "Jewish artistic bankrupters."[4]

Bruno Walter's concert was not canceled: Richard Strauss conducted it.[5] This, in turn, led Arturo Toscanini to announce in early June that, in protest, he would not conduct at the Bayreuth Festival. Propaganda Minister Joseph Goebbels noted laconically in his diary: "Toscanini canceled Bayreuth."[6]

The same public "mood" must have convinced the Dresden Opera House to hound out its music director, Fritz Busch, no Jew himself but accused of having too many contacts with Jews and of having invited too many Jewish artists to perform.[7] Other methods were also used: When the Hamburg Philharmonic Society published its program for the celebration of Brahms's hundredth birthday, it was informed that Chancellor Hitler would be ready to give his patronage to the celebrations on condition that all Jewish artists (among them the pianist Rudolf Serkin) disappear from the program. The offer was gladly accepted.[8]

The rush to de-Judaize the arts produced its measure of confusion. Thus, on April 1, a Lübeck newspaper reported that in the small town of Eutin, in nearby Schleswig-Holstein, the last concert of the winter season had offered a surprise: "In place of the Kiel City Orchestra's excellent cellist, John de J., Professor Hofmeier presented a piano recital. We are informed that it has been established that John de J. is Jewish." Soon after, however, there was a telegram from de J. to Hofmeier: "Claim false. Perfect documents." On May 5 the district party leader S. announced that the Dutch-born German citizen de J. was a Lutheran, as several generations of his forebears had been.[9]

The relief felt at not being Jewish must have been immense. In his (barely) fictionalized rendition of the career of the actor and later manager of the Berlin National Theater, Göring's protégé Gustav Gründgens, Klaus Mann described that very peculiar euphoria: "But even if the Nazis remained in power, what had he, Höfgen [Gründgens], to fear from them? He belonged to no party. And he wasn't a Jew. This fact above all others—that he wasn't a Jew—struck Hendrik all of a sudden as immensely comforting and important. He had never in the past estimated the true worth of this considerable and unsuspected advantage. He wasn't a Jew and so he could be forgiven everything."[10]

A few days after the Reichstag elections of March 5, all members of the Prussian Academy of the Arts received a confidential letter from the poet Gottfried Benn asking them whether they were ready, "in view of the 'changed political situation,'" to remain members of the parent Academy of Arts and Sciences, in which case they would have to abstain from any criticism of the new German regime. Moreover the members would have to manifest the right "national cultural" attitude by signing a declaration of loyalty. Nine of the twenty-seven members of the literature section replied negatively, among them the novelists Alfred Döblin, Thomas Mann, Jakob Wassermann, and Ricarda Huch. Mann's brother, the novelist Heinrich Mann, had already been expelled because of his left-wing political views.[11]

Max von Schillings, the new president of the Prussian Academy, put pressure on the "Aryan"* novelist Ricarda Huch not to resign. There was an exchange of letters, with Huch in her final retort alluding to Heinrich Mann's dismissal and to the resignation of Alfred Döblin, who was Jewish: "You mention the gentlemen Heinrich Mann and Dr. Döblin. It is true that I did not agree with Heinrich Mann, and I did not always agree with Dr. Döblin, although on some matters I did. In any case I can only wish that all non-Jewish Germans would seek as conscientiously to recognize and to do what is right, would be as open, honest and decent as I have always found him to be. In my judgment he could not have acted any differently than he did, in the face of the harassment of the Jews. That my resignation from the Academy is not motivated by sympathy for these gentlemen, in spite of the particular respect and sympathy I have for Dr. Döblin, is something everyone who knows me, either personally or from my books, will recognize. Herewith I declare my resignation from the academy."[12]

Living in Vienna, the novelist Franz Werfel, who was Jewish, perceived things differently. He was quite willing to sign the declaration, and on March 19 he wired Berlin for the necessary forms. On May 8 Schillings informed Werfel that he could not remain a member of the academy; two days later a number of his books were among those publicly burned. In the summer of 1933, after the establishment of the Reich Chamber of Culture (Reichskulturkammer, or RKK), and as part of it, of the Reich Association of German Writers, Werfel tried again: "Please note that I am a Czechoslovak citizen," he wrote, "and a resident of Vienna. At the same time, I wish to declare that I have always kept my distance from any political organization or activity. As a member of the German minority in Czechoslovakia, resident in Austria, I am subject to the laws of these states." Needless to say, Werfel never received an answer.[13] The novelist

---

*The Nazis gave a peculiar ideological twist to a great many words, such as "German" (as opposed to "Jewish"), "healthy" (often meaning racially healthy or not spoiled by Jews), "modernity," and so on. As the meanings are almost always recognizable, quotation marks will henceforth be avoided in most instances.

possibly wanted to ensure the German sale of his forthcoming novel, *The Forty Days of Musa Dagh,* a story based on the extermination of the Armenians by the Turks during the World War. The book was in fact published in the Reich at the end of 1933, but finally banned in February 1934.[14]

Albert Einstein was visiting the United States on January 30, 1933. It did not take him long to react. Describing what was happening in Germany as a "psychic illness of the masses," he ended his return journey in Ostend (Belgium) and never again set foot on German soil. The Kaiser Wilhelm Society dismissed him from his position; the Prussian Academy of Sciences expelled him; his citizenship was rescinded. Einstein was no longer a German. Prominence and fame shielded no one. Max Reinhardt was expelled from the directorship of the German Theater, which was "transferred to the German people," and fled the Reich. Max Liebermann, at eighty-six possibly the best-known German painter of the time, was too old to emigrate when Hitler came to power. Formerly president of the Prussian Academy of Arts, and in 1933 its honorary president, he held the highest German decoration, the Pour le Mérite. On May 7 Liebermann resigned from the academy. As the painter Oskar Kokoschka wrote from Paris in a published letter to the editor of the *Frankfurter Zeitung,* none of Liebermann's colleagues deemed it necessary to express a word of recognition or sympathy.[15] Isolated and ostracized, Liebermann died in 1935; only three "Aryan" artists attended his funeral. His widow survived him. When, in March 1943, the police arrived, with a stretcher, for the bedridden eighty-five-year-old woman to begin her deportation to the East, she committed suicide by swallowing an overdose of the barbiturate Veronal.[16]

As peripheral as it may seem in hindsight, the cultural domain was the first from which Jews (and "leftists") were massively expelled. Schillings's letter was sent immediately after the March 1933 Reichstag elections, and publication of Hinkel's interview preceded the promulagation of the Civil Service Law of April 7, which will be discussed further on. Thus, even before launching their first systematic anti-Jewish measures of exclusion, the new rulers of Germany had turned against the most visible representatives of the "Jewish spirit" that henceforth was to be eradicated. In general the major anti-Jewish measures the Nazis would take from then on in the various domains were not only acts of terror but also symbolic statements. This dual function expressed the pervasive presence of ideology within the system: Its tenets had to be ritually reasserted, with the persecution of chosen victims as part of the ongoing ritual. There was more. The double significance of the regime's initiatives engendered a kind of split consciousness in a great part of the population: For instance, people might not agree with the brutality of the dismissals of Jewish intellectuals from their positions, but they welcomed the cleansing of the "excessive influence" of Jews from German cultural life. Even some of the most celebrated German exiles, such as Thomas Mann, were not immune, at least for a time, from this kind of dual vision of the events.

A non-Jew, though married to one, Mann was away from Germany when the Nazis came to power, and he did not return. Writing to Einstein on May 15, he mentioned the painfulness to him of the very idea of exile: "For me to have been forced into this role, something thoroughly wrong and evil must surely have taken place. And it is my deepest conviction that this whole 'German Revolution' is indeed wrong and evil."[17] The author of *The Magic Mountain* was no less explicit months later, in a letter to his one-time friend, the ultranationalist historian of literature Ernst Bertram, who had become a staunch supporter of the new regime: "'We shall see,' I wrote to you a good while back, and you replied defiantly: 'Of course we shall.' Have you begun to see? No, for they are holding your eyes closed with bloody hands, and you accept the 'protection' only too gladly. The German intellectuals—forgive the word, it is intended as a purely objective term—will in fact be the very last to begin to see, for they have too deeply, too shamefully collaborated and exposed themselves."[18] But in fact much ambiguity remained in Mann's attitudes: To ensure the continuing publication and sale of his books in Germany, he carefully avoided speaking out against the Nazis for several years. And, at the outset, some Nazi organizations, such as the National Socialist Students Association, were careful about him as well: Thomas Mann's books were not included in the notorious May 10, 1933, auto-da-fé.[19]

Mann's ambivalence (or worse), particularly with regard to the Jews, surfaces in his diary entries during this first phase: "Isn't after all something significant and revolutionary on a grand style happening in Germany?" he wrote on April 4, 1933. "As for the Jews. . . . That Alfred Kerr's arrogant and poisonous Jewish garbling of Nietzsche is now excluded, is not altogether a catastrophe; and also the de-Judaization of justice isn't one."[20] He indulged in such remarks time and again, but it is perhaps in the diary entry of July 15, 1934, that Mann expressed his strongest resentments: "I was thinking about the absurdity of the fact, that the Jews, whose rights in

Germany are being abolished and who are being pushed out, have an important share in the spiritual issues which express themselves, obviously with a grimace, in the political system [Nazism] and that they can in good part be considered as the precursors of the anti-liberal turn."[21] As examples, Mann mentioned the poet Karl Wolfskehl, a member of the esoteric literary and intellectual circle around the poet Stefan George, and particularly the Munich eccentric Oskar Goldberg. There is some discrepancy between such expressions as "an important share," "in good part," and "the precursors of the anti-liberal turn" and these two marginal examples.[22] He went further: "In general I think that many Jews [in Germany] agree in their deepest being with their new role as tolerated guests who are not part of anything except, it goes without saying, as far as taxes are concerned."[23] Mann's anti-Nazi position was not to become clear, unambiguous, and public until early 1936.[24]

Mann's attitude illustrates the pervasiveness of split consciousness, and thus explains the ease with which Jews were expelled from cultural life. Apart from a few courageous individuals such as Ricarda Huch, there was no countervailing force in that domain—or, for that matter, in any other.

Hitler certainly had no split consciousness regarding anything Jewish. Yet, in 1933 at least, he deferred to Winifred Wagner (the English-born widow of Richard Wagner's son Siegfried, who was the guiding force at Bayreuth): "Amazingly," as Frederic Spotts puts it, that year Hitler even allowed the Jews Alexander Kipnis and Emanuel List to sing in his presence.[25]

II

Three days before the Reichstag elections of March, the Hamburg edition of the Jewish newspaper *Israelitisches Familienblatt* published a telling article under the headline HOW SHALL WE VOTE ON MARCH 5?: "There are many Jews," the article said, "who approve of the present-day right wing's economic program but who are denied the possibility of joining its parties, as these have, in a completely illogical way, associated their economic and political goals with a fight against Jewry."[26]

A benefit for Jewish handicrafts had taken place at Berlin's Café Leon on January 30, 1933. The news of Hitler's accession to the chancellorship became known shortly before the event began. Among the attending representatives of Jewish organizations and political movements, only the Zionist rabbi Hans Tramer referred to the news and spoke of it as a major change; all the other speakers kept to their announced subjects. Tramer's speech "made no impression. The entire audience considered it panic-mongering. There was no response."[27] The board of the Central Association of German Citizens of the Jewish Faith (Zentralverein deutscher Staatsbürger jüdischen Glaubens) on the same day concluded a public declaration in the same spirit: "In general, today more than ever we must follow the directive: wait calmly."[28] An editorial in the association's newspaper for January 30, written by the organization's chairman, Ludwig Holländer, was slightly more worried in tone, but showed basically the same stance: "The German Jews will not lose the calm they derive from their tie to all that is truly German. Less than ever will they allow external attacks, which they consider unjustified, to influence their inner attitude toward Germany."[29]

By and large there was no apparent sense of panic or even of urgency among the great majority of the approximately 525,000 Jews living in Germany in January 1933.[30] As the weeks went by, Max Naumann's Association of National German Jews and the Reich Association of Jewish War Veterans hoped for no less than integration into the new order of things. On April 4, the veterans' association chairman, Leo Löwenstein, addressed a petition to Hitler including a list of nationalistically oriented suggestions regarding the Jews of Germany, as well as a copy of the memorial book containing the names of the twelve thousand German soldiers of Jewish origin who had died for Germany during the World War. Ministerial Councillor Wienstein answered on April 14 that the chancellor acknowledged receipt of the letter and the book with "sincerest feelings." The head of the Chancellery, Hans Heinrich Lammers, received a delegation of the veterans on the twenty-eighth,[31] but with that the contacts ceased. Soon Hitler's office stopped acknowledging petitions from the Jewish organization. Like the Central Association, the Zionists continued to believe that the initial upheavals could be overcome by a reassertion of Jewish identity or simply by patience; the Jews reasoned that the responsibilities of power, the influence of conservative members of the government, and a watchful outside world would exercise a moderating influence on any Nazi tendency to excess.

Even after the April 1 Nazi boycott of Jewish businesses, some well-known German-Jewish figures, such as Rabbi Joachim Prinz, declared that it was unreasonable to take an anti-Nazi position. For Prinz, arguing against Germany's "reorganization," whose aim was "to give people bread and work . . . was neither intended nor possible."[32] The declaration may have been merely tactical, and it must be kept in mind that many Jews were at a loss how to react. Some eccentrics went much further. Thus, as late as the summer of 1933, in the opening statement of his lectures on the Roman poet Horace, the Kiel University historian Felix Jacoby declared: "As a Jew I find myself in a difficult situation. But as a historian I have learned not to consider historical events from a private perspective. Since 1927, I have voted for Adolf Hitler, and I consider myself lucky to be able to lecture on Augustus' poet in the year of the national revival. Augustus is the only figure of world history whom one may compare to Adolf Hitler."[33] This, however, was a rather exceptional case.

For some Jews the continuing presence of the old, respected President Paul von Hindenburg as head of state was a source of confidence; they occasionally wrote to him about their distress. "I was engaged to be married in 1914," Frieda Friedmann, a Berlin woman, wrote to Hindenburg on February 23: "My fiancé was killed in action in 1914. My brothers Max and Julius Cohn were killed in 1916 and 1918. My remaining brother, Willy, came back blind. . . . All three received the Iron Cross for their service to the country. But now it has gone so far that in our country pamphlets saying, 'Jews, get out!' are being distributed on the streets, and there are open calls for pogroms and acts of violence against Jews. . . . Is incitement against Jews a sign of courage or one of cowardice when Jews comprise only one percent of the German people?" Hindenburg's office promptly acknowledged receipt of the letter, and the president let Frieda Friedmann know that he was decidedly opposed to excesses perpetrated against Jews. The letter was transmitted to Hitler, who wrote in the margin: "This lady's claims are a swindle! Obviously there has been no incitement to a pogrom!"[34]

The Jews finally, like a considerable part of German society as a whole, were not sure—particularly before the March 5, 1933, Reichstag elections—whether the Nazis were in power to stay or whether a conservative military coup against them was still possible. Some Jewish intellectuals came up with rather unusual forecasts. "The prognosis," Martin Buber wrote to philosopher and educator Ernst Simon on February 14, "depends on the outcome of the imminent fight between the groups in the government. We must assume that no shift in the balance of power in favor of the National Socialists will be permitted, even if their parliamentary base vis-à-vis the German nationalists is proportionally strengthened. In that case, one of two things will happen: either the Hitlerites will remain in the government anyway; then they will be sent to fight the proletariat, which will split their party and render it harmless for the time being. . . . Or they will leave the government. . . . As long as the present condition holds, there can be no thought of Jew-baiting or anti-Jewish laws, only of administrative oppression. Anti-Semitic legislation would be possible only if the balance of power shifted in favor of the National Socialists, but as I have said above, this is hardly to be expected. Jew-baiting is only possible during the interval between the National Socialists' leaving the government and the proclamation of a state of emergency."[35]

III

The primary political targets of the new regime and of its terror system, at least during the first months after the Nazi accession to power, were not Jews but Communists. After the Reichstag fire of February 27, the anti-Communist hunt led to the arrest of almost ten thousand party members and sympathizers and to their imprisonment in newly created concentration camps. Dachau had been established on March 20 and was officially inaugurated by SS chief Heinrich Himmler on April 1.[36] In June, SS Group Leader Theodor Eicke became the camp's commander, and a year later he was appointed "inspector of concentration camps": Under Himmler's aegis he had become the architect of the life-and-death routine of the camp inmates in Hitler's new Germany.

After the mass arrests that followed the Reichstag fire, it was clear that the "Communist threat" no longer existed. But the new regime's frenzy of repression—and innovation—did not slacken; quite the contrary. A presidential decree of February 28 had already given Hitler emergency powers. Although the Nazis failed to gain an absolute majority in the March 5 elections, their coalition with the ultraconservative German National People's Party (Deutschnationale Volkspartei, or DNVP) obtained it. A few days later, on March 23, the Reichstag divested itself of its functions by passing the Enabling Act, which gave full legislative and executive powers to the chancellor (at the outset new legislation was discussed with the cabinet ministers, but the final decision was Hitler's). The rapidity of changes that followed was stunning: The states were brought into line; in May the trade unions were abolished and replaced by the German Labor Front; in July all political parties were dissolved with the sole exception of the National Socialist German Workers Party (Nationalsozialistische Deutsche

Arbeiterpartei, or NSDAP). Popular support for this torrential activity and constant demonstration of power snowballed. In the eyes of a rapidly growing number of Germans, a "national revival" was under way.[37]

It has often been asked whether the Nazis had concrete goals and precise plans. In spite of internal tensions and changing circumstances, short-term goals in most areas were systematically pursued and rapidly achieved. But the final objectives of the regime, the guidelines for long-term policies, were defined in general terms only, and concrete steps for their implementation were not spelled out. Yet these vaguely formulated long-term goals were essential not only as guidelines of sorts but also as indicators of boundless ambitions and expectations: They were objects of true belief for Hitler and his coterie; they mobilized the energies of the party and of various sectors of the population; and they were expressions of faith in the correctness of the way.

Anti-Jewish violence spread after the March elections. On the ninth, Storm Troopers (Sturmabteilung, or SA) seized dozens of East European Jews in the Scheunenviertel, one of Berlin's Jewish quarters. Traditionally the first targets of German Jew-hatred, these *Ostjuden* were also the first Jews as Jews to be sent off to concentration camps. On March 13 forcible closing of Jewish shops was imposed by the local SA in Mannheim; in Breslau, Jewish lawyers and judges were assaulted in the court building; and in Gedern, in Hesse, the SA broke into Jewish homes and beat up the inhabitants "with the acclamation of a rapidly growing crowd." The list of similar incidents is a long one.[38] There were also killings. According to the late March (bimonthly) report of the governing president of Bavaria, "On the 15th of this month, around 6 in the morning, several men in dark uniforms arrived by truck at the home of the Israelite businessman Otto Selz in Straubing. Selz was dragged from his house in his nightclothes and taken away. Around 9:30 Selz was shot to death in a forest near Wang, in the Landshut district. The truck is said to have arrived on the Munich-Landshut road and to have departed the same way. It carried six uniformed men and bore the insignia II.A. Several people claim to have noticed that the truck's occupants wore red armbands with a swastika."[39] On March 31 Interior Minister Wilhelm Frick wired all local police stations to warn them that Communist agitators disguised in SA uniforms and using SA license plates would smash Jewish shop windows and exploit the occasion to create disturbances.[40] This could have been standard Nazi disinformation or some remaining belief in possible Communist subversion. On April 1, the

Göttingen police station investigating the damage to Jewish stores and the local synagogue on March 28, reported having caught two members of the Communist Party and one Social Democrat in possession of parts of Nazi uniforms; headquarters in Hildesheim was informed that the men arrested were the perpetrators of the anti-Jewish action.[41]

Much of the foreign press gave wide coverage to the Nazi violence. The *Christian Science Monitor,* however, expressed doubts about the accuracy of the reports of Nazi atrocities, and later justified retaliation against "those who spread lies against Germany." And Walter Lippmann, the most prominent American political commentator of the day and himself a Jew, found words of praise for Hitler and could not resist a sideswipe at the Jews. These notable exceptions notwithstanding, most American newspapers did not mince words about the anti-Jewish persecution.[42] Jewish and non-Jewish protests grew. These very protests became the Nazis' pretext for the notorious April 1, 1933, boycott of Jewish businesses. Although the anti-Nazi campaign in the United States was discussed at some length during a cabinet meeting on March 24,[43] the final decision in favor of the boycott was probably made during a March 26 meeting of Hitler and Goebbels in Berchtesgaden. But in mid-March, Hitler had already allowed a committee headed by Julius Streicher, party chief of Franconia and editor of the party's most vicious anti-Jewish newspaper, *Der Stürmer,* to proceed with preparatory work for it.

In fact, the boycott had been predictable from the very moment the Nazis acceded to power. The possibility had often been mentioned during the two preceding years,[44] when Jewish small businesses had been increasingly harassed and Jewish employees increasingly discriminated against in the job market.[45] Among the Nazis much of the agitation for anti-Jewish economic measures was initiated by a motley coalition of "radicals" belonging either to the Nazi Enterprise Cells Organization (Nationalsozialistische Betriebszellenorganisation, or NSBO) headed by Reinhold Muchow or to Theodor Adrian von Renteln's League of Middle-Class Employees and Artisans (Kampfbund für den gewerblichen Mittelstand), as well as to various sections of the SA activated for that purpose by Otto Wagener, an economist and the SA's former acting chief of staff. Their common denominator was what former number two party leader Gregor Strasser once called an "anti-capitalist nostalgia";[46] their easiest way of expressing it: virulent anti-Semitism.

Such party radicals will be encountered at each major stage of anti-

Jewish policy up to and including the Kristallnacht pogrom of November 1938. In April 1933 they can be identified as members of the party's various economic interest groups, but also among them were jurists like Hans Frank (the future governor-general of occupied Poland) and Roland Freisler (the future president of the People's Tribunal) and race fanatics like Gerhard Wagner and Otto Gross, not to speak of Streicher, Goebbels, the SA leadership, and, foremost among them, Hitler himself. But specifically as a pressure group, the radicals consisted mainly of "old fighters"—SA members and rank-and-file party activists dissatisfied with the pace of the National Socialist revolution, with the meagerness of the spoils that had accrued to them, and with the often privileged status of comrades occupying key administrative positions in the state bureaucracy. The radicals were a shifting but sizable force of disgruntled party members seething for increased action and for the primacy of the party over the state.[47]

The radicals' influence should not be overrated, however. They never compelled Hitler to take steps he did not want to take. When their demands were deemed excessive, their initiatives were dismissed. The anti-Jewish decisions in the spring of 1933 helped the regime channel SA violence into state-controlled measures;[48] to the Nazis, of course, these measures were also welcome for their own sake.

Hitler informed the cabinet of the planned boycott of Jewish-owned businesses on March 29, telling the ministers that he himself had called for it. He described the alternative as spontaneous popular violence. An approved boycott, he added, would avoid dangerous unrest.[49] The German National ministers objected, and President Hindenburg tried to intervene. Hitler rejected any possible cancellation, but two days later (the day before the scheduled boycott) he suggested the possibility of postponing it until April 4—if the British and American governments were to declare immediately their opposition to the anti-German agitation in their countries; if not, the action would take place on April 1, to be followed by a waiting period until April 4.[50]

On the evening of the thirty-first, the British and American governments declared their readiness to make the necessary declaration. Foreign Minister Konstantin Freiherr von Neurath made it known, however, that it was too late to change course; he then mentioned Hitler's decision of a one-day action followed by a waiting period.[51] In fact, the possibility of resuming the boycott on April 4 was no longer being considered.

In the meantime Jewish leaders, mainly in the United States and

Palestine, were in a quandary: Should they support mass protests and a counterboycott of German goods, or should confrontation be avoided for fear of further "reprisals" against the Jews of Germany? Göring had summoned several leaders of German Jewry and sent them to London to intervene against planned anti-German demonstrations and initiatives. Simultaneously, on March 26, Kurt Blumenfeld, president of the Zionist Federation for Germany, and Julius Brodnitz, president of the Central Association, cabled the American Jewish Committee in New York: WE PROTEST CATEGORICALLY AGAINST HOLDING MONDAY MEETING, RADIO AND OTHER DEMONSTRATIONS. WE UNEQUIVOCALLY DEMAND ENERGETIC EFFORTS TO OBTAIN AN END TO DEMONSTRATIONS HOSTILE TO GERMANY.[52] By appeasing the Nazis the fearful German-Jewish leaders were hoping to avoid the boycott.

The leaders of the Jewish community in Palestine also opted for caution, the pressure of public opinion notwithstanding. They sent a telegram to the Reich Chancellery "offering assurances that no authorized body in Palestine had declared or intended to declare a trade boycott of Germany."[53] American Jewish leaders were divided; most of the Jewish organizations in the United States were opposed to mass demonstrations and economic action, mainly for fear of embarrassing the president and the State Department.[54] Reluctantly, and under pressure from such groups as the Jewish War Veterans, the American Jewish Congress finally decided otherwise. On March 27 protest meetings took place in several American cities, with the participation of church and labor leaders. As for the boycott of German goods, it spread as an emotional grass-roots movement that, over the months, received an increasing measure of institutional support, at least outside Palestine.[55]

Goebbels's excitement was irrepressible. In his diary entry for March 27, he wrote: "I've dictated a sharp article against the Jews' atrocity propaganda. At its mere announcement the whole *mischpoke* [*sic*, Yiddish for "family"] broke down. One must use such methods. Magnanimity doesn't impress the Jews." March 28: "Phone conversation with the Führer: the call for the boycott will be published today. Panic among the Jews!" March 29: "I convene my assistants and explain the organization of the boycott to them." March 30: "The organization of the boycott is complete. Now we merely need to press a button and it starts."[56] March 31: "Many people are going around with their heads hanging and seeing specters. They think the boycott will lead to war. By defending ourselves, we can only win respect.

A small group of us hold a last discussion and decide that the boycott should start tomorrow with fullest intensity. It will last one day and then be followed by an interruption until Wednesday. If the incitement in foreign countries stops, then the boycott will stop, otherwise a fight to the end will start."[57] April 1: "The boycott against the international atrocities propaganda broke out in the fullest intensity in Berlin and all over the Reich. The public," Goebbels added, "has everywhere shown its solidarity."[58]

In principle the boycott could have caused serious economic damage to the Jewish population as, according to Avraham Barkai, "more than sixty percent of all gainfully employed Jews were concentrated in the commercial sector, the overwhelming majority of these in the retail trade. . . . Similarly, Jews in the industrial and crafts sectors were active largely as proprietors of small businesses and shops or as artisans."[59] In reality, however, the Nazi action ran into immediate problems.[60]

The population proved rather indifferent to the boycott and sometimes even intent on buying in "Jewish" stores. According to the *Völkischer Beobachter* of April 3, some shoppers in Hannover tried to enter a Jewish-owned store by force.[61] In Munich repeated announcements concerning the forthcoming boycott resulted in such brisk business in Jewish-owned stores during the last days of March (the public did not yet know how long the boycott would last) that the *Völkischer Beobachter* bemoaned "the lack of sense among that part of the population which forced its hard-earned money into the hands of enemies of the people and cunning slanderers."[62] On the day of the boycott many Jewish businesses remained shut or closed early. Vast throngs of onlookers blocked the streets in the commercial districts of the city center to watch the unfolding event: They were passive but in no way showed the hostility to the "enemies of the people" the party agitators had expected.[63] A Dortmund rabbi's wife, Martha Appel, confirms in her memoirs a similarly passive and certainly not hostile attitude among the crowds in the streets of that city's commercial sector. She even reports hearing many expressions of discontent with the Nazi initiative.[64] This atmosphere seems to have been common in most parts of the Reich. The bimonthly police report in the Bavarian town of Bad Tölz, south of Munich, is succinct and unambiguous: "The only Jewish shop, 'Cohn' on the Fritzplatz, was not boycotted."[65]

The lack of popular enthusiasm was compounded by a host of unforeseen questions: How was a "Jewish" enterprise to be defined? By its name, by the Jewishness of its directors, or by Jewish control of all or part of its capital? If the enterprise were hurt, what, in a time of economic crisis, would happen to its Aryan employees? What would be the overall consequences, in terms of possible foreign retaliation, of the action on the German economy?

Although impending for some time, the April boycott was clearly an improvised action. It may have aimed at channeling the anti-Jewish initiatives of the SA and of other radicals; at indicating that, in the long run, the basis of Jewish existence in Germany would be destroyed; or, more immediately, at responding in an appropriately Nazi way to foreign protests against the treatment of German Jews. Whatever the various motivations may have been, Hitler displayed a form of leadership that was to become characteristic of his anti-Jewish actions over the next several years: He usually set an *apparent* compromise course between the demands of the party radicals and the pragmatic reservations of the conservatives, giving the public impression that he himself was above operational details.[66] Such restraint was obviously tactical; in the case of the boycott, it was dictated by the state of the economy and wariness of international reactions.[67]

For some Jews living in Germany, the boycott, despite its overall failure, had unexpected and unpleasant consequences. Such was the case of Arthur B., a Polish Jew who had been hired on February 1 with his band of "four German musicians (one of them a woman)" to perform at the Café Corso in Frankfurt. A month later B.'s contract was extended to April 30. On March 30, B. was dismissed by the café owner for being Jewish. B. applied to the Labor Court in Frankfurt to obtain payment of the money owed him for the month of April. The owner, he argued, had known when she hired him that he was a Polish Jew. She had been satisfied with the band's work and thus had no right to dismiss him without notice and payment. The court rejected his plea and charged him with the costs, ruling that the circumstances created by Jewish incitement against Germany—which had led some customers to demand the bandleader's dismissal and brought threats from the local *Gau* (main party district) leadership that the Café Corso would be boycotted as a Jewish enterprise if Arthur B. were to continue working there—could have caused severe damage to the defendant and was therefore sufficient reason for the dismissal. "Whether the defendant already knew when she hired him that the plaintiff was a Jew is irrelevant," the court concluded, "as the national revolution with its drastic consequences for the Jews took place after the plaintiff had been hired; the defendant could not have known at the time that the plaintiff's belonging to the Jewish race would later play such a significant role."[68]

The possibility of further boycotts remained open. "We hereby inform you," said a letter of August 31 from the Central Committee of the Boycott Movement (Zentralkomitee der Boykottbewegung) in Munich to the party district leadership of Hannover-South, "that the Central Committee for Defense Against Jewish Atrocities and Boycott Agitation . . . continues its work as before. The organization's activity will, however, be pursued quietly. We ask you to observe and inform us of any cases of corruption or other economic activities in which Jews play a harmful role. You may then wish to inform your district or local leadership in an appropriate way about such cases as just mentioned. As indicated in the last internal party instruction from the Deputy Führer [for Party affairs] Party Comrade [Rudolf] Hess, any public statements of the Central Committee must first be submitted to him."[69]

At the same time it was nonetheless becoming increasingly clear to Hitler himself that Jewish economic life was not to be openly interfered with, at least as long as the German economy was still in a precarious situation. A fear of foreign economic retaliation, whether orchestrated by the Jews or as an expression of genuine outrage at Nazi persecutions, was shared by Nazis and their conservative allies alike and dictated temporary moderation. Once Hjalmar Schacht moved from the presidency of the Reichsbank to become minister of the economy, in the summer of 1934, noninterference with Jewish business was quasi-officially agreed upon. A potential source of tension thus arose between party activists and the upper echelons of party and state.

According to the German Communist periodical *Rundschau*, by then published in Switzerland, only the smaller Jewish businesses—that is, the poorer Jews—were harmed by the Nazi boycott. Large enterprises such as the Berlin-based Ullstein publishing empire or Jewish-owned banks—Jewish big business—did not suffer at all.[70] What looks like merely an expression of Marxist orthodoxy was in part true, because harming a Jewish department-store chain such as Tietz could have put its fourteen thousand employees out of work.[71] For that very reason Hitler personally approved the granting of a loan to Tietz to ease its immediate financial difficulties.[72]

At Ullstein, one of the largest publishers in Germany (it had its own printing plant and issued newspapers, magazines, and books), the Nazi enterprise cell within the company itself addressed a letter to Hitler on

June 21, describing the disastrous consequences of a surreptitiously continuing boycott for the Jewish firm's employees: "Ullstein, which on the day of the official boycott was excluded from the action due to its being an enterprise of vital importance," the cell's leader wrote to Hitler, "is at present suffering acutely from the boycott movement. The great majority of the work force are party members and an even larger number are in the cell. With every passing day, this work force is increasingly upset by weekly and monthly dismissals, and it urgently requests me to petition the appropriate authorities in order that the livelihoods of thousands of good national comrades [members of the national-racial community, or *Volksgenossen*] not be endangered. Ullstein's publication numbers have gone down by more than half. I am daily informed of quite hair-raising boycott cases. For instance, for a long time now the party enrollment of the head of the Ullstein office in Freienwalde has been rejected on the grounds that as an employee of a Jewish publishing house he would actually cause harm to the party."[73]

This was complicated enough as it was, but the Communist *Rundschau* would have had even more to ponder if it had been aware of the many contradictions in the attitudes of major German banks and corporations toward anti-Jewish measures. First there were remnants of the past. Thus, in March 1933, when Hans Luther was replaced by Schacht as president of the Reichsbank, three Jewish bankers still remained on the bank's eight-member council and signed the authorization of his appointment.[74] This situation did not last much longer. As a result of Schacht's proddings and the party's steady pressure, the country's banks banished Jewish directors from their boards, as, for example, the dismissal of Oskar Wassermann and Theodor Frank from the board of the Deutsche Bank.[75] It is symptomatic of a measure of uneasiness with this step that the dismissals were linked to promises (obviously never fulfilled) of eventual reemployment.[76]

During the first years of the regime, however, there are indications of a somewhat unexpected moderation and even helpfulness on the part of big business in its dealings with non-Aryan firms. Pressure for business takeovers and other ruthless exploitation of the weakened status of Jews came mainly from smaller, midsized enterprises, and much less so, at least until the fall of 1937, from the higher reaches of the economy.[77] Some major corporations even retained the services of Jewish executives for years. But some precautions were taken. Thus, although most Jewish board members of the chemical industry giant I. G. Farben stayed on for a while, the closest Jewish associates of its president, Carl Bosch, such as Ernst Schwarz and Edmund Pietrowski, were reassigned to positions outside the Reich, the former in New York, the latter in Switzerland.[78]

Highly visible Jews had to go, of course. Within a few months, the banker Max Warburg was excluded from one corporate board after another. When he was banished from the board of the Hamburg-Amerika Line, the dignitaries assembled to bid him good-bye were treated to a strange scene. As, in view of the circumstances, no one else seemed ready with a valedictory, the Jewish banker himself delivered a farewell address: "To our regret," he began, "we have learned that you have decided to leave the board of the company and consider this decision irrevocable," and he ended no less appropriately: "And now I would like to wish you, dear Mr. Warburg, a *calm old age*, good luck and many blessings to your family."[79]

IV

When the Nazis acceded to power, they could in principle refer to the goals of their anti-Jewish policy as set down in the twenty-five-point party program of February 24, 1920. Points 4, 5, 6, and 8 dealt with concrete aspects of the "Jewish question." Point 4: "Only members of the nation may be citizens of the State. Only those of German blood, whatever their creed, may be members of the nation. Accordingly no Jew may be a member of the nation." Point 5: "Non-citizens may live in Germany only as guests and must be subject to laws for aliens." Point 6: "The right to vote on the state's government and legislation shall be enjoyed by the citizens of the state alone." Point 8: "All non-German immigration must be prevented. We demand that all non-Germans who entered Germany after 2 August 1914 shall be required to leave the Reich forthwith." Point 23 demanded that control of the German press be solely in the hands of Germans.[80]

Nothing in the program indicated ways of achieving these goals, and the failure of the April 1933 boycott is a good example of the total lack of preparation for their tasks among Germany's new masters. But, at least in their anti-Jewish policy, the Nazis soon became masters of improvisation; adopting the main points of their 1920 program as short-term goals, they learned how to pursue them ever more systematically.

On March 9 State Secretary Heinrich Lammers conveyed a request from the Reich chancellor to Minister of the Interior Frick. He was asked by Hitler to take into consideration the suggestion of State Secretary Paul Bang of the Ministry of the Economy about the application of "a racial

[*völkisch*] policy" toward East European Jews: prohibition of further immigration, cancellation of name changes made after 1918, and expulsion of a
certain number of those who had not yet been naturalized.[81] Within a week
Frick responded by sending instructions to all states (*Länder*):

> In order to introduce a racial policy (*völkische Politik*), it is
> necessary to:
>
> 1. Oppose the immigration of Eastern Jews.
> 2. Expel Eastern Jews living in Germany without a residence
>   permit.
> 3. Stop the naturalization of Eastern Jews.[82]

Bang's suggestions were in line with Points 5 (on naturalization) and 8
(on immigration) of the 1920 party program. As early as 1932, moreover,
both the German National Minister of the Interior Wilhelm Freiherr von
Gayl and the Nazi Helmut von Nicolai had formulated concrete proposals
regarding East European Jews,[83] and a month before Frick issued his guidelines the Prussian Ministry of the Interior had already taken the initiative to
cancel an order previously given to the police to avoid the expulsion of East
European Jews who had been accused by the police of "hostile activities" but
had lived in Germany for a long period.[84] On July 14, 1933, these measures
were enhanced by the Law for the Repeal of Naturalization and Recognition
of German Citizenship, which called for the cancellation of naturalizations
that had taken place between November 9, 1918, and January 30, 1933.[85]

The measures taken against the so-called Eastern Jews were overshad-
owed by the laws of April 1933.[86] The first of them— the most fundamental one because of its definition of the Jew—was the April 7 Law for the
Restoration of the Professional Civil Service. In its most general intent, the
law aimed at reshaping the entire government bureaucracy in order to
ensure its loyalty to the new regime. Applying to more than two million
state and municipal employees, its exclusionary measures were directed
against the politically unreliable, mainly Communists and other opponents
of the Nazis, and against Jews.[87] Paragraph 3, which came to be called the
"Aryan paragraph," reads: "1. Civil servants not of Aryan origin are to
retire. . . ." (Section 2 listed exceptions, which will be examined later.) On
April 11 the law's first supplementary decree defined "non-Aryan" as "anyone descended from non-Aryan, particularly Jewish, parents or grandparents. It suffices if one parent or grandparent is non-Aryan."[88]

For the first time since completion of the emancipation of the German Jews in 1871, a government, by law, had reintroduced discrimination against the Jews. Up to this point the Nazis had unleashed the most extreme anti-Jewish propaganda and brutalized, boycotted, or killed Jews on the assumption that they could somehow be identified as Jews, but no formal disenfranchisement based on an exclusionary definition had yet been initiated. The definition as such—whatever its precise terms were to be in the future—was the necessary initial basis of all the persecutions that were to follow.[89]

Wilhelm Frick was at the immediate origin of the Civil Service Law; he had already proposed the same legislation to the Reichstag as far back as May 1925. On March 24, 1933, he submitted the law to the cabinet. On March 31 or April 1, Hitler probably intervened to support the proposal. The atmosphere surrounding the boycott undoubtedly contributed to the rapid drafting of the text. Although the scope of the law was general, the anti-Jewish provision represented its very core.[90]

The definition of Jewish origin in the Civil Service Law was the broadest and most comprehensive, and the provisions for assessment of each doubtful case the harshest possible. In the elaboration of the law we find traces of the anti-Semitic and racial zeal of Achim Gercke, the specialist for race research at the Ministry of the Interior,[91] a man who during his student days at Göttingen had started, with some help from faculty and staff, to set up a card index of all Jews—as defined by racial theory; that is, in terms of Jewish ancestry—living in Germany.[92] For Gercke the anti-Jewish laws were not limited to their immediate and concrete object; they also had an "educational" function: Through them "the entire national community becomes enlightened about the Jewish question; it learns that the national community is a community of blood; for the first time it understands race thinking and, instead of an overly theoretical approach to the Jewish question, it is confronted with a concrete solution."[93]

In 1933 the number of Jews in the civil service was small. As a result of Hindenburg's intervention (following a petition by the Association of Jewish War Veterans that was also supported by the elderly Field Marshal August von Mackensen), combat veterans and civil servants whose fathers or sons had been killed in action in World War I were exempted from the law. Civil servants, moreover, who had been in state service by August 1, 1914, were also exempt.[94] All others were forced into retirement.

Legislation regarding Jewish lawyers illustrates, even more clearly than the economic boycott, how Hitler maneuvered between contradictory demands from Nazi radicals on the one hand and from his DNVP allies on the other. By the end of March, physical molestation of Jewish jurists had spread throughout the Reich. In Dresden, Jewish judges and lawyers were dragged out of their offices and even out of courtrooms during proceedings, and, more often than not, beaten up. According to the *Vossische Zeitung* (quoted by the *Jüdische Rundschau* of March 28), in Gleiwitz, Silesia, "a large number of young men entered the court building and molested several Jewish lawyers. The seventy-year-old legal counselor Kochmann was hit in the face and other lawyers punched all over. A Jewish woman assessor was taken to jail. The proceedings were interrupted. Finally, the police had to occupy the building in order to put an end to the disturbances."[95] There were dozens of similar events throughout Germany. At the same time local Nazi leaders such as the Bavarian justice minister, Hans Frank, and the Prussian justice minister, Hanns Kerrl, on their own initiative announced measures for the immediate dismissal of all Jewish lawyers and civil servants.

Franz Schlegelberger, state secretary of the Ministry of Justice, reported to Hitler that these local initiatives created an entirely new situation and demanded rapid legislation to impose a new, unified legal framework. Schlegelberger was backed by his minister, DNVP member Franz Gürtner. The Justice Ministry had prepared a decree excluding Jewish lawyers from the bar on the same basis—but also with the same exemptions regarding combat veterans and their relatives, and longevity in practice, as under the Civil Service Law. At the April 7 cabinet meeting Hitler unambiguously opted for Gürtner's proposal. In Hitler's own words: "For the moment . . . one has to deal only with what is necessary."[96] The decree was confirmed the same day and made public on April 11.

Because of the exemptions, the initial application of the law was relatively mild. Of the 4,585 Jewish lawyers practicing in Germany, 3,167 (or almost 70 percent) were allowed to continue their work; 336 Jewish judges and state prosecutors, out of a total of 717, were also kept in office.[97] In June 1933 Jews still made up more than 16 percent of all practicing lawyers in Germany.[98] These statistics should, however, not be misinterpreted. Though still allowed to practice, Jewish lawyers were excluded from the national association of lawyers and listed not in its annual directory but in a separate guide; all in all, notwithstanding the support of some Aryan institutions and individuals, they worked under a "boycott by fear."[99]

Nazi rank-and-file agitation against Jewish physicians did not lag far behind the attacks on Jewish jurists. Thus, for example, according to the March 2 *Israelitisches Familienblatt*, an SS physician, Arno Hermann, tried to dissuade a woman patient from consulting a Jewish physician named Ostrowski. The Physicians' Honor Tribunal that heard Ostrowski's complaint condemned Hermann's initiative. Thereupon Leonardo Conti, the newly appointed Nazi commissioner for special affairs in the Prussian Ministry of the Interior, violently attacked the Honor Tribunal's ruling in an article published in the *Völkischer Beobachter*. In the name of the primacy of "inner conviction" and "world view," Conti argued that "every nondegenerate woman must and will internally shrink from being treated by a Jewish gynecologist; this has nothing to do with racial hatred, but belongs to the medical imperative according to which a relation of mutual understanding must grow between spiritually related physicians and patients."[100]

Hitler was even more careful with physicians than with lawyers. At the April 7 cabinet meeting, he suggested that measures against them be postponed until an adequate information campaign could be organized.[101] At this stage, after April 22, Jewish doctors were merely barred de facto from clinics and hospitals run by the national health insurance organization, with some even allowed to continue to practice there. Thus, in mid-1933, nearly 11 percent of all practicing German physicians were Jews. Here is another example of Hitler's pragmatism in action: Thousands of Jewish physicians meant tens of thousands of German patients. Disrupting the ties between these physicians and a vast number of patients could have caused unnecessary discontent. Hitler preferred to wait.

On April 25 the Law Against the Overcrowding of German Schools and Universities was passed. It was aimed exclusively against non-Aryan pupils and students.[102] The law limited the matriculation of new Jewish students in any German school or university to 1.5 percent of the total of new applicants, with the overall number of Jewish pupils or students in any institution not to exceed 5 percent. Children of World War I veterans and those born of mixed marriages contracted before the passage of the law were exempted from the quota. The regime's intention was carefully explained in the press. According to the *Deutsche Allgemeine Zeitung* of April 27: "A self-respecting nation cannot, on a scale accepted up to now, leave its higher activities in the hands of people of racially foreign origin. . . . Allowing the presence of too high a percentage of people of foreign origin in relation to their percentage in the general population could be interpreted as an acceptance of the superiority of other races, something decidedly to be rejected."[103]

The April laws and the supplementary decrees that followed compelled at least two million state employees and tens of thousands of lawyers, physicians, students, and many others to look for adequate proof of Aryan ancestry; the same process turned tens of thousands of priests, pastors, town clerks, and archivists into investigators and suppliers of vital attestations of impeccable blood purity; willingly or not these were becoming part of a racial bureaucratic machine that had begun to search, probe, and exclude.[104]

Often enough the most unlikely cases surfaced to be caught in the bizarre but unrelenting bureaucratic process triggered by the new legislation. Thus, for the six following years, the April 7 law would create havoc in the life of one Karl Berthold, an employee of the social benefits office (*Versorgungsamt*) in Chemnitz, Silesia.[105] According to a June 17, 1933, letter sent from the Chemnitz office to the main social benefits office in Dresden, the "suspicion exists that he [Karl Berthold] is *possibly* of non-Aryan origin on his father's side."[106] The letter indicated that Berthold was most probably the illegitimate son of a Jewish circus "artiste," Carl Blumenfeld, and of an Aryan mother who had died sixteen years earlier. On June 23 the Dresden office submitted the case to the Ministry of Labor, with the comment that unequivocal documentary proof was unavailable, that Berthold's outward appearance did not dispel the suspicion of a non-Aryan origin, but that, on the other hand, the fact that he was raised in the house of his maternal grandfather "in a Christian, strongly militaristic-national spirit, worked in his favor, so that the characteristics of the non-Aryan race, in case he was burdened on his father's side, would be compensated for by his upbringing."[107]

On July 21 the Ministry of Labor forwarded Berthold's file (which by then included seventeen appended documents) to the Ministry of the Interior with a request for speedy evaluation. On September 8, the ministry's specialist for racial research, Achim Gercke, gave his opinion: Carl Blumenfeld's paternity was confirmed, but Gercke could not avoid mentioning that, according to all available dates, Blumenfeld must have been only thirteen years old when Karl Berthold was conceived: "The impossibility of such a fact cannot be taken for granted," Gercke wrote, "as among Jews sexual maturity comes earlier, and similar cases are known."[108]

It did not take long for the main office in Dresden to be informed of

Gercke's computations and to do some simple arithmetic of its own. On September 26 the Dresden office wrote to the Ministry of Labor pointing out that, as Berthold had been born on March 23, 1890—when Blumenfeld was still under thirteen—the baby had to have been conceived "when the artist Carl Blumenfeld was only eleven and a half. It is difficult to assume," the Dresden letter continued, "that a boy of eleven and a half could have fathered a child with a woman of twenty-five." The Dresden office demanded that the obvious be recognized: Karl Berthold was not Carl Blumenfeld's child.[109] Needless to say, that opinion was rejected.

Berthold's story, which with its ups and downs would continue to unfold until 1939, is in many ways a parable; it will reappear sporadically until the paradoxical decision that settled Berthold's fate.

As denunciations poured in, investigations came to be conducted at all levels of the civil service. It took Hitler's personal intervention to put an end to an inquiry into the ancestry of Leo Killy, a member of the Reich Chancellery staff accused of being a full Jew. Killy's family documents cleared him of any suspicion, at least in Hitler's eyes.[110] The procedures varied: Fräulein M., who merely wished to marry a civil servant, wanted to be reassured about her Aryan ancestry, as her grandmother's name, Goldmann, could raise some doubts. The examination was performed in Professor Otmar von Verschuer's genetics department in the Kaiser Wilhelm Institute for Anthropology, Human Genetics, and Eugenics in Berlin. One of the questions Verschuer's specialists had to solve was: "Can Fräulein M. be described as a non-Aryan in the sense that she can be recognized as such by a layman on the basis of her mental attitude, her environment, or her outward appearance?" The "genetic examination," based on photographs of Fräulein M.'s relatives and on aspects of her own physical appearance, led to most positive results. The report excluded any signs of Jewishness. Although Fräulein M. had "a narrow, high and convexly projecting nose," it concluded that she had inherited the nose from her father (not from the grandmother burdened with the name Goldmann) and thus was a pure Aryan.[111]

In September 1933 Jews were forbidden to own farms or engage in agriculture. That month the establishment, under the control of the Propaganda Ministry, of the Reich Chamber of Culture, enabled Goebbels to limit the participation of Jews in the new Germany's cultural life. (Their systematic expulsion, which would include not only writers and artists but also owners of important businesses in the cultural domain, was for that reason delayed until 1935.)[112] Also under the aegis of Goebbels's Propaganda Ministry, Jews were barred from belonging to the Journalists' Association and, on October 4, from being newspaper editors. The German press had been cleansed. (Exactly a year later, Goebbels recognized the right of Jewish editors and journalists to work, but only within the framework of the Jewish press.)[113]

In Nazi racial thinking, the German national community drew its strength from the purity of its blood and from its rootedness in the sacred German earth. Such racial purity was a condition of superior cultural creation and of the construction of a powerful state, the guarantor of victory in the struggle for racial survival and domination. From the outset, therefore, the 1933 laws pointed to the exclusion of the Jews from all key areas of this utopian vision: the state structure itself (the Civil Service Law), the biological health of the national community (the physicians' law), the social fabric of the community (the disbarring of Jewish lawyers), culture (the laws regarding schools, universities, the press, the cultural professions), and, finally, the sacred earth (the farm law). The Civil Service Law was the only one of these to be fully implemented at this early stage, but the symbolic statements they expressed and the ideological message they carried were unmistakable.

Very few German Jews sensed the implications of the Nazi laws in terms of sheer long-range terror. One who did was Georg Solmssen, spokesman for the board of directors of the Deutsche Bank and son of an Orthodox Jew. In an April 9, 1933, letter addressed to the bank's board chairman, after pointing out that even the non-Nazi part of the population seemed to consider the new measures "self-evident," Solmssen added: "I am afraid that we are merely at the beginning of a process aiming, purposefully and according to a well-prepared plan, at the economic and moral annihilation of all members, without any distinctions, of the Jewish race living in Germany. The total passivity not only of those classes of the population that belong to the National Socialist Party, the absence of all feelings of solidarity becoming apparent among those who until now worked shoulder to shoulder with Jewish colleagues, the increasingly more obvious desire to take personal advantage of vacated positions, the hushing up of the disgrace and the shame disastrously inflicted upon people who, although innocent, witness the destruction of their honor and their existence from one day to the next—all of this indicates a situation so hopeless that it would be wrong not to face it squarely without any attempt at prettification."[114]

*   *   *

There was some convergence between the expressions of the most extreme anti-Semitic agenda of German conservatives at the beginning of the century and the Nazi measures during the early years of the new regime. In his study of the German Civil Service, Hans Mommsen pointed to the similarity between the "Aryan paragraph" of the Civil Service Law of April 1933 and the Conservative Party's so-called Tivoli program of 1892. [115] The program's first paragraph declared: "We combat the widely obtrusive and subversive Jewish influence on our popular life. We demand a Christian authority for the Christian people and Christian teachers for Christian pupils."[116]

The Conservatives, in other words, demanded the exclusion of Jews from any government position and from any influence on German education and culture. As for the main thrust of the forthcoming 1935 Nuremberg laws—segregation of the Jews according to racial criteria and placing of the Jewish community as such under "alien status"—this had already been demanded by radical Conservative anti-Semites, particularly by Heinrich Class, president of the Pan-Germanic League, in a notorious pamphlet, entitled *Wenn ich der Kaiser wär (If I Were the Kaiser)*, published in 1912. Thus, although what was to become the Nazi program of action was a Nazi creation, the overall evolution of the German right-wing parties during the Weimar years gave birth to a set of anti-Jewish slogans and demands that the extreme nationalist parties (the DNVP in particular) shared with the Nazis.

The conservative state bureaucracy had sometimes anticipated Nazi positions on Jewish matters. The Foreign Ministry, for instance, tried, well before the Nazis came to power, to defend Nazi anti-Semitism. After January 1933, with the blessings of State Secretary Bernhard Wilhelm von Bülow and Foreign Minister Neurath, senior officials of the Ministry intensified these efforts.[117] In the spring of 1933, anti-Jewish propaganda work in the Foreign Ministry was bolstered by the establishment of a new Department Germany (Referat Deutschland), to which this task was specifically given.

At the Prussian Ministry of the Interior, State Secretary Herbert von Bismarck of the DNVP participated in the anti-Jewish crusade with no less vehemence than Frick, the Nazi minister. Apparently stung by the recently published biography of his great-uncle Otto, the Iron Chancellor, by Emil Ludwig (his real name was Emil Ludwig Kohn), Bismarck demanded prohibition of the use of pseudonyms by Jewish authors. Moreover, as Bismarck put it, "national pride is deeply wounded by those cases in which Jews with Eastern Jewish names have adopted particularly nice German surnames, such as, for example, Harden, Olden, Hinrichsen, etc. I consider a review of name changes urgently necessary in order to revoke changes of that kind."[118]

On April 6, 1933, an ad hoc committee—following an initiative that probably originated in the Prussian Interior Ministry—started work on a draft for a Law Regulating the Position of the Jews. Again the German Nationals were heavily represented on the eight-member drafting committee. A copy of this draft proposal, sent in July 1933 to the head of Department Germany of the Foreign Ministry, remained in the archives of the Wilhelmstrasse. The draft suggests the appointment of a "national guardian" (*Volkswart*) for dealing with Jewish affairs and employs the term "Jewish council" (*Judenrat*) in defining the central organization that is to represent the Jews of Germany in their dealings with the authorities, particularly with the *Volkswart*. Already in the draft are many of the discriminatory measures that were to be taken later,[119] although at the time, nothing came of this initiative. Thus for part of the way at least, Nazi policies against the Jews were identical with the anti-Semitic agenda set by the German Conservatives several decades before Hitler's accession to power.[120]

And yet the curtailment of the economic measures against the Jews was also a conservative demand, and whatever exceptions were introduced into the April laws were instigated by the most prominent conservative figure of all, President Hindenburg. Hitler understood perfectly how essentially different his own anti-Jewish drive was from the traditional anti-Semitism of the old field marshal, and in his answer to Hindenburg's request of April 4, regarding exceptions to the exclusion of Jews from the civil service, limited himself to the regular middle-of-the-road anti-Jewish arguments of the moderate breed of conservatives to which Hindenburg belonged. It was in fact Hitler's first lengthy statement on the Jews since he became chancellor.

In his April 5 letter, Hitler started by using the argument of a Jewish "inundation." With regard to the civil service, the Nazi leader argued that the Jews, as a foreign element and as people with ability, had entered governmental positions and "were sowing the seed of corruption, the extent of which no one today has any adequate appreciation." The international

Jewish "atrocity and boycott agitation" precipitated measures that are intrinsically defensive. Hitler nonetheless promised that Hindenburg's request regarding Jewish veterans would be implemented. Then he moved to a strangely premonitory finale: "In general, the first goal of this cleansing process is intended to be the restoration of a certain healthy and natural relationship; and second, to remove from specified positions important to the state those elements that cannot be entrusted with the life or death of the Reich. Because in the coming years we will inevitably have to take precautions to ensure that certain events that cannot be disclosed to the rest of the world for higher reasons of state really remain secret."[121]

Again, Hitler was utilizing some of the main tenets of conservative anti-Semitism to the full: the over-representation of Jews in some key areas of social and professional life, their constituting a nonassimilated and therefore foreign element in society, the nefarious influence of their activities (liberal or revolutionary), particularly after November 1918. Weimar, the conservatives used to clamor, was a "Jewish republic." Hitler had not forgotten to mention, for the special benefit of a field marshal and Prussian landowner, that in the old Prussian state the Jews had had little access to the civil service and that the officer corps had been kept free of them. There was some irony in the fact that a few days after Hitler's letter to Hindenburg, the old field marshal himself had to answer a query from Prince Carl of Sweden, president of the Swedish Red Cross, about the situation of the Jews in Germany. The text of Hindenburg's letter to Sweden was in fact dictated by Hitler, with the early draft prepared by Hindenburg's office significantly changed (any admission of acts of violence against Jews was omitted, and the standard theme of the invasion of the Reich by Jews from the East strongly underlined).[122] Thus, over his own signature, the president of the Reich sent a letter not very different from the one Hitler had addressed to him on April 4. But soon Hindenburg would be gone, and this source of annoyance would disappear from Hitler's path.

V

The city of Cologne forbade the use of municipal sports facilities to Jews in March 1933.[123] Beginning April 3 requests by Jews in Prussia for name changes were to be submitted to the Justice Ministry, "to prevent the covering up of origins."[124] On April 4 the German Boxing Association excluded all Jewish boxers.[125] On April 8 all Jewish teaching assistants at universities in the state of Baden were to be expelled immediately.[126] On April 18 the party district chief (Gauleiter) of Westphalia decided that a Jew would be allowed to leave prison only if the two persons who had submitted the request for bail, or the doctor who had signed the medical certificate, were ready to take his place in prison."[127] On April 19 the use of Yiddish was forbidden in cattle markets in Baden.[128] On April 24 the use of Jewish names for spelling purposes in telephone communications was forbidden.[129] On May 8 the mayor of Zweibrücken prohibited Jews from leasing places in the next annual town market.[130] On May 13 the change of Jewish to non-Jewish names was forbidden.[131] On May 24 the full Aryanization of the German gymnastics organization was ordered, with full Aryan descent of all four grandparents stipulated.[132] Whereas in April Jewish doctors had been excluded from state-insured institutions, in May privately insured institutions were ordered to refund medical expenses for treatment by Jewish doctors only when the patients themselves were non-Aryan. Separate lists of Jewish and non-Jewish doctors would be ready by June.[133]

On April 10 the president of the state government and minister for religious affairs and education of Hesse had demanded of the mayor of Frankfurt that the Heinrich Heine monument be removed from its site. On May 18 the mayor replied that "the bronze statue was thrown off its pedestal on the night of April 26–27. The slightly damaged statue has been removed and stored in the cellar of the ethnological museum."[134]

In fact, according to the Stuttgart city chronicle, in the spring of 1933 hardly a day went by without some aspect of the "Jewish question" coming up in one way or another. On the eve of the boycott, several well-known local Jewish physicians, lawyers, and industrialists left the country.[135] On April 5 the athlete and businessman Fritz Rosenfelder committed suicide. His friend, the World War I ace Ernst Udet, flew over the cemetery to drop a wreath.[136] On April 15 the Nazi Party demanded the exclusion of Berthold Heymann, a Socialist (and Jewish) former cabinet minister in Württemberg, from the electoral list.[137] On April 20 the Magistrate's Court of Stuttgart tried the chief physician of the Marienspital (Saint Mary's Hospital), Caesar Hirsch, in absentia. Members of his staff testified that he had declared he would not return to Nazi Germany, "as he refused to live in such a homeland."[138] On April 27 three hundred people demonstrated on the Königsstrasse against the opening of a local branch of the Jewish-owned shoe company Etam.[139] On April 29 a Jewish veterinarian who wanted to resume his service at the slaughterhouse was threatened by several butchers and taken "into custody."[140] And so it continued, day in and day out.

In his study of the Nazi seizure of power in the small city of Northeim (renamed Thalburg), near Hannover, William Sheridan Allen vividly describes the changing fate of the town's 120 Jews. Mostly small businessmen and their families, they were well assimilated and for several generations had been an integral part of the community. In 1932 a Jewish haberdasher had celebrated the 230th anniversary of the establishment of his shop.[141] Allen tells of a banker named Braun, who tried hard to maintain his German nationalist stance and to disregard the increasingly insulting measures introduced by the Nazis: "To the solicitous advice that was given to him to leave Thalburg, he replied, 'Where should I go? Here I am the Banker Braun; elsewhere I would be the Jew Braun.'"[142]

Other Jews in Thalburg were less confident. Within a few months the result was the same for all. Some withdrew from the various clubs and social organizations to which they had belonged; others received letters of dismissal under various pretexts. "Thus," as Allen expresses it, "the position of the Jews in Thalburg was rapidly clarified, certainly by the end of the first half-year of Hitler's regime. . . . The new state of affairs became a fact of life; it was accepted. Thalburg's Jews were simply excluded from the community at large."[143]

For young Hilma Geffen-Ludomer, the only Jewish child in the Berlin suburb of Rangsdorf, the Law Against the Overcrowding of German Schools meant total change. The "nice, neighborly atmosphere" ended "abruptly. . . . Suddenly, I didn't have any friends. I had no more girlfriends, and many neighbors were afraid to talk to us. Some of the neighbors that we visited told me: 'Don't come anymore because I'm scared. We should not have any contact with Jews.'" Lore Gang-Salheimer, eleven in 1933 and living in Nuremberg, could remain in her school as her father had fought at Verdun. Nonetheless "it began to happen that non-Jewish children would say, 'No I can't walk home from school with you anymore. I can't be seen with you anymore.'"[144] "With every passing day under Nazi rule," wrote Martha Appel, "the chasm between us and our neighbors grew wider. Friends with whom we had had warm relations for years did not know us anymore. Suddenly we discovered that we were different."[145]

On the occasion of the general census of June 1933, German Jews, like everyone else, were defined and counted in terms of their religious affiliation and nationality, but their registration cards included more details than those of other citizens. According to the official *Statistik des deutschen Reiches*, these special cards "allowed for an overview of the biological and social situation of Jewry in the German Reich, insofar as it could be recorded on the basis of religious affiliation." A census "of Jewry living in the Reich on the basis of race" was not yet possible.[146]

VI

The Law for the Prevention of Genetically Diseased Offspring (Gesetz zur Verhütung erbkranken Nachwuchses) was adopted on July 14, 1933, the day on which all political parties with the exception of the NSDAP were banned and the laws against Eastern Jews (cancellation of citizenship, an end to immigration, and so on) came into effect. The new law allowed for the sterilization of anyone recognized as suffering from supposedly hereditary diseases, such as feeble-mindedness, schizophrenia, manic-depressive insanity, genetic epilepsy, Huntington's chorea, genetic blindness, genetic deafness, and severe alcoholism.[147]

The evolution leading to the July 1933 law was already noticeable during the Weimar period. Among eugenicists, the promoters of "positive eugenics" were losing ground, and "negative eugenics"—with its emphasis on the exclusion, that is, mainly the sterilization, of carriers of incapacitating hereditary diseases—was gaining the upper hand even within official institutions: A trend that had appeared on a wide scale in the West before World War I was increasingly dominating the German scene.[148] As in so many other domains, the war was of decisive importance: Weren't the young and the physically fit being slaughtered on the battlefield while the incapacitated and the unfit were being shielded? Wasn't the reestablishment of genetic equilibrium a major national-racial imperative? Economic thinking added its own logic: The social cost of maintaining mentally and physically handicapped individuals whose reproduction would only increase the burden was considered prohibitive.[149] This way of thinking was widespread and by no means a preserve of the radical right. Although the draft of a sterilization law submitted to the Prussian government in July 1932 still emphasized *voluntary* sterilization in case of hereditary defects,[150] the idea of *compulsory* sterilization seems to have been spreading.[151] It was nonetheless with the Nazi accession to power that the decisive change took place.

The new legislation was furthered by tireless activists such as Arthur

Gütt, who, after January 1933 besieged the Nazi Party's health department with detailed memoranda. Before long Leonardo Conti had Gütt nominated to a senior position at the Reich Ministry of the Interior.[152] The cardinal difference between the measures proposed by Gütt and included in the law and any previous legislation on sterilization was indeed the element of compulsion. Paragraph 12, section 1, of the new law stated that once sterilization had been decided upon, it could be implemented "against the will of the person to be sterilized."[153] This distinction is true for most cases, and on the official level. It seems, though, that even before 1933, patients in some psychiatric institutions were being sterilized without their own or their families' consent.[154] About two hundred thousand people were sterilized between mid–1933 and the end of 1937.[155] By the end of the war, the number had reached four hundred thousand.[156]

From the outset of the sterilization policies to the apparent ending of euthanasia in August 1941—and to the beginning of the "Final Solution" close to that same date—policies regarding the handicapped and the mentally ill on the one hand and those regarding the Jews on the other followed a simultaneous and parallel development. These two policies, however, had different origins and different aims. Whereas sterilization and euthanasia were exclusively aimed at enhancing the purity of the *Volksgemeinschaft*, and were bolstered by cost-benefit computations, the segregation and the extermination of the Jews—though also a racial purification process—was mainly a struggle against an active, formidable enemy that was perceived endangering the very survival of Germany and of the Aryan world. Thus, in addition to the goal of racial cleansing, identical to that pursued in the sterilization and euthanasia campaign and in contrast to it, the struggle against the Jews was seen as a confrontation of apocalyptic dimensions.

CHAPTER 2

Consenting Elites, Threatened Elites

I

About thirty SA men from Heilbronn arrived in Niederstetten, a small town in southwest Germany, on Saturday, March 25, 1933. Breaking into the few Jewish homes in the area, they took the men to the town hall and savagely beat them while local policemen kept watch at the building entrance. The scene was repeated that morning in neighboring Creglingen, where the eighteen male Jews found in the synagogue were also herded into the town hall. There the beatings led to the deaths of sixty-seven-year-old Hermann Stern and, a few days later, fifty-three-year-old Arnold Rosenfeld.

At the Sunday service the next day, Hermann Umfried, pastor of Niederstetten's Lutheran church, spoke up. His sermon was carefully phrased: It began with standard expressions of faith in the new regime and some negative remarks about Jews. But Umfried then turned to what had happened the previous day: "Only authorities are allowed to punish, and all authorities lie under divine authority. Punishment can be meted out only against those who are evil and only when a just sentence has been handed down. What happened yesterday in this town was unjust. I call on all of you to help see to it that the German people's shield of honor may remain unsullied!" When the attacks against Pastor Umfried started, no local, regional, or national church institution dared to come to his support or to express even the mildest opposition to violence against Jews. In January 1934 the local district party leader (*Kreisleiter*) ordered Umfried to resign. Increasingly anguished by the possibility that not only he but also his wife and their four daughters would be shipped off to a concentration camp, the pastor committed suicide.

Eight years and eight months later, at 2:04 P.M. on November 28, 1941, the first transport of Jews left the Niederstetten railroad station. A second batch boarded the train in April 1942, and the third and last in August of that year. Of the forty-two Jews deported from Niederstetten, only three survived.[1]

The boycott of Jewish businesses was the first major test on a national scale of the attitude of the Christian churches toward the situation of the Jews under the new government. In historian Klaus Scholder's words, "during the decisive days around the first of April, no bishop, no church dignitaries, no synod made any open declaration against the persecution of the Jews in Germany."[2] In a radio address broadcast to the United States on April 4, 1933, the most prominent German Protestant clergyman, Bishop Otto Dibelius, justified the new regime's actions, denying that there was any brutality even in the concentration camps and asserting that the boycott—which he called a reasonable defensive measure—took its course amid "calm and order."[3] His broadcast was no momentary aberration. A few days later Dibelius sent a confidential Easter message to all the pastors of his province: "My dear Brethren! We all not only understand but are fully sympathetic to the recent motivations out of which the *völkisch* movement has emerged. Notwithstanding the evil sound that the term has frequently acquired, I have always considered myself an anti-Semite. One cannot ignore that Jewry has played a leading role in all the destructive manifestations of modern civilization."[4]

The Catholic Church's reaction to the boycott was not fundamentally different. On March 31, at the suggestion of the Berlin cleric Bernhard Lichtenberg, the director of the Deutsche Bank in Berlin and president of the Committee for Inter-Confessional Peace, Oskar Wassermann, asked Adolf Johannes Cardinal Bertram, chairman of the German Conference of Bishops, to intervene against the boycott. Himself reticent about intervening, Bertram set about asking other senior German prelates for their opinions by stressing that the boycott was part of an economic battle that had nothing to do with immediate church interests. From Munich, Michael Cardinal Faulhaber wired Bertram: HOPELESS. WOULD MAKE THINGS WORSE. IN ANY CASE ALREADY DYING DOWN. For Archbishop Conrad Gröber of Freiburg, the problem was merely that converted Jews among the boycotted merchants were also being damaged.[5] Nothing was done.

In a letter addressed at approximately the same time to the Vatican's secretary of state, Eugenio Cardinal Pacelli, the future Pope Pius XII, Faulhaber wrote: "We bishops are being asked why the Catholic Church, as often in its history, does not intervene on behalf of the Jews. This is not possible at this time because the struggle against the Jews would then, at the same time, become a struggle against the Catholics, and because the Jews can help themselves, as the sudden end of the boycott shows. It is especially unjust and painful that by this action the Jews, even those who have been baptized for ten and twenty years and are good Catholics, indeed even those whose parents were already Catholics, are legally still considered Jews, and as doctors or lawyers are to lose their positions."[6]

To the clergyman Alois Wurm, founder and editor of the periodical *Seele* (Soul), who asked why the church did not state openly that people could not be persecuted because of their race, the Munich cardinal answered in less guarded terms: "For the higher ecclesiastical authorities, there are immediate issues of much greater importance; schools, the maintaining of Catholic associations, sterilization are more important for Christianity in our homeland. One must assume that the Jews are capable of helping themselves." There is no reason "to give a pretext to the government to turn the incitement against the Jews into incitement against the Jesuits."[7]

Archbishop Gröber was no more forthcoming when he stated to Robert Leiber, a Jesuit who was to become the confessor of Pius XII: "I immediately intervened on behalf of the converted Jews, but so far have had no response to my action. . . . I am afraid that the campaign against Judah will prove costly to us."[8]

The main issue for the churches was one of dogma, particularly with regard to the status of converted Jews and to the links between Judaism and Christianity. The debate had become particularly acute within Protestantism, when, in 1932, the pro-Nazi German Christian Faith Movement published its "Guidelines." "The relevant theme was a sort of race conscious belief in Christ; race, people and nation as part of a God-given ordering of life."[9] Point 9 of "Guidelines," for example, reads: "In the mission to the Jews we see a serious threat to our people [*Volkstum*]. That mission is the entry way for foreign blood into the body of our *Volk*. . . . We reject missions to the Jews in Germany as long as Jews possess the right of citizenship and hence the danger of racial fraud and bastardization exists. . . . Marriage between Germans and Jews particularly is to be forbidden."[10]

The German Christian Movement had grown in nurturing soil, and it was not by chance that, in the 1932 church elections, it received a third of the vote. The traditional alliance between German Protestantism and German nationalist authoritarianism went too deep to allow a decisive and immediately countervailing force to arise against the zealots intent on purifying Christianity of its Jewish heritage. Even those Protestant theologians who, in the 1920s, had been ready to engage in dialogue with Jews—participating, for example, in meetings organized under the aegis of Martin Buber's periodical, Der Jude—now expressed, more virulently than before, the standard accusations of "Pharisaic" and "legalistic" manifestations of the Jewish spirit. As Buber wrote in response to a particularly offensive article by Oskar A. H. Schmitz published in Der Jude in 1925 under the title "Desirable and Undesirable Jews": "I have once again . . . noted that there is a boundary beyond which the possibility of encounter ceases and only the reporting of factual information remains. I cannot fight against an opponent who is thoroughly opposed to me, nor can I fight against an opponent who stands on a different plane than I."[11] As the years went by, such encounters became less frequent, and German Protestantism increasingly opened itself to the promise of national renewal and positive Christianity heralded by National Socialism.

The German Christian Movement's ideological campaign seemed strongly bolstered by the election, on September 27, 1933, of Ludwig Müller, a fervent Nazi, as Reich bishop—that is, as some sort of Führer's coordinator for all major issues pertaining to the Protestant churches. But precisely this election and a growing controversy regarding pastors and church members of Jewish origin caused a widening rift within the Evangelical Church.

In an implementation of the Civil Service Law, the synod governing the Prussian Evangelical Church demanded the forced retirement of pastors of Jewish origin or married to Jews. This initiative was quickly followed by the synods of Saxony, Schleswig-Holstein, Braunschweig, Lübeck, Hesse-Nassau, Tübingen, and Württemberg.[12] By the early fall of 1933, general adoption of the so-called Aryan paragraph throughout the Reich appeared to be a foregone conclusion. A contrary trend, however, simultaneously made its appearance, with a group of leading theologians issuing a statement on "The New Testament and the Race Question," which clearly rejected any theological justification for adoption of the paragraph[13] and, on Christmas 1933, Pastors Dietrich Bonhoeffer and Martin Niemöller (a widely admired World War I hero), founded an oppositional organization, the Pastors' Emergency League (Pfarrernotbund), whose initial thirteen hundred adherents grew within a few months to six thousand. One of the league's first initiatives was to issue a protest against the Aryan paragraph: "As a matter of duty, I bear witness that with the use of 'Aryan laws' within the Church of Christ an injury is done to our common confession of faith."[14] The Confessing Church was born.

But the steadfastness of the Confessing Church regarding the Jewish issue was limited to support of the rights of non-Aryan Christians. And even on this point Martin Niemöller made it abundantly clear, for example in his "Propositions on the Aryan Question" ("Sätze zur Arierfrage"), published in November 1933, that only theological considerations prompted him to take his position. As he was to state at his 1937 trial for criticism of the regime, defending converted Jews "was uncongenial to him."[15] "This perception [that the community of all Christians is a matter to be taken with utter seriousness]," wrote Niemöller in the "Propositions," "requires of us, who as a people have had to carry a heavy burden as a result of the influence of the Jewish people, a high degree of self-denial, so that the desire to be freed from this demand [to maintain one single community with the converted Jews] is understandable. . . . The issue can only be dealt with . . . if we may expect from the officials [of the Church] who are of Jewish origin . . . that they impose upon themselves the restraint necessary in order to avoid any scandal. It would not be helpful if today a pastor of non-Aryan origin was to fill a position in the government of the church or had a conspicuous function in the mission to the people."[16]

Dietrich Bonhoeffer's attitude changed over the years, but even in him a deep ambivalence about the Jews as such would remain. "The state's measures against the Jewish people are connected . . . in a very special way with the Church," he declared with regard to the April boycott. "In the Church of Christ, we have never lost sight of the idea that the 'Chosen People,' who nailed the Saviour of the world to the cross, must bear the curse of the action through a long history of suffering."[17] Thus it is precisely a theological view of the Jews that seems to have molded some of Bonhoeffer's pronouncements. Even his friend and biographer Eberhard Bethge could not escape the conclusion that in Bonhoeffer's writings "a theological anti-Judaism is present."[18] Theological anti-Judaism" was not uncommon within the Confessing Church, and some of its most respected personalities, such as Walter Künneth, did not hesitate to equate Nazi and Jewish interpretations of the "Jewish election," as based on race, blood, and *Volk*, in opposition to the Christian view of election by God's grace.[19] Such comparisons were to reappear in Christian anti-Nazi polemics in the mid-thirties and later.

The "Aryan paragraph" applied to only twenty-nine pastors out of eighteen thousand; among these, eleven were excluded from the list because they had fought in World War I. The paragraph was never centrally enforced; its application depended on local church authorities and local Gestapo officials.[20] From the churches' viewpoint, the real debate was about principle and dogma, which excluded unconverted Jews. When, in May 1934, the first national meeting of the Confessing Church took place in Barmen, not a word was uttered about the persecutions: This time not even the converted Jews were mentioned.[21]

On the face of it the Catholic Church's attitude toward the new regime should have been firmer than that of the Protestants. The Catholic hierarchy had expressed a measure of hostility to Hitler's movement during the last years of the republic, but this stance was uniquely determined by church interests and by the varying political fortunes of the Catholic Center Party. The position of many German Catholics toward Nazism before 1933 was fundamentally ambiguous: "Many Catholic publicists . . . pointed to the anti-Christian elements in the Nazi program and declared these incompatible with Catholic teaching. But they went on to speak of the healthy core of Nazism which ought to be appreciated—its reassertion of the values of religion and love of fatherland, its standing as a strong bulwark against atheistic Bolshevism."[22] The general attitude of the Catholic Church regarding the Jewish issue in Germany and elsewhere can be defined as a "moderate anti-Semitism" that supported the struggle against "undue Jewish influence" in the economy and in cultural life. As Vicar-General Mayer of Mainz expressed it, "Hitler in *Mein Kampf* had 'appropriately described' the bad influence of the Jews in press, theater and literature. Still, it was un-Christian to hate other races and to subject the Jews and foreigners to disabilities through discriminatory legislation that would merely bring about reprisals from other countries."[23]

Soon after he took power, and intent on signing a Concordat with the

Vatican, Hitler tried to blunt possible Catholic criticism of his anti-Jewish policies and to shift the burden of the arguments onto the church itself. On April 26 he received Bishop Wilhelm Berning of Osnabrück as delegate from the Conference of Bishops, which was meeting at the time. The Jewish issue did not figure on Berning's agenda, but Hitler made sure to raise it on his own. According to a protocol drafted by the bishop's assistant, Hitler spoke warmly and quietly, now and then emotionally, without a word against the church and with recognition of the bishops: "I have been attacked because of my handling of the Jewish question. The Catholic Church considered the Jews pestilent for fifteen hundred years, put them in ghettos, etc., because it recognized the Jews for what they were. In the epoch of liberalism the danger was no longer recognized. I am moving back toward the time in which a fifteen-hundred-year-long tradition was implemented. I do not set race over religion, but I recognize the representatives of this race as pestilent for the state and for the church and perhaps I am thereby doing Christianity a great service by pushing them out of schools and public functions."[24] The protocol does not record any response by Bishop Berning.

On the occasion of the ratification of the Concordat, in September 1933, Cardinal Secretary of State Pacelli sent a note to the German chargé d'affaires defining the church's position of principle: "The Holy See takes this occasion to add a word on behalf of those German Catholics who themselves have gone over from Judaism to the Christian religion or who are descended in the first generation, or more remotely, from Jews who adopted the Catholic faith, and who for reasons known to the Reich government are likewise suffering from social and economic difficulties."[25] In principle this was to be the consistent position of the Catholic and the Protestant churches, although in practice both submitted to the Nazi measures against converted Jews when they were racially defined as Jews.

The dogmatic confrontation the Catholic hierarchy took up was mainly related to the religious link between Judaism and Christianity. This position found an early expression in five sermons preached by Cardinal Faulhaber during Advent of 1933. Faulhaber rose above the division between Catholics and Protestants when he declared: "We extend our hand to our separated brethren, to defend together with them the holy books of the Old Testament." In Scholder's words: "Faulhaber's sermons were not directed against the practical, political anti-Semitism of the time, but against its principle, the racial anti-Semitism that was attempting to enter the Church."[26] Undoubtedly this was the intention of the sermons and the main thrust of Faulhaber's argumentation, but the careful distinctions established by the cardinal could mislead his audience about his and the church's attitude toward the Jews living among them.

"So that I may be perfectly clear and preclude any possible misunderstanding," Faulhaber declared, "let me begin by making three distinctions. We must first distinguish between the people of Israel before and after the death of Christ. Before the death of Christ, during the period between the calling of Abraham and the fullness of time, the people of Israel were the vehicle of Divine Redemption. . . . It is only with this Israel and the early biblical period that I shall deal in my Advent sermons." The cardinal then described God's dismissal of Israel after Israel had not recognized Christ, adding words that may have sounded hostile to the Jews who did not recognize Christ's revelation: "The daughters of Zion received their bill of divorce and from that time forth, Ahasuerus wanders, forever restless, over the face of the earth." Faulhaber's second theme now followed:

"We must distinguish between the Scriptures of the Old Testament on the one hand and the Talmudic writings of post-Christian Judaism on the other. . . . The Talmudic writings are the work of man; they were not prompted by the spirit of God. It is only the sacred writings of pre-Christian Judaism, not the Talmud, that the Church of the New Testament has accepted as her inheritance.

"Thirdly, we must distinguish in the Old Testament Bible itself between what had only transitory value and what had permanent value. . . . For the purpose of our subject, we are concerned only with those religious, ethical and social values of the Old Testament which remain as values also for Christianity."[27]

Cardinal Faulhaber himself later stressed that, in his Advent sermons, he had wished only to defend the Old Testament and not to comment on contemporary aspects of the Jewish issue.[28] In fact, in the sermons he was using some of the most common clichés of traditional religious anti-Semitism. Ironically enough, a report of the security service of the SS interpreted the sermons as an intervention in favor of the Jews, quoting both foreign newspaper comments and the Jewish Central Association's newspaper, in which Rabbi Leo Baerwald of Munich had written: "We take modest pride that it is through us that revelation was given to the world."[29]

Discussion of the Concordat with the Vatican was item 17 on the agenda of the July 14 cabinet meeting. According to the minutes, the Reich chancellor dismissed any debate about the details of the agreement. "He expressed the opinion that one should only consider it as a great achievement. The Concordat gave Germany an opportunity and created an area of trust which was particularly significant in the developing struggle against international Jewry."[30]

This remark can hardly be interpreted as merely a political ploy aimed at convincing the other members of the government of the necessity of accepting the Concordat without debate, as the fight against world Jewry was certainly not a priority on the conservative ministers' agenda. Thus a chance remark opens an unusual vista on Hitler's thoughts, again pointing toward the trail of his obsession: the "developing struggle" against a global danger—world Jewry. Hitler, moreover, did indeed consider the alliance with the Vatican as being of special significance in this battle. Is it not possible that the Nazi leader believed that the traditional anti-Jewish stance of the Christian churches would also allow for a tacit alliance against the common enemy, or at least offer Nazism the advantage of an "area of trust" in the "developing struggle"? Did Hitler not in fact say as much to Bishop Berning? For a brief instant there appears to be an ominous linkage between the standard procedures of politics and the compulsions of myth.

II

The questionnaire addressed to university professors (in Germany they were civil servants) reached Hermann Kantorowicz, professor of the philosophy and history of law at the University of Kiel, on April 23, 1933. To the question about the racial origins of his grandparents, he replied: "Since there is no time to inquire as to which sense of the term 'race' is being utilized, I shall limit myself to the following declaration: as all four of my grandparents died a long time ago and the necessary measurements, etc., were never made, I am unable to ascertain scientifically (anthropologically) what racial group they belonged to. Understood in its common significance, their race was German, as they all spoke German as their mother tongue, which means that it was Indo-European or Aryan. Their race in the sense of the first supplementary decree to the Law of April 7, 1933, section 2, paragraph 1, sentence 3 was the Jewish religion."[31] One may wonder what made a greater impression on the official who received the filled-out form: the sarcasm or the thoroughness?

It was somewhat gratuitous to send the questionnaire to Kantorowicz, since Minister of Education Bernhard Rust, citing paragraph 3 of the Law for the Restoration of the Professional Civil Service, had already dismissed him on April 14, along with a number of other, mainly Jewish, professors. Sixteen prominent names among them were published in the *Deutsche Allgemeine Zeitung* on that same day.[32] During the year 1933, about twelve hundred Jews holding academic positions would be dismissed.[33]

In Göttingen, where some of the most illustrious members of the theoretical physics and mathematics faculties were Jews (or, in one instance, married to one), each of the three main figures chose a different response: Nobel laureate James Franck sent a public letter of resignation (published in the *Göttinger Zeitung*) but planned to stay in Germany, Max Born (who, after the war, would also receive a Nobel Prize in physics) left the university quietly, and Richard Courant decided to utilize the exception clauses of the law in order to keep his position. Within a few months, however, all three emigrated.[34] In his letter Franck rejected the exemption granted to him as a war veteran because, in his words, "we Germans of Jewish origin are being treated like foreigners and like enemies of our country." Franck's letter led to a public declaration by forty-two of his Göttingen colleagues describing the Jewish physicist's statement as an "act of sabotage" and expressing the hope that "the government would speed up the necessary cleansing measures."[35]

At Tübingen old traditions and new impulses neatly converged. The number of Jewish faculty members dismissed was distinctly low—for a simple reason: No Jew had ever been appointed to a full professorship at this institution, and there were very few Jews among the lower-ranking appointees. Nonetheless, whoever could be expelled was expelled. Hans Bethe, a future Nobel Prize winner in physics, was told to go because of his Jewish mother; the philosophy professor Traugott Konstantin Oesterreich was dismissed on the pretext that he was not politically reliable, but in reality because his wife was of Jewish origin. The same fate almost befell the non-Jewish art historian Georg Weise. The suspicion that Weise's wife was Jewish led to his dismissal, until unimpeachable documentary evidence of Frau Weise-Andrea's Aryan origins was produced and led to Weise's reinstatement.[36]

What happened in Freiburg seems paradigmatic. On April 1 the local Nazi paper, *Der Allemanne,* published lists of Jewish physicians, dentists, and so on, who were to be boycotted; some days later the same paper ran a list of Jewish members of the university medical faculty (the list had been provided by the head of psychiatry). In the meantime, on April 6, the Reich governor of Baden, Robert Wagner, moving ahead of decisions about to be taken in Berlin, ordered the dismissal of Jewish civil servants. On April 10 a delegation of Freiburg University deans and professors traveled to Karlsruhe to plead on behalf of the mayor of Freiburg, who was being threatened with dismissal on political grounds. During their meeting at the ministry, the delegation was reminded that dismissals of Jewish faculty members had to be carried out promptly. According to notes taken by the official in charge of university matters, "The professors promised that the decree would be loyally implemented." It was. On the same day the rector instructed the deans of all schools to dismiss all faculty members of Jewish religion or origin and, for verification, to obtain their signatures on the notices of dismissal. On April 12 the ministry in Karlsruhe was informed that "by 10 A.M. the order had been completely fulfilled." The notification to the Jewish members of the medical school faculty read in its entirety: "According to the order of the academic rectorate, I inform you that, with reference to Ministry Order No. A 7642, you are placed on indefinite leave. Signed: the Dean, Rehn."[37]

In Heidelberg, where the number of professors of Jewish origin was particularly significant, there were attempts at procrastination by the academic senate and the rector, but to no avail. At the beginning of the summer semester of 1933, forty-five "non-Aryans" were still teaching; by August of the same year, only twenty-four were left (those who benefited from the various exception clauses).[38] No organized or individual protests were recorded.

The attitude of some of the privileged non-Aryan scholars was often ambiguous—or worse. On April 25 the Kaiser Wilhelm Society administration in Berlin had been notified by the Ministry of the Interior that all Jewish and half-Jewish department heads and staff members had to be dismissed; directors of institutes were exempted from this measure. Fritz Haber, a Jew and a Nobel laureate, who would have had to dismiss three of his four department heads and five of his thirteen staff members, resigned on April 30. "The other directors (including those who were themselves Jewish) reported their Jewish employees according to instructions."[39] Among those who thus conformed, the Jews Jakob Goldschmidt and Otto Meyerhof, and the half-Jewish Otto Warburg, were the most prominent. For the geneticist Goldschmidt, "Nazism was preferable to Bolshevism," and Otto Warburg, it seems, thought the regime would not last beyond

1934,[40] a belief that did not hamper his retaining his position throughout the whole Nazi period. Warburg's case was strange indeed. His cancer research was so highly valued by the Nazis—apparently even by Hitler himself—that in 1941, when the possibility of dismissal arose because of his half-Jewish origins, he was turned into a quarter Jew on Göring's instructions.[41] As for Meyerhof, he apparently tried to shield some of his Jewish employees, only to be denounced by his codirector, Professor Richard Kuhn.[42] He emigrated in 1938.

It seems, therefore, safe to suggest that when, in January 1934, on the anniversary of the foundation of the German Empire, the Göttingen professor of ancient history Ulrich Karstedt declared that "one should not grumble . . . because in a Jewish shop a window pane has been smashed or because the daughter of the cattle dealer Levi was refused admission to a student corporation,"[43] he was making something of an understatement, not only with regard to the general situation of the Jews in Germany but in the universities as well.[44]

There were a few mild petitions in favor of Jewish colleagues, such as the praise bestowed in May 1933 by the Heidelberg medical faculty on its Jewish members: "We cannot overlook the fact that German Jewry is contributing to great scientific achievements and that major medical personalities come from its midst. Precisely as physicians we feel the duty, keeping in mind the requirements of people and state, to represent the viewpoint of true humanity and to express our worries, as the danger threatens that all sense of responsibility is being pushed aside by emotional or impulsive violence. . . ."[45] This careful declaration was in fact atypical, as the medical schools of German universities showed a much higher percentage of party members than other disciplines.[46] And in its attitude toward Jews, Heidelberg was not basically different from the other German universities.[47]

In April 1933 twelve professors from various fields expressed support for their Jewish colleague, the Munich University philosopher Richard Hönigswald; addressed to the Bavarian Ministry of Education, their letter was backed by the dean of the Munich philosophical faculty. The ministry solicited additional advice and received a set of negative answers, including one from Martin Heidegger, and Hönigswald was dismissed.[48]

Some individual interventions have become well known. There was, for example, Max Planck's (unsuccessful) intervention with Hitler in favor of Fritz Haber's reinstatement[49] and, paradoxically, Heidegger's intervention against the dismissal of Siegfried Tannhäuser and Georg von Hevesy. The dismissal of such eminent scientists, Heidegger explained to the Baden authorities, would have negative consequences abroad and harm Germany's foreign policy.[50]

Heidegger had become rector of Freiburg University in April 1933. He was already on record regarding the presence of Jews in German academic life. In a letter of October 20, 1929, to Victor Schwörer, acting president of the Emergency Fund, established to support needy scholars, the philosopher had stated that the only existing option was either the systematic strengthening of "our" German intellectual life or its definitive abandonment "to growing Judaization in the wider and narrower sense."[51] When Heidegger's mathematics professor, Alfred Löwy, was compelled as a Jew to take early retirement in April 1933, the newly appointed rector wished him "the strength to overcome the hardships and difficulties carried by such times of change."[52] Elfride Heidegger used almost exactly the same words in her letter of April 29, 1933, to Malvine Husserl, the wife of her husband's Jewish mentor, the philosopher Edmund Husserl; she added, however, that although the Civil Service Law was hard, it was reasonable from a German point of view.[53]

Shortly before her departure from Germany in the summer of 1933, Hannah Arendt had written in what was possibly her strongest letter to Heidegger, her teacher and lover, that rumors had reached her about his ever more distant, even hostile attitude toward Jewish colleagues and students. The tone of his answer, as paraphrased by Elzbieta Ettinger, in what would be his last letter to Arendt until after the war, is revealing enough: "To Jewish students . . . he generously gave of his time, disruptive though it was to his own work, getting them stipends and discussing their dissertations with them. Who comes to him in an emergency? A Jew. Who insists on urgently discussing his doctoral degree? A Jew. Who sends him voluminous work for urgent critique? A Jew. Who asks him for help in obtaining grants? Jews!!"[54]

On November 3, 1933, Heidegger announced that economic support would be denied to "Jewish or Marxist" students, or to anyone else defined as a "non-Aryan" according to the new laws.[55] On December 13 he sought financial aid for a volume of pro-Hitler speeches by German professors to be distributed worldwide; he concluded his request with an assurance: "Needless to say, non-Aryans shall not appear on the title page."[56] On the sixteenth of that month, he wrote to the head of the Nazi Professors

Association at Göttingen about Eduard Baumgarten, an ex-student and colleague of his: Baumgarten "frequented, very actively, the Jew Fränkel, who used to teach at Göttingen and was just recently fired from here." Simultaneously Heidegger refused to continue the supervision of doctoral dissertations by Jewish students and referred them to Martin Honecker, a professor of church philosophy.[57]

Heidegger's attitude toward Husserl remains unclear. Although, according to his biographer Rüdiger Safranski, it is untrue that Heidegger forbade Husserl access to the philosophy department, he actually broke all contact with him (as he did with all other Jewish colleagues and disciples) and did nothing to alleviate Husserl's growing isolation. When Husserl died, Heidegger was ill. Would he otherwise have attended the funeral, along with the single "Aryan" faculty member who thought fit to do so, the historian Gerhard Ritter?[58] The dedication of his magnum opus *Being and Time* to Husserl was omitted from the 1944 edition at the publisher's demand, but Heidegger's footnote expression of gratitude to his Jewish mentor was left in. Contradictions abound, with possibly the strangest of them all being Heidegger's praise, in the mid-thirties, for Spinoza, and his declaration that "if Spinoza's philosophy was Jewish, then all philosophy from Leibniz to Hegel was also Jewish."[59]

On April 22, 1933, Heidegger sent an entreaty to Carl Schmitt, by far the most prestigious German political and legal theorist of the time, pleading with him not to turn his back on the new movement. The entreaty was superfluous, as Schmitt had already made his choice. Like Heidegger—and this seems to have been the first rule to follow—he had stopped answering letters from Jewish students, colleagues, and other scholars with whom he had previously been in close touch (in Schmitt's case, one of the striking examples is the abrupt end he put to his correspondence with the Jewish political philosopher Leo Strauss).[60] And, to make sure that there was no misunderstanding about where he stood, Schmitt introduced some outright anti-Semitic remarks into the new (1933) edition of his *Concept of the Political*.[61] In any event, Schmitt's anti-Jewish positions were to be definitely more outspoken, extreme, and virulent than those of the Freiburg philosopher.

During the summer semester of 1933, both Schmitt and Heidegger took part in a lecture series organized by Heidelberg students. Heidegger spoke on "The University in the New Reich"; Schmitt's theme was "The New Constitutional Law." They were preceded in the same series by Dr.

Walter Gross, head of the racial policy office of the Nazi Party, who spoke on "The Physician and the Racial Community." On May 1, in Freiburg, Heidegger had become party member 3-125-894; on the same day, in Cologne, Schmitt joined the party as member 2-098-860.[62]

Hannah Arendt left the country and, by way of Prague and Geneva, reached Paris; there she soon started working for the Zionist Youth Emigration Organization. The main reason for her early emigration, she later said, was more than anything else the behavior of her Aryan friends, such as Benno von Wiese, who—subject to no outside pressure whatso-ever—adhered enthusiastically to the new system's ideals and norms.[63] Yet, in general, her criticism of Heidegger remained muted.

The responses of Jewish academics to the new regime's measures and to the new attitudes of colleagues and friends varied from one individual to another. Within that broad spectrum, a peculiar situation was that of Jews who had been long-standing militant German nationalists but, unlike Felix Jacoby, did not opt for total blindness to the regime's actions. On April 20 Ernst Kantorowicz, a medieval historian at Frankfurt University, sent a letter to the minister of science and education of Hesse that tellingly expresses the great slowness, hesitancy, and regretfulness—despite the harsh new Nazi policies—of the retreat of such Jews from their former positions. "Although," Kantorowicz wrote, "as a war volunteer from August 1914 on, as a frontline soldier throughout the war, as a post-war fighter against Poland, against the Spartacists, and against the Republic of the Councils [of workers and soldiers] in Posen, Berlin, and Munich, I am not obliged to expect dismissal because of my Jewish origins; although in view of my publications on the Hohenstaufen Emperor Frederick II, I do not need any attestation from the day before yesterday, yesterday, or today regarding my attitude toward a nationally oriented Germany; although beyond all immediate trends and occurrences, my fundamentally positive attitude toward a nationally governed Reich has not been undermined even by the most recent events; and although I certainly need not expect student disturbances to interrupt my teaching—so that the issue of unhampered teaching at the level of the entire university need not be considered in my case—as a Jew, I see myself compelled nonetheless to draw the consequences of what has happened and give up my teaching for the coming summer semester."[64] Kantorowicz was not tendering his resignation; he was merely withdrawing from the next semester. The implication was that he would wait for the policies of the new national Germany to change.

Whereas the attitude of the majority of "Aryan" university professors could be defined as "cultured Judeophobia,"[65] among the students a radical brand of Judeophobia had taken hold. At the end of the nineteenth century, some Austrian student corporations, followed by German ones, had already excluded Jews on a racial basis—that is, even baptized Jews were not accepted.[66] Michael Kater attributes a portion of extreme student anti-Semitism to competition—mainly in the remunerative fields of law and medicine, in which the percentage of Jewish students was indeed high, as was the percentage of Jews in these professions. In any case, in the early years of the Weimar Republic the majority of German student fraternities joined the German University League (Deutscher Hochschulring), an organization with openly *völkisch* and anti-Semitic aims, which soon came to control student politics.[67] Membership in the league was conditional on fully Aryan origin, with racial Germans from Austria or the Sudetenland accepted despite their not being German citizens. The league dominated the universities until the mid-1920s, when it was replaced by the National Socialist Students Association (*Nationalsozialistischer Deutscher Studentenbund*).[68] And demonstrations and acts of physical aggression by right-wing students against their enemies became common on German campuses from the late twenties on.[69]

Soon professors who were too explicitly pacifist or anti-nationalist, such as Theodor Lessing, Günther Dehn, Emil Julius Gumbel, Hans Nawiasky, and Ernst Cohn, came under attack.[70] Gumbel was driven out of Heidelberg even before the Nazis came to power. In 1931 Nazis gained a majority in the German Student Association (Deutsche Studentenschaft); this was the first national association to come under their control. Within a short time a whole cohort of young intellectuals would put its energy and ability at the disposal of the party and its policies.[71]

After January 1933 student groups took matters into their own hands, not unlike the SA. The national leader of the Nazi student organization, Oskar Stabel, announced shortly before the April 1 boycott that student pickets would be posted that day at the entrances to Jewish professors' lecture halls and seminar rooms in order to "dissuade" anyone from entering.[72] Such was the case, for example, at the Technical University in Berlin. Later on Nazi students with cameras positioned themselves on the podiums of lecture halls so as to take pictures of students attending classes taught by Jews.[73] This kind of student agitation was strongly encouraged by a violently anti-Jewish speech delivered on May 5 by Education Minister Rust in the Berlin university auditorium, and by such comments on the speech as these in the official *Preussische Zeitung*: "Science for a Jew does not mean a task, an obligation, a domain of creative organization, but a business and a way of destroying the culture of the host people. Thus the most important chairs of so-called German universities were filled with Jews. Positions were vacated to allow them to pursue their parasitic activities, which were then rewarded with Nobel Prizes."[74]

In early April 1933, the National Socialist Student Association established a press and propaganda section. Its very first measure, decided on April 8, was to be "the public burning of destructive Jewish writing" by university students as a reaction to world Jewry's "shameless incitement" against Germany. An "information" campaign was to be undertaken between April 12 and May 10; the public burnings were scheduled to start on university campuses at 6:00 P.M. on the last day of the campaign.

The notorious twelve theses the students prepared for ritual declamation during the burnings were not exclusively directed against Jews and the "Jewish spirit": Among the other targets were Marxism, pacifism, and the "overstressing of the instinctual life" (that is, "the Freudian School and its journal *Imago*"). It was a rebellion of the German against the "un-German spirit." But the main thrust of the action remained essentially anti-Jewish; in the eyes of the organizers, it was meant to extend anti-Jewish action from the economic domain (the April 1 boycott) to the entire field of German culture.

On April 13 the theses were affixed to university buildings and billboards all over Germany. Thesis 7 read: "When the Jew writes in German, he lies. He should be compelled, from now on, to indicate on books he wishes to publish in German: 'translated from the Hebrew.'"[75]

On the evening of May 10, rituals of exorcism took place in most of the university cities and towns of Germany. More than twenty thousand books were burned in Berlin, and from two to three thousand in every other major German city.[76] In Berlin a huge bonfire was lit in front of the Kroll Opera House, and Goebbels was one of the speakers. After the speeches, in the capital as in the other cities, slogans against the banned authors were chanted by the throng as the poisonous books (by Karl Marx, Ferdinand Lassalle, Sigmund Freud, Maximilian Harden, and Kurt Tucholsky, among many others) were hurled, batch after batch, into the flames. "The great searchlights on the Opera Square," wrote the *Jüdische Rundschau*, "also threw their light onto the swallowing up of our existence and our fate. Not only Jews have been accused, but also men of pure German blood. The latter are being judged only for their deeds. For Jews, however, there is no need for a specific reason; the old saying holds: 'The Jew will be burnt.'"[77]

The Nazi students did not limit their activities to disrupting the lectures of Jewish professors and burning dangerous books. They attempted to impose their will at every level when it came to the hiring of teachers or their reinstatement as war veterans. On May 6 the leader of the Nazi student association of the Technical University in Hildburghausen, Thuringia, sent an anything but subservient letter to the Thuringian education minister in Weimar. The students had been told that a Jewish teacher named Bermann was to be reinstated. After casting doubt on the validity of Bermann's claim to frontline service during World War I, the student leader went on: "Agitation among the students is very strong, as some forty percent are members of the National Socialist Student Association, and to be taught by a racially alien teacher is incompatible with their convictions. The National Socialist Student Association addresses the urgent demand to the National Socialist government of Thuringia not to reinstate the Jewish teacher."[78] Whether Bermann was reinstated or not is not known, but even seasoned Nazis considered the student activism something of an embarrassment. "I have been informed by State Minister of the Interior, Party member Fritsch," wrote one of the district leaders for central Germany to Manfred von Killinger, prime minister of Saxony, on August 12, "that the State Ministry is not pleased with the situation at the University of Leipzig. . . . Over the last three months I have fought rigorously and consistently against any radicalization of the university. According to your wishes, I have therefore forbidden the National Socialist students to boycott any professors."[79]

Sometimes students themselves perceived that they had gone too far: They had even blacklisted H. G. Wells and Upton Sinclair. The Foreign Ministry was up in arms because among the authors whose works had been burned in front of the Kroll Opera House on May 10 was the then famous promoter of European union, Count Richard Cudenhove-Kalergi. The student leader Gerhard Gräfe confided to a correspondent that he was denying that Cudenhove's writings were burned, but that precautions would have to be taken in the future.[80] Such reservations also took other forms: In his diary entries for 1933, Victor Klemperer, a Jewish professor of Romance literature at Dresden's Technical University who had been exempted from dismissal owing to wartime combat service, several times mentioned that the most assiduous participant in his seminar was the female leader of the university's Nazi student cell.[81]

A comparison between the attitudes of the churches and those of the universities toward the regime's anti-Jewish measures of 1933 reveals basic similarities along with some (very) minor differences. Although outright supporters of National Socialism as a whole were a small minority both in the churches and in the universities, those in favor of the national revival heralded by the new regime were definitely a majority. That majority shared a conservative-nationalist credo that easily converged with the main ideals proclaimed by the regime at its beginning. But what distinguished the churches' attitudes was the existence of certain specific interests involving the preservation of some basic tenets of Christian dogma. The Jews as Jews were abandoned to their fate, but both the Protestant and Catholic churches attempted to maintain the preeminence of such fundamental beliefs as the supersession of race by baptism and the sanctity of the Old Testament. (Later, at times, the private attitudes of Catholics and of members of the Confessing Church toward the persecution of the Jews would even be critical, mainly because of growing tension between them and the regime.) Nothing of the kind hampered acceptance by university professors of the regime's anti-Jewish acts. In principle the German academic elite was committed to pursuit of learning unimpeded by state intervention, but, as has been seen, other values and beliefs weighed far more heavily with it in the twenties and early thirties. The enthusiastic "self-coordination" (*Selbstgleichschaltung*) of the universities demonstrated that there was no fundamental opposition but rather a substantial measure of convergence between the inner core of the mandarins' faith and National Socialism's public stance as it appeared at the outset. In such a context, motivation for taking a stand in favor of Jewish colleagues and students was minimal. The consequences of such an overall moral collapse are obvious. In many ways elite groups were a bridge between National Socialist extremism and the wider reaches of German society; thus, their ready abandonment of the Jews sets their attitudes and responses in a fateful historical light.

When Pastor Umfried criticized the attack on the Jews of his town, no church authority supported him; when Jewish businesses were boycotted, no religious voice was heard; when Hitler launched his diatribe against the Jews, Bishop Berning did not respond. When Jewish colleagues were dismissed, no German professor publicly protested; when the number of Jewish students was drastically reduced, no university committee or faculty member expressed any opposition; when books were burned throughout the Reich, no intellectual in Germany, or for that matter anyone else within the country, openly expressed any shame. Such total collapse is more than unusual. As the first months of 1933 went by, Hitler must have seen that he could count on the genuine support of church and university; whatever opposition may have existed, it would not be expressed as long as direct institutional interests and basic dogmatic tenets were not threatened. The concrete situation of the Jews was a litmus test of how far any genuine moral principle could be silenced; although the situation was to become more complex later on, during this early period the result of the test was clear.

III

While Germany's intellectual and spiritual elites were granting their explicit or tacit support to the new regime, the leading figures of the Jewish community were trying to hide their distress behind a façade of confidence: Despite all difficulties, the future of Jewish life in Germany was not being irretrievably endangered. Ismar Elbogen, one of the most prominent Jewish historians of the time, expressed what was probably the most common attitude when he wrote: "They can condemn us to hunger but they cannot condemn us to starvation."[82] This was the spirit that presided over the establishment of the National Representation of German Jews (Reichsvertretung der deutschen Juden), formally launched in 1933, on the initiative of the president and the rabbi of the Essen community.[83] It would remain the umbrella organization of local and national Jewish associations until 1938, headed throughout by the Berlin rabbi Leo Baeck, the respected chairman of the Association of German Rabbis and a scholar of repute,[84] and by the lay leader Otto Hirsch. Despite opposition from "national German Jews," ultra-Orthodox religious groups, and, sporadically, from the Zionist movement, the National Representation played a significant role in the affairs of German Jewry until its transformation, after a transition period in 1938–39, into the National Association of Jews in Germany (Reichsvereinigung der Juden in Deutschland), an organization very closely controlled by the Gestapo.

There was not any greater sense of urgency at the National Representation than there was among most individual Jews in Germany. In early 1934 Otto Hirsch would still be speaking out against "hasty" emigration: He believed in the possibility of maintaining a dignified Jewish life in the new Germany.[85] That Alfred Hirschberg, the most prominent personality of the Central Association, denied "any need at all to enlarge upon the utopia of resettlement [in Palestine]" was true to type, but that a publication of the Zionist Pioneer organization defined unprepared immigration to Eretz Israel as "a crime against Zionism" comes as a surprise, perhaps because of the vehemence of its tone.[86]

Not all German Jewish leaders displayed such nonchalance. One who insistently demanded immediate emigration was Georg Kareski, head of the right-wing [Revisionist] Zionist Organization. A vocal but marginal personality even within German Zionism, Kareski was ready to organize the exodus of the Jews from Germany by cooperating, if need be, with the Gestapo and the Propaganda Ministry. He may indeed have maneuvered to establish his own authority within German Jewry by exploiting his collaboration with the Nazis,[87] but his sense of urgency was real and premonitory.

Even as the months went by, the leaders of German Jewry did not, in general, gain much insight into the uncompromisingly anti-Jewish stance of the Nazis. Thus, in August 1933, Werner Senator, who had returned to Germany from Palestine in order to become a director of the newly established Central Committee for Help and Reconstruction (Zentralausschuss für Hilfe und Aufbau), suggested, in a memorandum sent to the American Joint Distribution Committee, that a dialogue be established between the Jews and the Nazis. In his opinion, such a dialogue "should lead to a kind of Concordat, like the arrangements between the Roman Curia and European States."[88]

No Roman Curia and no Concordat were mentioned as examples in the "Memorandum on the Jewish Question" that the representatives of Orthodox Jewry sent to Hitler on October 4. The signatories brought to the Reich chancellor's attention the injustice of the identification of Jewry with Marxist materialism, the unfairness of the attribution to an entire community of the mistakes of some of its members, and the tenuousness of the connection between the ancient Jewish race and the modern, uprooted, ultrarationalistic Jewish writers and journalists. Orthodox Jewry disavowed the atrocity propaganda being directed against Germany, and its delegates reminded Hitler of the Jewish sacrifices during World War I. The authors of the letter were convinced that the new government did not have in mind the annihilation of German Jewry, but in case they were wrong on this point, they demanded to be told so. Again, on the assumption that such was not the aim of the regime, the representatives of Orthodox Jewry demanded that the Jews of Germany be granted a living space within the living space of the German people, where they could practice their religion and follow their professions "without being endangered and insulted." The memorandum was filed before it even reached Hitler's desk.[89]

Thirty-seven thousand of the approximately 525,000 Jews in Germany left the country in 1933; during the four following years, the annual number of emigrants remained much lower than that (23,000 in 1934, 21,000 in 1935, 25,000 in 1936, 23,000 in 1937).[90] In 1933 about 73 percent of the emigrants left for countries in Western Europe, 19 percent for Palestine, and 8 percent chose to go overseas.[91] Such seeming lack of enthusiasm for leaving a country where segregation, humiliation, and a whole array of persecutory measures were becoming steadily worse was due, first of all, to the inability of most of the Jewish leadership and mainly of ordinary German Jews to grasp an essentially unpredictable course of events. "I do not believe," Klaus Mann wrote in his autobiography, "that the insights of shopkeeper Moritz Cohn differ basically from those of his neighbor, the shopkeeper Friedrich Müller."[92] Most of the Jews expected to weather the storm in Germany. In addition, the material difficulty of emigrating was considerable, especially in a period of economic uncertainty; it entailed an immediate and heavy material loss: Jewish-owned property was sold at ever lower prices, and the emigration tax (the Brüning government's 1931 "tax on capital flight," which was levied on assets of two hundred thousand Reichsmarks and up, was raised by the Nazis to a levy on assets of fifty thousand Reichsmarks and up) was prohibitive. The Reichsbank's purely arbitrary exchange rate for the purchase of foreign currency by emigrants further depleted steadily shrinking assets: Thus, until 1935, Jewish emigrants exchanged their marks at 50 percent of their value, then at 30 percent, and finally, on the eve of the war, at 4 percent.[93] Although the Nazis wanted to get rid of the Jews of Germany, they were intent on dispossessing them first by increasingly harsh methods.

In one instance only were the economic conditions of emigration somewhat facilitated. Not only did the regime encourage Zionist activities on the territory of the Reich[94], but concrete economic measures were taken to ease the departure of Jews for Palestine. The so-called Haavarah (Hebrew: Transfer) Agreement, concluded on August 27, 1933, between the German Ministry of the Economy and Zionist representatives from Germany and Palestine, allowed Jewish emigrants indirect transfer of part of their assets and facilitated exports of goods from Nazi Germany to Palestine.[95] As a result, some one hundred million Reichsmarks were transferred to Palestine, and most of the sixty thousand German Jews who arrived in that country during 1933–39 could thereby ensure a minimal basis for their material existence.[96]

Economic agreement and some measure of cooperation in easing Jewish emigration from Germany (and in 1938 and 1939) from post-Anschluss Austria and German-occupied Bohemia-Moravia) to Palestine, were of course purely instrumental. The Zionists had no doubts about the Nazis' evil designs on the Jews, and the Nazis considered the Zionists first and foremost Jews. About Zionism itself, moreover, Nazi ideology and Nazi policies were divided from the outset: while favoring, like all other European extreme anti-Semites, Zionism as a means of enticing the Jews to leave Europe, they also considered the Zionist Organization established in Basel in 1897 as a key element of the Jewish world conspiracy—a Jewish state in Palestine would be a kind of Vatican coordinating Jewish scheming all over the world. Such necessary but unholy contacts between Zionists and Nazis nonetheless continued up to the beginning (and into) the war.

One of the main benefits the new regime hoped to reap from the Haavarah was a breach in the foreign Jewish economic boycott of Germany. The Nazi fears of a significant Jewish boycott were, in fact, basically unreal, but Zionist policy responded to what the Germans hoped to achieve. The Zionist organizations and the leadership of the Yishuv (the Jewish community in Palestine) distanced themselves from any form of mass protest or boycott to avoid creating obstacles to the new arrangements. Even before the conclusion of the Haavarah Agreement, such "cooperation" sometimes took bizarre forms. Thus, in early 1933, Baron Leopold Itz Edler von Mildenstein, a man who a few years later was to become chief of the Jewish section of the SD (the Sicherheitsdienst, or security service, the SS intelligence branch headed by Reinhard Heydrich), was invited along with his wife to tour Palestine and write a series of articles for Goebbels's *Der Angriff.* And so it was that the Mildensteins, accompanied by Kurt Tuchler, a leading member of the Berlin Zionist Organization, and his wife, visited Jewish settlements in Eretz Israel. The highly positive articles, entitled "A Nazi Visits Palestine," were duly published, and, to mark the occasion, a special medallion cast, with a swastika on one side and a Star of David on the other.[97]

Seen from the perspective of 1933 and in the light of Nazi interests at the time, the *Angriff* series may have looked less strange than they appear today. The same can be said about the memorandum sent to Hitler by the leaders of the Zionist Organization for Germany on June 22, 1933. In Francis Nicosia's words, "It seemed to profess a degree of sympathy for the *völkisch* principles of the Hitler regime and argued that Zionism was compatible with these principles."[98] This compatibility was clearly defined: "Zionism believes that the rebirth of the national life of a people, which is now occurring in Germany through the emphasis on its Christian and national character, must also come about among the Jewish people. For the Jewish people, too, national origin, religion, common destiny and a sense of its uniqueness must be of decisive importance to its existence. This demands the elimination of the egotistical individualism of the liberal era, and its replacement with a sense of community and collective responsibility."[99] It further demanded for the Jews a place in the overall structure, based on the race principle, established by National Socialism, so that they too, in the sphere allocated to them, could make a fruitful contribution to the life of the fatherland.[100]

In the summer of 1933, one of the main Zionist leaders in Palestine, the German-born Arthur Ruppin, paid a visit to the Nazi race theoretician Hans F. K. Günther at the University of Jena. "The Jews," Günther reassured him, "were not inferior to the Aryans, they were simply different. This meant that a 'fair solution' had to be found for the Jewish problem. The professor was extremely friendly, Ruppin recorded with satisfaction."[101] Thus, despite rapid awareness of the Nazis' unmitigated hatred of Jews, some Zionist leaders' early responses to the new German situation were not negative. There was a widespread hope that the Nazi policy of furthering Jewish emigration from Germany offered great opportunities for the Yishuv. A stream of important visitors came from Palestine to observe conditions in Germany. The Labor Zionist leader Moshe Belinson reported to Berl Katzenelson, the editor of the main Labor daily, *Davar*: "The streets are paved with more money than we have ever dreamed of in the history of our Zionist enterprise. Here is an opportunity to build and flourish like none we have ever had or ever will have."[102]

Zionist hopes were moderated by practical worries about excessive numbers of immigrants. "In order that the immigration not flood the existing settlement in Palestine like lava," Ruppin declared at the Zionist Congress held in Prague in the summer of 1933, "it must be proportionate to a certain percentage of that settlement."[103] This remained the policy for several years to come, and well after the passage of the 1935 Nuremberg racial laws, both the German Zionists and the leaders of the Yishuv were still envisaging an annual rate of fifteen to twenty thousand German-Jewish emigrants, extending over a period of twenty to thirty years.[104]

Whatever the practical steps that were envisioned, Zionist rhetoric was clear: Palestine was the only possible haven and solution. This was not obvious to some of the German Jews, who, on arrival in the land of Israel, were suddenly faced with a new and unexpected reality. The novelist Arnold Zweig, a left-wing Zionist of long standing who had arrived in the summer of 1933, summed up his feelings about his new homeland in a diary entry on December 31: "In Palestine. In foreign parts."[105]

Some leaders of German Jewry still believed in 1933 that the Nazis would be duly impressed by an objective presentation of Jewish contributions to German culture. A few months after the change of regime, and with Max Warburg's and Leo Baeck's encouragement, Leopold Ullstein, a younger member of the publishing family, launched the preparation of a wide-ranging study to that effect. Within a year a hefty volume was ready, but in December 1934 its publication was prohibited. "The naïve reader of this study," the Gestapo report pronounced, "would get the impression that the whole of German culture up to the National Socialist revolution was carried by Jews. The reader would receive an entirely false picture of the real activity, particularly of the decomposing action of the Jews on German culture. Moreover, well-known Jewish crooks and speculators are presented to the reader as victims of their time and their dirty dealings glossed over. . . . In addition, Jews generally known as enemies of the state . . . are presented as remarkable carriers of German culture."[106] Jewish culture for Jews, however, was another matter, and whereas Ullstein had set his sights on an untimely enterprise, another Berlin Jew, Kurt Singer, the former deputy director of the Berlin City Opera, came up with a different kind of idea: the establishment of a Jewish cultural association (Kulturbund deutscher Juden).

Singer's Kulturbund fitted Nazi needs. When Singer's project of autonomous cultural activities by Jews and for Jews (only) was submitted to the new Prussian authorities, it received Göring's approval. For all practical purposes, it was controlled on the Nazi side by the same Hans Hinkel who was already in charge of the de-Judaization of cultural life in Prussia. On the face of it the Kulturbund appeared to be a perfectly functional initiative to solve the problems created both for the regime and for the Jews by the expulsion from German cultural life of approximately eight thousand Jewish writers, artists, musicians, and performers of all kinds, as well as their coworkers and agents.[107] Apart from the work it provided and the soothing psychological function it filled for part of the Jewish community, the Kulturbund also offered to the surrounding society an easy way to dismiss any potential sense of embarrassment: "Aryans who found the regime's anti-Semitic measures distasteful could reassure themselves that Jewish artists were at least permitted to remain active in their chosen professions."[108]

The Kulturbund also played another role, unseen but no less real, which pointed to the future: As the first Jewish organization under the direct supervision of a Nazi overlord, it foreshadowed the Nazi ghetto, in which a pretense of internal autonomy camouflaged the total subordination of an appointed Jewish leadership to the dictates of its masters. The Kulturbund was hailed by an array of Jewish intellectuals as offering the opportunity for a new Jewish cultural and spiritual life to a community under siege.[109] This ongoing misunderstanding of the true meaning of the situation was compounded by the ambition of some of its founders: to create a cultural life of such quality that it would teach the Germans a lesson. The literary critic Julius Bab summed it all up with extraordinary naïveté when he wrote in a letter of June 1933: "It remains a bitter fact—it is a ghetto enterprise, but one that we certainly want to accomplish so well that the Germans will have to be ashamed."[110] Bab's statement could also mean that the Germans would feel ashamed to be treating the carriers of such high culture so shabbily.

Sporadically Hinkel would inform his wards of works Jews were no longer allowed to perform. In the theater Germanic legends and performances of works from the German Middle Ages and German romanticism were prohibited. For a time the classical period was allowed, but Schiller was forbidden in 1934 and Goethe in 1936. Among foreign writers Shakespeare was allowed, but Hamlet's "To be or not to be" soliloquy was forbidden: In a Jewish theater in the Third Reich, "the oppressor's wrong, the proud man's contumely" could have sounded subversive, hence that line led to the exclusion of the entire speech.[111] Needless to say, despite the attachment of German Jews to the works of Richard Wagner and Richard Strauss, these composers were not to be performed by Jews. Beethoven was forbidden them in 1937, but Mozart had to wait until the next year, after the Anschluss.[112]

Such growing constraints notwithstanding, the activity of the Kulturbund, both in Berlin and, soon after, in all major German cities, was remarkable. More than 180,000 Jews from all parts of Germany became active members of the association. In its first year the Kulturbund staged 69 opera performances and 117 concerts, and, from mid-1934 to mid-1935, 57 opera performances and 358 concerts.[113] The opera repertory included works by Mozart, Offenbach, Verdi, Johann Strauss, Donizetti, Rossini, Tchaikovsky, and Saint-Saëns, among others. Although, apart from the ideological and financial constraints, the choice of works performed was mainly traditional, in 1934 the Frankfurt Kulturbund organized a concert in honor of Arnold Schoenberg's sixtieth birthday, and the Cologne branch organized a performance of Paul Hindemith's children's opera *Wir bauen eine Stadt* (We're building a town)—locating it in Palestine.[114]

In principle the Jews were increasingly to be fed on "Jewish works." But even this principle did not always satisfy the Nazi mind. On October 26, 1933, Rainer Schlösser, the Reich director of theaters of the Ministry of Propaganda, recommended to Hinkel that performances of Emil Bernhard's (Emil Cohn's) *Die Jagd Gottes* (God's hunt) be forbidden, as the play was "a kind of 'consolation for the Jews,' a kind of 'heartening' for the Jews." Moreover, the action took place against a background of mistreatment of Jews by Cossacks: "It is easy to imagine with whom these Cossacks would be identified."[115]

Jewish audiences must have been partly aware that Kulturbund activities were intended to have a soothing effect on them. Nonetheless, Kulturbund theaters like the one on Berlin's Charlottenstrasse became a spiritual lifeline. The tram conductors knew their public; "Charlottenstrasse," they would call out. "Jewish culture—everybody off!"[116]

"The goal of our stage," declared the director of theater activities of the Rhine-Ruhr Kulturbund in the November/December 1933 issue of its periodical, "is to bring to all the joy and courage to face life by letting them participate in the eternal values of poetry or by discussing the problems of our time, but also by showing lighthearted pieces and not rejecting them. We intend to keep the connection with the German *Heimat* [homeland]

and to form at the same time a connecting link with our great Jewish past and with a future that is worth living for."[117]

IV

By the end of 1933, tens of millions of people inside and outside Germany were aware of the systematic policy of segregation and persecution launched by the new German regime against its Jewish citizens. Yet, as already noted at the outset, it may have been impossible for most people, Jews and non-Jews alike, to have a clear idea of the goals and limits of this policy. There was anxiety among the Jews of Germany but no panic or any widespread sense of urgency. It is hard to evaluate how much importance German society at its various levels granted to an issue that was not on any priority list. Political stabilization, the dismantling of the Left, economic improvement, national revival, and international uncertainties were certainly more present in the minds of many than the hazy outlines of the Jewish issue; for most Germans the issues and challenges of daily life in times of political change and of economic turmoil were the paramount focus of interest, whatever their awareness of other problems may have been. It is against this background that Hitler's own obsession with the Jewish issue must be considered.

In a remarkable dispatch sent to Foreign Minister Sir John Simon on May 11, 1933, the British ambassador in Berlin, Sir Horace Rumbold, described the course taken by an interview with Hitler once he had alluded to the persecution of the Jews: "The allusion to the treatment of the Jews resulted in the Chancellor working himself up into a state of great excitement. 'I will never agree,' he shouted, as if he were addressing an open-air meeting, 'to the existence of two kinds of law for German nationals. There is an immense amount of unemployment in Germany, and I have, for instance, to turn away youths of pure German stock from higher education. There are not enough posts for the pure-bred Germans, and the Jews must suffer with the rest. If the Jews engineer a boycott of German goods from abroad, I will take care that this hits the Jews in Germany.' These remarks were delivered with great ferocity. Only his excitement, to which I did not wish to add, prevented me from pointing out that there were, in fact, two standards of treatment of German nationals, inasmuch as those of Jewish race were being discriminated against." At the end of the dispatch, Rumbold returned to the issue: "My comment on the foregoing is that Herr Hitler is himself responsible for the anti-

Jewish policy of the German government and that it would be a mistake to believe that it is the policy of his wilder men whom he has difficulty in controlling. Anybody who has had the opportunity of listening to his remarks on the subject of Jews could not have failed, like myself, to realize that he is a fanatic on the subject."[118]

The American consul general in Berlin reached the same conclusion." One of the most unfortunate features of the situation," George S. Messersmith wrote to Secretary of State Cordell Hull on November 1, 1933, "is that, as I have already pointed out in previous dispatches and again in this one, Mr. Hitler himself is implacable and *unconvinced* and is the real head of the anti-Jewish movement. He can be reasonable on a number of subjects, but on this he can only be passionate and prejudiced."[119]

Hitler did not express his obsession with the Jewish peril in major public utterances during 1933, but its lurking presence can be perceived in his remarks about the Concordat, in the last part of the letter to Hindenburg, in the discussion with Bishop Berning, as well as in outbursts such as those reported by foreign diplomats. It is no less apparent, however, that the new chancellor was not yet sure of the leeway granted him by the shifting political and economic situation. International reactions did concern him. As he put it in his meeting with the Reich district governors, on July 6, 1933, for Germany the most dangerous front at the time was the external one: "One should not irritate it, when it is not necessary to deal with it. To reopen the Jewish question would mean to start a world-wide uproar again."[120] Clearly the shaky economic circumstances of the Reich were also a major factor in his decisions, as already noted. Once the bumbling minister of the Economy, Alfred Hugenberg, and his ineffective successor Kurt Schmitt, had been eased out, Hitler, on July 30, 1934, appointed Hjalmar Schacht, the conservative "wizard," as minister and overlord of the economy of the Reich. For practical economic reasons, Schacht insisted that no major interference with Jewish business would be allowed.[121] In general terms Hitler backed Schacht's position until the new transition period of 1936–37. Finally, on some matters such as the issue of Jewish physicians, Hitler certainly took into account German public opinion: In other words he understood the need for tactical pragmatism regarding immediate anti-Jewish measures, and thus his policy had to remain, for a time at least, close to the preexisting anti-Jewish agenda of the conservatives.

The extent to which Hitler was torn between hatred of the Jews and desire for radical action on the one hand, and the need for tactical restraint on the other, emerged clearly in the July 14 cabinet meeting, at which he declared that the Concordat with the Vatican would help the Reich in its struggle against world Jewry. When, during discussion of criteria for the continued exercise of the legal profession by Jews, several ministers suggested that the identification of frontline veterans should be based on membership in combat units, Hitler protested: "The Jewish nation in its totality was being rejected. Therefore all Jews had to be dismissed [from the professions]. One could make an exception only for those who had taken part in direct combat. Only participation in combat, and not mere presence in the combat zone, was decisive. A commission to check the rolls of the various units was necessary."[122] But how was this need for an ongoing struggle against the Jews to find its expression in the economic sphere, for example, without leading to the dangerous results of which Hitler was well aware? When the topic was raised at the same cabinet meeting, the Reich chancellor launched into an explanation that clearly laid bare the dilemma facing him. "The Jews were continuing their silent boycott of Germany," Hitler explained, "and their aim was to bring about the downfall of the present regime. Therefore it was only just that the Jews in Germany be the first to feel the effects of this boycott. There were too many enterprises in Germany, and clearly some would have to disappear. In this situation the opponent, Jewish enterprises, had to be the first to go—equal treatment in this domain would be wrong."[123] In other words, Jewish enterprises had to be discriminated against—to a point: Within the category of those enterprises that had to disappear, the Jewish ones were to be the first on the list. Such a statement could be read in many ways.

That Hitler also manipulated the Jewish issue in order to achieve some general political goals is not impossible. Although the economic boycott of Jewish businesses had to stop, at least officially, the menacing party rhetoric clearly indicated that henceforward the Jews were considered potential hostages whose fate would depend upon the outside world's attitude toward the new Germany. Such use of the Jews would, incidentally, remain as a threatening theme throughout the thirties and find its most violent expression after the Kristallnacht pogrom of November 1938 and during the last months of peace, particularly in Hitler's Reichstag speech of January 1939. Moreover, the April 1933 boycott and the other early anti-Jewish measures allowed for some release of the pent-up violence simmering among the "party radicals." Throughout the coming months and years, but particularly in 1935, Hitler would use his anti-Jewish policies as a safety valve against a buildup of ideological or material resentment among the party rank and file and the more extreme of his underlings.

Finally, as far as the regime's first year is concerned, were there any indications that—beyond general ideological obsession and immediate tactics—Hitler was already considering further systematic steps against the Jews of Germany? It seems indeed that the idea of establishing a fundamental legal distinction between German (Aryan) citizens and the Jews living in Germany, a staple demand of many conservatives in the past and an item of the Nazi Party program, was on both the conservative civil service's agenda and Hitler's mind from the very outset of his government.

The first draft of a new citizenship law appears to have emerged in the Ministry of the Interior at the end of May 1933, and was submitted to the Committee of Experts for Population and Racial Policy of the Ministry in the following month.[124] No immediate results came of these efforts, but Hitler, it seems, was considering similar plans for the future. Thus, at a September 28 meeting with the minister of the interior and the Reich district governors, "Hitler explained that he would have preferred to take a step-by-step approach toward sharpening anti-Jewish measures; this could have been achieved had a citizenship law been established which would then have allowed him to take further, sharper steps. However, the boycott started by the Jews had called for an immediate and very sharp reaction."[125]

As will be seen, even in the atmosphere of uncertainty following his accession to power, Hitler did not lose sight of his ideological goals with regard to the Jews, as well as in relation to the other issues that formed the core of his worldview. Although he avoided public statements on the Jewish issue, he could not restrain himself entirely. In his closing speech at the September 1933 Nuremberg party rally, called (for the occasion) the Congress of Victory, he launched into disparaging comments about the Jews in his expostulations on the racial foundations of art: "It is a sign of the horrible spiritual decadence of the past epoch that one spoke of styles without recognizing their racial determinants. . . . Each clearly formed race has its own handwriting in the book of art, insofar as it is not, like Jewry, devoid of any creative artistic ability."[126] As for the function of a worldview, Hitler defined it in his address: "Worldviews," he declared, "consider the achievement of political power only as the precondition for the beginning of the fulfillment of their true mission. In the very term 'worldview' there lies the solemn commitment to make all enterprise dependent upon a specific initial conception and a visible direction. Such a conception can be right or wrong; it is the starting point for the attitude to be taken toward all manifestations and occurrences of life and thereby a compelling and obligatory rule for all action."[127] In other words a worldview as defined by Hitler was a quasi-religious framework encompassing immediate political goals. Nazism was no mere ideological discourse; it was a political religion commanding the total commitment owed to a religious faith.[128]

The "visible direction" of a worldview implied the existence of "final goals" that, their general and hazy formulation notwithstanding, were supposed to guide the elaboration and implementation of short-term plans. Before the fall of 1935 Hitler did not hint either in public or in private what the final goal of his anti-Jewish policy might be. But much earlier, as a fledgling political agitator, he had defined the goal of a systematic anti-Jewish policy in his notorious first political text, the letter on the "Jewish question" addressed on September 16, 1919, to one Adolf Gemlich. In the short term the Jews had to be deprived of their civil rights: "The final aim however must be the uncompromising removal of the Jews altogether."[129]

CHAPTER 3

Redemptive Anti-Semitism

I

On the afternoon of November 9, 1918, Albert Ballin, the Jewish owner of the Hamburg-Amerika shipping line, took his life. Germany had lost the war, and the Kaiser, who had befriended him and valued his advice, had been compelled to abdicate and flee to Holland, while in Berlin a republic was proclaimed. On the thirteenth, two days after the Armistice, Ballin was buried at Ohlsdorf, a suburb of Hamburg. "In the midst of revolution," writes Ballin's biographer, "the city paused to pay tribute to its most distinguished citizen, and from Amerongen the ex-Kaiser telegraphed his condolences to Frau Ballin."[1]

Ballin's life and death were but one last illustration of the paradoxical existence of the Jews of Germany during the Second Reich. Some had achieved remarkable success but were held at arm's length; many felt "at home in Germany" but were perceived as strangers; almost all were loyal citizens but engendered suspicion. Thus, two years before the collapse, on October 11, 1916, by which time the military situation had reached a complete stalemate, the Prussian war minister signed a decree ordering a census of all Jews in the armed forces "to determine . . . how many Jews subject to military duty were serving in every unit of the German armies."[2] The War Ministry explained that it was "continually receiving complaints from the population that large numbers of men of the Israelite faith who are fit for military service are either exempt from military duties or are evading their obligation to serve under every conceivable pretext."[3] The census was held on November 1, 1916.

From the beginning of the war, the Jews of Germany had, like all other Germans, joined the army; very soon a number of them became officers.

For the castelike Prussian officer corps in particular, this was a bitter pill to swallow, and officer organizations turned to anti-Semitic groups to find ways of putting an end to these promotions.[4] A wave of rumors, originating both within and outside the army, described Jewish soldiers as lacking in ability and courage, and accused Jews in great numbers of shirking front-line duty, settling into rear-echelon office jobs or flocking into the "war economy corporations" established for the acquisition of raw materials and food supplies.[5]

The industrialist Walther Rathenau, who was Jewish, had in fact become the head of the new War Resources Department in the War Ministry, and on the initiative of Ballin, the bankers Max Warburg and Carl Melchior (also Jewish), the Central Purchasing Company was established for acquiring foreign food products through a network of war corporations. According to extreme nationalist Germans, these corporations were becoming instruments of Jewish speculation and exploitation of the nation in its time of peril: "The war profiteers were first of all essentially Jews," wrote Gen. Erich Ludendorff in his memoirs. "They acquired a dominant influence in the 'war corporations' . . . which gave them the occasion to enrich themselves at the expense of the German people and to take possession of the German economy, in order to achieve one of the power goals of the Jewish people."[6] Hitler, in *Mein Kampf*, wrapped it all up in his own typical style: "The general mood [in the army] was miserable. . . . The offices were filled with Jews. Nearly every clerk was a Jew and nearly every Jew was a clerk. . . . As regards economic life, things were even worse. Here the Jewish people had become really 'indispensable.' The spider was slowly beginning to suck the blood out of the people's pores. Through the war corporations, they had found an instrument with which, little by little, to finish off the nation's free economy."[7]

Due to the professional structure of the Jewish population, approximately 10 percent of the directors of the war corporations were Jews.[8] Continuous anti-Jewish attacks induced a Catholic Center deputy, Matthias Erzberger, to demand a Reichstag inquiry.[9] He was supported by a coalition of liberals and conservatives. Even some Social Democrats joined in.[10] It was in this atmosphere that the Prussian War Ministry announced its decision to conduct its census of Jews (*Judenzählung*).

The Jews reacted, but only meekly. Warburg, then already one of the most influential Jews in imperial Germany, met with War Minister Stein in March 1917 to ask for the release of a statement that Jews were fighting as bravely as other Germans. Stein refused, and in order to underline the Jewish traits he most disliked, lectured Warburg about Heinrich Heine.[11]

The results of the census were not published during the war, ostensibly out of consideration for the Jews, as they were termed "devastating" by officials of the War Ministry.[12] Immediately after the Armistice, pseudo results were leaked to the radical anti-Semitic Völkischer Schutz- und Trutzbund by the Jew-hating General Wrisberg and used as anti-Jewish propaganda on a massive scale.[13] Only at the beginning of the 1920s did a systematic study of the material show it to be "the greatest statistical monstrosity of which an administration had ever been responsible."[14] Detailed analysis indicated that Jewish participation in frontline service was equivalent to that of the general population, with a minimal deviation due to age and occupational structure. The damage had nonetheless been done.

Ernst Simon, who had volunteered for the army to find a sense of community with the German nation, perceived that the *Judenzählung* was more than the initiative of some malevolent officials. It was the "real expression of a real mood: that we were strangers, that we did not belong, that we had to be specially tagged, counted, registered and dealt with."[15] Walther Rathenau wrote to a friend in the summer of 1916: "The more Jews are killed [in action] in this war, the more obstinately their enemies will prove that they all sat behind the front in order to deal in war speculation. The hatred will grow twice and threefold."[16]

After almost two decades of relative latency, the Jewish issue had resurfaced in full force in German political life during the 1912 Reichstag elections, which were soon dubbed the "Jewish elections" (*Judenwahlen*).[17] The real political issue was the growth of the Left. However, as the Jews—opposed to (and by) the Conservatives and disappointed by the stand taken toward them by the National Liberals—turned to the Progressives and, in particular, to the Social Democrats, they became identified with the left-wing peril.[18]

The elections marked the disappearance of the anti-Semitic splinter parties and represented a significant setback for the Conservative right. The Social Democrats emerged as the strongest single party on the German scene, more than doubling their number of seats in the Reichstag from 53 to 110. Of the 300 candidates favored by organizations in which Jews were prominent, 88 were elected.[19] These results proved that the majority of the voters did not manifestly harbor intense anti-Jewish feelings, but the reaction of the Right was different and immediate. It had become obvious to the right-wing press that Jewish money and the Jewish spirit were in control of the "gold" and the "red" internationals, those two most dangerous enemies of the German nation. Even for a publication as close to the Lutheran Church as the *Christlichsoziale Reichsbote*, the workers who voted for the Social Democrats were "driven by the Jewish whip" held by "the manipulators of international Jewish capitalism."[20]

Frantic activity now spread throughout the extreme right, with approximately twenty new ultra-nationalist and racist organizations springing up on the political scene. Some of them, such as the Reichshammerbund and the Germanenorden, were coalitions of previously existing groups.[21] Among larger groups, the evolution of the Pan-German League is particularly telling. In his previously mentioned 1912 pamphlet, *If I Were the Kaiser*, league president Heinrich Class fully spelled out a program for the complete expulsion of the Jews from German public life—that is, from public office, from the liberal professions, and from banks and newspapers. Jews would lose the right to own land. Jewish immigration would be banned, and all Jewish noncitizens deported. Those who were citizens would be subject to "alien Status" (*Fremdenrecht*). A Jew would be defined as a person belonging to the Jewish religious community on January 18, 1871, the day the German Empire was proclaimed, as would all the descendants of such persons, even if only one grandparent was Jewish.[22]

A few months later a memorandum was submitted to the crown prince, Wilhelm II's eldest son, by another member of the league, Konstantin von Gebsattel; it proposed the same measures against the Jews as well as a "coup d'état" to put an end to parliamentarianism in Germany. The crown prince—who later would become a member of the SS—was "captivated" by Gebsattel's memorandum and transmitted it to his father and to Chancellor Theobald von Bethmann-Hollweg. Himself a strange mix of traditional conservatism and radical right-wing opinions,[23] the Kaiser was dismissive. He considered Gebsattel an "oddball," the Pan-Germans who supported such plans "dangerous people," and the idea of excluding the Jews from public life "downright childish"; Germany would be cutting itself off from civilized nations. The chancellor was more deferential to the crown prince, but no less negative.[24]

The Association Against Jewish Arrogance (Verband gegen die Überhebung des Judentums) was established on February 11, 1912, by the remnants of the old anti-Semitic parties and various other anti-Semitic organizations. Its aim was the creation, under nationalist auspices, of a mass movement to achieve political change. "One of their top priorities was to exclude the Jewish 'race' from the nation's public life. The founding of the association, clearly linked to the 1912 elections, was but one more manifestation of the new right's determined 'defense' against *Juda*."[25]

II

Jews never represented more than approximately 1 percent of Germany's overall population in the late nineteenth and early twentieth centuries. Between the beginning of the century and 1933, that percentage slightly declined. The Jewish community, however, gained in visibility by gradually concentrating in the large cities, keeping to certain professions, and absorbing an increasing number of easily identifiable East European Jews.[26]

The general visibility of the Jews in Germany was enhanced by their relative importance in the "sensitive" areas of business and finance, journalism and cultural activities, medicine and the law, and, finally, by their involvement in liberal and left-wing politics. The social discrimination to which the Jews were subjected, and their own striving for advancement and acceptance, easily explain their patterns of activity. Interpreted as Jewish subversion and domination, these patterns in turn led, at least in parts of German society, to further hostility and rejection.

Of the fifty-two private banks in Berlin at the beginning of the nineteenth century, thirty were Jewish-owned. Later on Bismarck asked the Rothschilds to recommend a private banker (it was to be Gerson Bleichröder), and Kaiser Wilhelm I chose for himself the banker Moritz Cohn. When, at the turn of the century, many private banks became shareholding companies, Jews frequently held a controlling percentage of the shares or served as directors of the new enterprises. Add the banking aristocracy of the Warburgs, the Arnholds, the Friedländer-Fulds, the Simons, the Weinbergs, and so on, to such financial potentates as chainstore owners Abraham Wertheim and Leonhard and Oskar Tietz, founding electrical industrialist Emil Rathenau, publisher Rudolf Mosse, and shipping magnate Albert Ballin, and it becomes obvious that Jews held an eminent and visible place in the financial world of imperial Germany.[27]

The Jewish economic elite's particular function during the nineteenth century had been its decisive role in capital mobilization and concentration through development of the Berlin stock market,[28] and linkage of the still relatively parochial German economy with world markets.[29] The centrality of "Jewish" banking during the Weimar period did not decrease,[30] contrary to what has sometimes been argued. But there was no correlation between Jewish economic activity and any kind of lasting political influence in German society.

Culture was possibly the most sensitive domain. In March 1912 a telling exchange was triggered by an article written by a young Jewish intellectual, Moritz Goldstein, and published in the arts journal *Kunstwart* under the title "Deutsch-jüdischer Parnass" (German-Jewish Parnassus). As Goldstein put it, "We Jews administer the spiritual possessions of a people that denies us the right and the capability of doing so."[31] After admitting to Jewish influence on the press and in the literary world, Goldstein reemphasized the insuperable rift between the Jewish "administrators" of German culture, who believed they were speaking for and to the Germans, and the Germans themselves, who considered such presumption insufferable. What, then, was the way out? Zionism, Goldstein thought, was no option for people of his background and generation. In an emotional and most emphatic fashion, he called instead for an act of courage on the part of the Jews of Germany: that, in spite of their deep feelings for Germany and all things German, in spite of their centuries-long presence in the land, they must turn their backs on the host society and stop vowing ever-renewed and ever-unrequited love.[32] On the cultural level Jews should now turn to Jewish issues, not only for their own sake but to create "a new type of Jew, new not in life but in literature."[33] Goldstein's closing was on an emotional par with the rest: "We demand recognition of a tragedy that, with a heavy heart, we have exposed to all."[34]

Goldstein's sharp diagnosis/tearful lament induced the editor of *Kunstwart*, Ferdinand Avenarius, to produce in the August issue a long comment entitled "Aussprachen mit Juden" (Debates with Jews). "We are not anti-Semites," he wrote. "We know that there are domains in which the Jews are more able than we are, and that we have greater ability in others; we hope that with good will on both sides, peaceful co-operation will be possible, but we are convinced that relations cannot continue much longer in their present form." Avenarius called for some sort of "negotiation" between "leaders" of "both sides in order to avoid bitter cultural battles [*Kulturkämpfe*]. . . . Given the growing excitement [Avenarius did not specify whose]," he did not believe that success could easily be achieved.[35] The argument was clear, the "we" and "they" even clearer. But as to the basic facts (though obviously not their interpretation), both Goldstein and (implicitly) Avenarius were not entirely wrong.

As for the press—excluding the great number of conservative and specifically Christian newspapers and periodicals, as well as most of the regional papers—there was, on the national level, a strong Jewish presence in ownership, editorial responsibility, and major cultural or political commentary. Rudolf Mosse's publishing empire included the *Berliner Tageblatt*, the *Morgenzeitung*, the *Volkszeitung*, and the *Börsenblatt*. The Ullstein family owned the *Neues Berliner Tageblatt*, the *Abendpost*, the *Illustrierte Zeitung*, and *B.Z. am Mittag*, "the first German paper based completely on street sales."[36] The paper with the largest circulation, the *Morgenpost*, also belonged to Ullstein, as eventually did the *Vossische Zeitung*, "Berlin's oldest newspaper."[37] Among the three most prominent publishers who took the largest share of the pre-1914 daily press—Mosse, Ullstein, and Scherl—the first two were Jews.[38] The relative importance of these three publishers would be altered somewhat in the twenties by the acquisition of Scherl by the ultra-right-wing Alfred Hugenberg and by the consequent rapid expansion of his press holdings.

The editors in chief and main editorial writers of many of the most influential newspapers (such as Theodor Wolff, editor of the *Berliner Tageblatt*; Georg Bernhard, editor of the *Vossische Zeitung*; and Bernhard Guttmann, the influential Berlin correspondent of the *Frankfurter Zeitung*), were Jews, as were dozens of other political commentators, cultural critics, and satirists in a wide array of dailies and periodicals.[39]

In book publishing Mosse and Ullstein were major figures, as was Samuel Fischer, who founded his publishing house in Berlin in 1886. Fischer, as important in the history of modern German literature as, for example, Random House or Scribner's in the United States, published Thomas Mann, Gerhart Hauptmann, and Hermann Hesse, among others.[40]

Along with Jewish publishers and editors in chief, there was a solid group of Jewish readers and theater- and concertgoers. A striving for *Bildung* (culture/education) had turned the Jewish bourgeoisie into the self-appointed (and ecstatic) carrier of German culture. Writing in December 1896 about the first performance of Gerhart Hauptmann's play *Die versunkene Glocke* (The Sunken Bell), Baroness Hildegard von Spitzemberg noted in her diary: "The house was packed with Jews and

Jew-companions and with the representatives of press and literature: Maximilian Harden, Hermann Sudermann, Erich Schmidt, Theodor Fontane, Ludwig Pietsch, the last [two] of whom, however, shook their heads disapprovingly and did not join in the frenetic applause of the poet's [playwright's] supporters."[41] Fontane and Pietsch were non-Jews.

The situation was possibly even more extreme in Austria-Hungary. At the end of the nineteenth century, Jews owned more than 50 percent of the major banks in the Austrian part of the empire, and occupied nearly 80 percent of the key positions in the banking world.[42] In the Hungarian part, the Jewish economic presence, which benefited from the full support of the Hungarian aristocracy, was even more widespread. "Above all, Jews were prominent among the great press tycoons. They owned, edited, and very extensively contributed to most of the leading newspapers of Vienna. Though his words were somewhat exaggerated, it was nonetheless telling that Harry Wickham Steed, the London *Times* correspondent in the Austrian capital, could write that 'economically, politically and in point of general influence they [the Jews] are . . . the most significant element in the Monarchy.'"[43]

During the early decades of the nineteenth century, the harmonious assimilation of the Jews into German society, as in other countries of Western and Central Europe—later made formally possible by the full emancipation of 1869 and 1871—could appear to many as a reasonable prospect.[44] More than anything else, the Jews themselves wanted to join the ranks of the German bourgeoisie; this collective "project" was undoubtedly their overriding goal.[45] Lay leaders and enlightened rabbis never tired of stressing the importance of *Bildung* and *Sittlichkeit* (manners and morals).[46] Although the great majority of Jews did not abandon Judaism entirely, the collective effort of adaptation led to deep reshapings of Jewish identity in the religious domain as well as in a variety of secular pursuits and attitudes.[47] The modern German Jew, however, did create—consciously or not—a specific subculture that, although aiming at integration, resulted in a new form of separation.[48] Religious-cultural distinctiveness was reinforced by the increasingly negative reactions of society in general to the very rapidity of the Jews' social and economic ascent. Economic success and growing visibility without political power produced, in part at least, their own nemesis. In his biography of Bismarck's banker, Gerson Bleichröder, Fritz

Stern alluded to the shift in attitudes from the 1870s on: "[Bleichröder's] middle years described the moment of the least troubled amalgamation of German and Jewish society; his declining years [he died in 1893] marked the first organized repudiation of that amalgamation, and his very success was taken as a warrant for repudiation."[49]

One may readily agree with German historian Thomas Nipperdey that in comparison to that of France, Austria, or Russia, German anti-Semitism on the eve of World War I was certainly not the most extreme. One may also agree with his statement that pre-1914 anti-Semitism should be evaluated both within its own historical context and from the perspective of later events ("under the sign of Auschwitz").[50] However, his related statement that the Jews of Germany themselves considered the anti-Semitism of those years a marginal issue, a remnant of prior discrimination that would disappear in due time is less convincing.[51] Any perusal of contemporary testimonies indicates that Jews held diverse views regarding the attitudes of society in general toward them. It needs only Moritz Goldstein's lament to show that some German Jews were quite aware of the fact that the chasm between them and the surrounding society was growing.

This was true not only in Germany. Two equally remarkable literary representations of Austria before the Great War, Stefan Zweig's *The World of Yesterday* and Arthur Schnitzler's *The Road into the Open*, provide contrary assessments of how the Jews perceived their own situation. For Zweig anti-Semitism was practically nonexistent; for Schnitzler it was at the center of his characters' consciousness and existence. In any event, whatever the relative strength of prewar anti-Semitism may have been, its presence was a necessary condition for the massive anti-Jewish hostility that spread throughout Germany during the war years and increasingly after the defeat of 1918. Moreover, the prewar scene also provided some of the ideological tenets, political demands, and institutional frameworks that endowed postwar anti-Semitism with its early structures and immediate goals.

When one considers the wider European scene, the achievements, political attitudes, and cultural options of Jews at the end of the nineteenth century appear as those of members of an identifiable minority, stemming in part from the peculiar historical development of this minority. But these achievements and options were first and foremost those of individuals whose goal was the kind of success that led to integration into society in general. For the anti-Semite, however, the situation looked entirely different: Jewish striving and Jewish success, real or imaginary, were perceived as the behavior of a foreign and hostile minority group acting collectively to exploit and dominate the majority.

As long as merely a few Jews, under the patronage of kings and princes, managed to climb the social ladder, their limited number, the function they fulfilled, and the protection they were granted checked the spread of hostility. When, as Hannah Arendt pointed out in somewhat different terms,[52] emancipation allowed for the social advancement of a large number of Jews within a context in which their social function was losing its specificity and in which political power no longer backed them, they increasingly became the targets of various forms of social resentment. Modern anti-Semitism was fueled by this conjunction of increasing visibility and increasing weakness.

A common trigger of various forms of nonracial anti-Jewish resentment was undoubtedly the very existence of a Jewish difference. Liberals demanded that, in the name of universalist ideals, the Jews should accept the complete disappearance of their particular group identity; nationalists, on the other hand, demanded such disappearance for the sake of a higher particularist identity, that of the modern nation-state. Although the majority of Jews were more than eager to travel a long way down the road to cultural and social assimilation, most of them rejected total collective disappearance. Thus, as moderate as Jewish particularism may have been, it antagonized its liberal supporters and incensed its nationalist opponents. Jewish visibility in highly sensitive domains exacerbated the irritant inherent in difference.

Racial anti-Semites also claimed that their anti-Semitic campaign was based on the Jews' difference. However, whereas for the nonracial anti-Semite such difference could and *should* have been totally effaced by the complete assimilation and disappearance of the Jews as such, the racial anti-Semite argued that the difference was indelible, that it was inscribed in the blood. For the nonracial anti-Semite, a solution to the "Jewish question" was possible within society in general; for the racial anti-Semite, because of the dangerous racial impact of Jewish presence and equality, the only solution was exclusion (legal and possibly physical) from society in general. This well-known basic picture should be completed by two aspects of the modern anti-Jewish scene that are either barely mentioned by many historians or considered all-encompassing by others: the survival of traditional religious anti-Semitism and the related proliferation of conspiracy theories in which the Jews always played a central role.

Whether or not Christian hostility toward the Jews was intermittent, whether or not the Jews themselves contributed to the exacerbation of this hostility,[53] does not alter the fact that, in dogma, ritual, and practice, Christianity branded the Jews with what appeared to be an indelible stigma. That stigma had been effaced neither by time nor by events, and throughout the nineteenth and the early decades of the twentieth centuries, Christian religious anti-Semitism remained of central importance in Europe and in the Western world in general.

In Germany, apart from the general motives of Christian anti-Semitism, Christian anti-Jewish attitudes also stemmed from the particular situation of the churches throughout the imperial era. German Catholics were antagonized by Jewish support for the National Liberals, who were Bismarck's allies during his anti-Catholic campaign of the 1870s, the Kulturkampf;[54] conservative Protestants were firmly committed to the Christian nature of the Second Reich, and even liberal Protestants, in their attempt to rationalize Christianity, entered into confrontations with liberal Jews keen on demonstrating the pagan core of the Christian religion.[55] Finally, in Germany, France, and Austria, political use of Christian anti-Jewish themes proved successful, at least for a time, in appealing to lower-middle-class voters.

For some historians the rootedness and the very permanence of Christian anti-Judaism has been the only basis of all forms of modern anti-Semitism. Jacob Katz, for example, sees modern anti-Semitism as but "a continuation of the premodern rejection of Judaism by Christianity, even when it [modern anti-Semitism] renounced any claim to be legitimized by it or even professed to be antagonistic to Christianity." In Katz's view any claims for an anti-Semitism that would be beyond "the Jewish Christian division" were but "a mere declaration of intent. No anti-Semite, even if he himself was anti-Christian, ever forwent the use of those anti-Jewish arguments rooted in the denigration of Jews and Judaism in earlier Christian times."[56] This interpretation is excessive, but the impact of religious anti-Judaism on other modern forms of anti-Semitism is apparent in several ways. First, a vast reservoir of almost automatic anti-Jewish reactions continued to accumulate as a result of early exposure to Christian religious education and liturgy, and to everyday expressions drawn from the pervasive and ongoing presence of the various denominations of the Christian creed. Second, the very notion of "outsider" applied by modern anti-Semitism to the Jew owed its tenacity not only to Jewish difference as such but also to the depth of its religious roots. Whatever else could be said about the Jew, he was first and foremost the "other," who had rejected Christ and revelation. Finally, perhaps the most powerful effect of religious anti-Judaism was the dual structure of the anti-Jewish image inherited from Christianity. On the one hand, the Jew was a pariah, the despised witness of the triumphal onward march of the true faith; on the other, from the Late Middle Ages onward, an opposite image appeared in popular Christianity and in millennarian movements, that of the demonic Jew, the perpetrator of ritual murder, the plotter against Christianity, the herald of the Antichrist, the potent and occult emissary of the forces of evil. It is this dual image that reappears in some major aspects of modern anti-Semitism. And, its threatening and occult dimension became the recurrent theme of the main conspiracy theories of the Western world.

The Christian phantasm of a Jewish plot against the Christian community may itself have been a revival of the pagan notion that the Jews were enemies of humanity acting in secret against the rest of the world. According to a popular medieval Christian legend, "a secret rabbinical synod convened periodically from all over Europe to determine which community was in turn to commit ritual murder."[57] From the eighteenth century on, new conspiracy theories also pointed to threats from a number of non-Jewish occult groups: Freemasons, Illuminati, Jesuits. In the landscape of modernity, paranoid political thought was acquiring a permanence of sorts. "What is the distinguishing thing about the paranoid style," wrote Richard Hofstadter, "is not that its exponents see conspiracies or plots here and there in history, but that they regard a 'vast' or 'gigantic' conspiracy as *the motive force* in historical events. History *is* a conspiracy, set in motion by demonic forces of almost transcendent power, and what is felt to be needed to defeat it is not the usual methods of political give-and-take, but an all-out crusade."[58]

Within this array of occult forces, the Jews were the plotters par excellence, the manipulators hidden behind all other secret groups that were merely their instruments. In the notorious two-pronged secret threat of "Jews and Freemasons," the latter were perceived as instruments of the former.[59] Jewish conspiracies, in other words, were at the very top of the conspiratorial hierarchy, and their aim was nothing less than total domination of the world. The centrality of the Jews in this phantasmic universe can be explained only by its roots in the Christian tradition.

Like any other national anti-Semitism at the end of the nineteenth century and during the years preceding World War I, anti-Semitism in imperial Germany was determined, as I have already indicated, both by dominant Christian and modern European trends and by the impact of specific historical circumstances, among which several further aspects should be stressed:

In general terms a structural dimension needs to be emphasized in distinguishing, for example, between French and German modes of national integration, with the relevance of such a distinction in terms of anti-Jewish attitudes becoming clearly apparent. Since the French Revolution, the French model of national integration had been that of a process fostered and implemented by the state on the basis of universal principles, those of the Enlightenment and the Revolution. Since the romantic revolution, the German model of national integration had been derived from and predicated upon the idea of the nation as a closed ethnocultural community independent of and sometimes opposed to the state. Whereas the French model implied the *construction* of national identity by way of a centralized educational system and all other means of socialization at the disposal of the state, the German model often posited the existence of inherited characteristics belonging to a preexisting organic community.[60]

By way of state-directed socialization and in the name of the secular republic's universal values, a Jew could become French, and not merely on a purely formal level. (This despite intensely hostile reactions from that substantial part of French society that rejected the Revolution, the republican state, and thus the Jews, identified as foreigners allied with the state and as carriers of the secular, subversive values of social upheaval and modernity.) Regardless of formal emancipation and equality of civic rights, the Jew was often kept at a distance by a German national community fundamentally closed to a group whose recognizable difference seemed to society in general to be rooted in alien ethnocultural—and, increasingly, racial—soil. A somewhat different (but not incompatible) interpretation has pointed to the fact that in France legal emancipation carried a prime expectation of gradual Jewish assimilation (also by way of the French educational system and its universalist values), whereas in Germany a widely shared position was that the process of assimilation should be imposed and monitored by bureacratic means, and that full emancipation should be granted only at the end of the process. As time went by, in Germany the success of Jewish assimilation was increasingly questioned. Therefore, even after the Jews of Germany were granted full emancipation, anti-Semites of all hues—and even liberals—could argue that total assimilation had not really been achieved and that the results of emancipation were problematic.[61]

The situation in Germany was further exacerbated by developments specific to the second half of the nineteenth century, mainly the various aspects of an extremely rapid process of modernization. By entirely transforming the country's social structures and by threatening its existing hierarchies, the onrush of German modernization seemed to endanger hallowed cultural values and the organic links of the community;[62] at the same time it seemed to allow the otherwise incomprehensible social ascent of the Jews, who were thus perceived as the promoters, carriers, and exploiters of that modernization. The Jewish threat now appeared to be both penetration by a foreign element into the innermost texture of the national community and furthering, by way of that penetration, not of modernity as such (enthusiastically embraced by the majority of German society) but of the evils of modernity.

It is within this context that other developments peculiar to Germany acquire their full significance. First, after the rise and fall of the German anti-Semitic parties between the mid-1870s and the late 1890s, anti-Jewish hostility continued to spread in German society at large through a variety of other channels—economic and professional associations, nationalistic political organizations, widely influential cultural groups. The rapid increase of such institutionalized infusions of anti-Jewish attitudes into the very heart of society did not take place—or at least not on such a scale—in other major Western or Central European countries. Second, in Germany a full-blown anti-Semitic ideology was systematically elaborated; it allowed more or less diffuse anti-Jewish resentment to adopt ready-made intellectual frameworks and formulas that in turn were to foster more extreme ideological constructs during the coming years of crisis. Such specific ideologization of German anti-Semitism was particularly visible, in two different ways, with regard to racial anti-Semitism. In its mainly biological form, racial anti-Semitism used eugenics and racial anthropology to launch a "scientific" inquiry into the racial characteristics of the Jew. The other strand of racial anti-Semitism, in its particularly German, mystical form, emphasized the mythic dimensions of the race and the sacredness of Aryan blood. This second strand fused with a decidedly religious vision, that of a German (or Aryan) Christianity, and led to what can be called "redemptive anti-Semitism."

III

Whereas ordinary racial anti-Semitism is one element within a wider racist worldview, in redemptive anti-Semitism the struggle against the Jews is the dominant aspect of a worldview in which other racist themes are but secondary appendages.

Redemptive anti-Semitism was born from the fear of racial degeneration and the religious belief in redemption. The main cause of degeneration was the penetration of the Jews into the German body politic, into German society, and into the German bloodstream. Germanhood and the Aryan world were on the path to perdition if the struggle against the Jews was not joined; this was to be a struggle to the death. Redemption would come as liberation from the Jews—as their expulsion, possibly their annihilation.

This new anti-Semitism has been depicted as part and parcel of the revolutionary fervor of the early nineteenth century, particularly of the revolutionary spirit of 1848. But it should be pointed out that the main bearers of the new anti-Jewish mystique had all turned against their revolutionary pasts; when Judaism was mentioned in their revolutionary writings, it was in a purely metaphorical sense (mainly as representing Mammon or "the Law"), and whatever revolutionary terminology remained in their new anti-Semitism was meant as "radical change," as "redemption" in a strongly religious sense, or, more precisely, in a racial-religious sense.[63]

Various themes of redemptive anti-Semitism can be found in *völkisch* ideology in general, but the run-of-the-mill *völkisch* obsessions were usually too down-to-earth in their goals to belong to the redemptive sphere. Among the *völkisch* ideologues, only the philosopher Eugen Dühring and the biblical scholar Paul de Lagarde came close to this sort of anti-Semitic eschatological worldview. The source of the new trend has to be sought elsewhere, in that meeting point of German Christianity, neoromanticism, the mystical cult of sacred Aryan blood, and ultraconservative nationalism: the Bayreuth circle.

I intentionally single out the Bayreuth circle rather than Richard Wagner himself. Although redemptive anti-Semitism derived its impact from the spirit of Bayreuth, and the spirit of Bayreuth would have been nonexistent without Richard Wagner, the depth of his personal commitment to this brand of apocalyptic anti-Semitism remains somewhat contradictory. That Wagner's anti-Semitism was a constant and growing obsession after the 1851 publication of his *Das Judentum in der Musik* (*Judaism in Music*) is unquestionable. That the maestro saw Jewish machinations hidden in every nook and cranny of the new German Reich is notorious. That the redemption theme became the leitmotiv of Wagner's ideology and work during the last years of his life is no less generally accepted. Finally, that the disappearance of the Jews was one of the central elements of his vision of redemption seems also well established. But what, in Wagner's message, was the concrete meaning of such a disappearance? Did it mean the abolition of the Jewish spirit, the vanishing of the Jews as a separate and identifiable cultural and ethnic group, or did redemption imply the actual physical elimination of the Jews? This last interpretation has been argued by, among others, historians such as Robert W. Gutman, Hartmut Zelinsky, and Paul Lawrence Rose.[64] The last in particular identifies Wagner's "revolutionary anti-Semitism" and its supposedly exterminatory streak with the composer's revolutionary ardor of 1848.

In *Judaism in Music*, the annihilation of the Jew (and the pamphlet's notorious final words: "the redemption of Ahasuerus—going under!") most probably means the annihilation of the Jewish spirit. In this finale the maestro heaps dithyrambic praise upon the political writer Ludwig Börne, a Jew who in his eyes exemplified the redemption from Jewishness into "genuine manhood" by "ceasing to be a Jew."[65] Börne's example is manifestly the path to be collectively followed. But Wagner's writings of the late 1870s and the 1880s and the redemptive symbolism of the *Ring* and especially of *Parsifal*, are indeed extraordinarily ambiguous whenever the Jewish theme directly or indirectly appears. Whether redemption from erotic lust, from worldly cravings, from the struggles for power is achieved, as in the *Ring*, by way of self-annihilation or, as in *Parsifal*, by mystical purification and the rebirth of a sanctified Germanic Christendom, the Jew remains the symbol of the worldly lures that keep humanity in shackles. Thus the redemptive struggle had to be a total struggle, and the Jew, like the evil and unredeemable Klingsor in *Parsifal*, had to disappear. In *Siegfried* the allusion is even more direct: The Germanic hero Siegfried kills the repulsive Nibelung dwarf Mime, whom Wagner himself identifies, according to Cosima Wagner's diaries, as a "*Jüdling*."[66] All in all the relation between Siegfried and Mime, overloaded with the most telling symbolism, was probably meant as a fierce anti-Semitic allegory of the relation between German and Jew—and of the ultimate fate of the Jew.[67]

Even the Master's jokes, like his "wish" that all Jews be burned at a performance of Lessing's *Nathan the Wise*,[68] expressed the underlying intensity of his exterminatory fantasies. And yet, Wagner's ideas about the Jews remained inconsistent, and the number of Jews in his entourage, from the pianists Carl Tausig and Josef Rubinstein to the conductor Hermann Levi and the impresario Angelo Neumann, is well known. Indeed, Wagner's behavior toward Levi was often overtly sadistic, and Rubinstein was a notoriously self-hating Jew. Yet these Jews belonged to the maestro's close entourage, and, more significant, Wagner gave Neumann considerable leeway regarding the handling of contracts and performances of his works: No consistently fanatical anti-Semite would have allowed such a massive compromise.

Although Wagner himself embraced the theoretical racism of the French essayist Arthur de Gobineau, the intellectual foundations of redemptive anti-Semitism were mainly fostered and elaborated by the other Bayreuthians, especially after the composer's death, during the reign of his widow, Cosima: Hans von Wolzogen, Ludwig Scheemann, and, first and foremost, the Englishman Houston Stewart Chamberlain. In a classic study of the Bayreuth Circle, Winfried Schüler defined Bayreuth's special significance within the anti-Semitic movement and Chamberlain's own decisive contribution: "It is in the nature of anti-Semitic ideologies to use a more or less prominent friend-foe model. What nonetheless gives Bayreuth's anti-Semitism an unmistakably particular aspect is the resoluteness with which the opposition between Germandom and Jewry is raised to the position of the central theme of world history. In Chamberlain's *Foundations* [his 1899 magnum opus, *The Foundations of the Nineteenth Century*] this dualistic image of history finds its tersest formulation."[69]

In line with Bayreuth's oft-repeated leitmotiv, Chamberlain called for the birth of a German-Christian religion, a Christianity cleansed of its Jewish spirit, as the sole basis for regeneration. In other words, the redemption of Aryan Christianity would be achieved only through the elimination of the Jew. But even here it is not entirely clear whether or not the redemptive struggle against the Jews was to be waged against the Jewish spirit only. In the closing lines of volume 1, after stating that in the nineteenth century, amid a chaos of mixed breeds, the two "pure" races that stood facing each other were the Jews and the Germans, Chamberlain writes: "No arguing about 'humanity' can alter the fact that this means a struggle. Where the struggle is not waged with cannon-balls, it goes on silently in the heart of society. . . . But this struggle, silent though it be, is above all others a struggle for life and death."[70] Chamberlain probably did not know himself what he meant by this in terms of concrete action, but he undoubtedly offered the most systematic formulation of what he considered the fundamental struggle shaping the course of world history.

Three years after the publication of Chamberlain's *Foundations*, the *Frankfurter Zeitung* had to admit that it "has caused more of a ferment than any other appearance on the book market in recent years."[71] By 1915 the book had sold more than one hundred thousand copies and was being widely referred to. As the years went by, Chamberlain, who in 1908 had married Richard and Cosima Wagner's daughter, Eva, became ever more obsessed with the "Jewish question." In nightmares, he reported, he saw himself kidnapped by Jews and sentenced to death.[72] "My lawyer friend in Munich," he informed an old acquaintance, "tells me that there is no living being whom the Jews hate more than me."[73] The war, and even more so the early years of the Weimar Republic, drove his obsession to its utmost limits. Hitler visited him in Bayreuth in 1923: The by now paralyzed prophet of redemptive anti-Semitism was granted the supreme happiness of meeting—and recognizing as such—Germany's savior from the Jews.[74]

IV

The impact of the Great War and the Bolshevik Revolution on the European imagination was stronger than that of any other event since the French Revolution. Mass death, shattering political upheavals, and visions of catastrophes to come fueled the pervasive apocalyptic mood that settled over Europe.[75] Beyond nationalist exacerbation in several countries, the hopes, fears, and hatreds of millions crystallized along the main political divide that would run through the history of the following decades: fear of revolution on one side, demand for it on the other. Those who feared the revolution frequently identified its leaders with the Jews. Now the proof for the Jewish world conspiracy was incontrovertible: Jewry was about to destroy all established order, annihilate Christianity, and impose its dominion. In her 1921 book, *World Revolution*, the English historian Nesta Webster asked, "who are . . . the authors of the Plot?. . . . What is their ultimate object in wishing to destroy civilization? What do they hope to gain by it? It is this apparent absence of motive, this seemingly aimless campaign of destruction carried on by the Bolsheviks of Russia, that has led many people to believe in the theory of a Jewish conspiracy to destroy

Christianity."[76] Webster was among these believers, and so, in his own way, at the time, was Thomas Mann. "We also spoke of the type of Russian Jew, the leader of the world revolutionary movement," Mann wrote in his diary on May 2, 1918, recording a conversation with Ernst Bertram, "that explosive mixture of Jewish intellectual radicalism and Slavic Christian enthusiasm." He added: "A world that still retains an instinct of self-preservation must act against such people with all the energy that can be mobilized and with the swiftness of martial law."[77]

The most explosive ideological mixture present in postwar Germany was a fusion of constant fear of the Red menace with nationalist resentment born of defeat. The two elements seemed to be related, and the chaotic occurrences that marked the early months of the postimperial regime seemed to confirm the worst suspicions and fuel the fires of hatred.

Two months after Germany's defeat, the extreme left-wing revolutionary Spartacists attempted to seize power in Berlin. The uprising failed, and on the evening of January 15, 1919, its main leaders, Karl Liebknecht and Rosa Luxemburg, probably having been betrayed, were arrested at their hiding place in Berlin-Wilmersdorf.[78] They were brought to the Eden Hotel, the headquarters of the Garde-Kavallerie-Schützen-Division, where they were interrogated by a Captain Pabst. Liebknecht was led out first, taken by car to the Tiergarten, and "shot while trying to escape." Luxemburg, already brutally beaten at the Eden, was dragged out half dead, moved from one car to another, and then shot. Her body was thrown into the Landwehrkanal, where it remained until March. A military tribunal acquitted most of the officers directly involved in the murders (sentencing only two of them to minimal imprisonment), and Defense Minister Gustav Noske, a Social Democrat, duly signed these unlikely verdicts. Rosa Luxemburg and her closest companions among the Berlin Spartacists, Leo Jogisches and Paul Levi, were Jews.

The prominence of Jews among the leaders of the revolution in Bavaria added fuel to the already passionate anti-Semitic hatred of the Right as did their role among the Berlin Spartacists. It was Kurt Eisner, the Jewish leader of the Independent Socialist Party (USPD) in Bavaria, who toppled the Wittelsbach dynasty, which for centuries had given Bavaria its kings. During his short term as prime minister, Eisner added enemies by publishing incriminating archives regarding Germany's responsibility for the outbreak of the war and appealing to the German people to help in rebuilding devastated areas of enemy territory, which was simply interpreted as a call for the enslavement of Germans "from children to old people, [who would] be obliged to carry stones for the war-torn areas."[79]

On February 21, 1919, Eisner was assassinated by Count Anton Arco-Valley, a right-wing law student. After a brief interim government of majority Socialists, the first of two Republics of the Councils was established. In fact only a minority among the leaders of the Bavarian republics were of Jewish origin, but some of their most visible personalities could be identified as such.[80]

Exacerbated right-wing opinion accused these Jewish leaders of being responsible for the main atrocity committed by the Reds: the shooting of hostages in the cellar of the Luitpold Gymnasium in Munich. To this day the exact sequence of events is unclear. Apparently, on April 26, 1919, seven activists of the radical anti-Semitic Thule Society, among them its secretary, Countess Heila von Westarp, were detained at the organization's office. Two officers of the Bavarian Army and a Jewish artist named Ernst Berger were added to the seven Thule members. On April 30, after news reached Munich, that the counterrevolutionary volunteer units, the Free Corps of Franz Freiherr Ritter von Epp, had killed Red prisoners in the town of Starnberg, the commander of the Red forces, a former navy man named Rudolf Egelhofer, ordered the shooting of the hostages. These executions, an isolated atrocity, became the quintessential illustration of Jewish Bolshevik terror in Germany; in the words of British historian Reginald Phelps, this "murder of hostages goes far to explain . . . the passionate wave of anti-Semitism that spread because the deed was alleged to represent the vengeance of 'Jewish Soviet leaders' . . . on anti-Semitic foes." Needless to say, the fact that Egelhofer and "all those directly connected with the shooting" were not Jews, and that one of the victims was Jewish, did not change these perceptions in the least.[81]

The impact of the situation in Berlin and Bavaria was amplified by revolutionary agitation in other parts of Germany. According to the pro-Nazi French historian Jacques Benoist-Méchin, revolutionaries of Jewish background were no less active in various other regional upheavals: "In Magdeburg, it is Brandes; in Dresden, Lipinsky, Geyer, and Fleissner; in the Ruhr, Markus and Levinsohn; in Bremerhaven and Kiel, Grünewald and Kohn; in the Palatinate, Lilienthal and Heine."[82] What is important here is not the accuracy of every detail but the widespread attitude it expressed.

These events in Germany were perceived in relation to simultaneous upheavals in Hungary: the establishment of Béla Kun's Soviet Republic and the fact that the "Jewish" presence was even more massive there than in Berlin and Munich. The British historian of Central Europe R. W. Seton-Watson noted in May 1919: "Anti-Semitic feeling is growing steadily in Budapest (which is not surprising, considering that not only the whole Government, save 2, and 28 out of the 36 ministerial commissioners are Jews, but also a large proportion of the Red officers)."[83] Some of these revolutionaries, such as the notorious Tibor Szamuely, were indeed downright sinister figures.[84] Finally, the massive disproportion of leaders of Jewish origin among the Bolsheviks themselves seemed to give cogency to what had become a pervasive myth that spread and resonated throughout the Western world.[85]

There was no mystery in the fact that Jews joined the revolutionary left in large numbers. These men and women belonged to the generation of newly emancipated Jews who had abandoned the framework of religious tradition for the ideas and ideals of rationalism and, more often than not, for socialism (or Zionism). Their political choices derived both from the discrimination to which they had been subjected, mainly in Russia but also in Central Europe, and from the appeal of the socialist message of equality. In the new socialist world, all of suffering humanity would be redeemed, and with that, the Jewish stigma would disappear: It was, for at least some of these "non-Jewish Jews,"[86] a vision of a secularized messianism, which may have sounded like a distant echo of the message of the Prophets they no longer recognized. In fact, almost all of them were actually hostile, in the name of revolutionary universalism, to anything Jewish. In no way did they represent the political tendencies of the great majority of the Central and Western European Jewish populations, which were politically liberal or close to the Social Democrats; only a fraction was decidedly conservative. For example, the German Democratic Party, favored by most German Jews, was the very epitome of the liberal center of the political scene.[87] Much of this was ignored by the non-Jewish public. Particularly in Germany the nationalist camp's accumulated hatred needed a pretext and a target for its outpourings. And so it pounced on the revolutionary Jews.

Rosa Luxemburg and the Jewish leaders in Bavaria represented the threat of Jewish revolution. For the nationalists the appointment of a number of Jewish cabinet ministers and other high officials proved that the hated republic was indeed in Jewish hands; the Right could point to Hugo Haase,

Otto Landsberg, Hugo Preuss, Eugen Schiffer, Emanuel Wurm, Oskar Cohn, and to the most visible Jewish minister of all, Walther Rathenau.[88] Rosa Luxemburg had been murdered on January 15, 1919; Walther Rathenau, appointed foreign minister barely six months before, was assassinated on June 25, 1922.

Rathenau's murderers—Erwin Kern (aged twenty-four) and Hermann Fischer (twenty-six), both members of a Free Corps unit called Naval Brigade Ehrhardt, and their accomplices Ernst Werner Techow (twenty-one), his brother Gerd (sixteen), and Ernst von Salomon, also a former Free Corps member—were, in Salomon's words, "young men from good families."[89] At their trial Techow declared that Rathenau was one of the Elders of Zion.[90]

The canonical text of the Jewish-conspiracy theorists, *Protocols of the Elders of Zion*, was secretly fabricated in the mid–1890s by order of Piotr Rachkovsky, chief of the Paris office of the Okhrana, the czarist secret police.[91] The *Protocols* comprised elements of two works from the 1860s, a French anti–Napoleon III pamphlet and a German anti-Semitic novel, *Biarritz*, by one Hermann Gödsche.[92] The entire concoction was meant to fight the spread of liberalism inside the Russian Empire. Rachkovsky was merely following the rich tradition of attributing worldwide conspiracies to Jews.

The *Protocols* remained obscure until the outbreak of the Russian Revolution. But the crumbling of the czarist regime and the disappearance of the Romanovs and then of the Hohenzollern and Habsburg dynasties suddenly endowed this mysterious text, which was carried westward by fleeing White Russians, with an entirely new significance. In Germany, where the *Protocols* was excerpted in 1919 in the *völkisch* publication *Auf Vorposten*, it came to be considered concrete proof of the existence of dark forces responsible for the nation's defeat in the war and for its postwar revolutionary chaos, humiliation, and bondage at the hands of the victors. Thirty-three German editions appeared in the years before Hitler's accession to power, and countless others after 1933.[93]

The various versions of the *Protocols* published over the decades in a variety of languages share a basically identical core consisting of purported discussions held among the "Elders of Zion" at twenty-four secret meetings. In the immediate future the elders are not to shy away from any violent means to achieve control of the world. Oddly enough total power is not intended to lead to some harsh despotism aimed only at benefiting the

Jews. The ultimate goal is described as the establishment of a just and socially oriented global regime. The people would rejoice at such beneficent government, and their satisfaction would ensure the survival of the Kingdom of Zion for centuries and centuries.

The last part of the *Protocols* reads like a prescription for some totalitarian utopia, precisely what many people longed for in that period of economic uncertainty and political crisis. Why, then, did this booklet inspire such fear and loathing? The hate effect of the *Protocols* was due simply to the very idea of *Jewish* domination over the Christian world. The elders were plotting the disintegration of Christendom. In the same vein the destruction of traditional elites and the very idea of revolution were terrifying to the upper- and middle-class majority of the *Protocols'* readers. A 1920 American edition, for instance, clearly linked the machinations of the Elders of Zion to the Bolshevik peril.[94]

In an article headlined "The Jewish Peril, a Disturbing Pamphlet: Call for Inquiry," the London *Times* of May 8, 1920, asked, "What are these 'Protocols'? Are they authentic? If so, what malevolent assembly concocted these plans and gloated over their exposition? Are they forgery? If so, whence comes the uncanny note of prophecy, prophecy in part fulfilled, in part far gone in the way of fulfillment?"[95] A year later the *Times* reversed itself, declaring that the *Protocols* was indeed a forgery. Nonetheless the May 1920 article had pointed to a fear buried deep in many minds: of falling victim to secret forces lurking in the dark. The *Protocols* thus exacerbated to the most extreme degree the paranoia prevalent in those years of crisis and disaster. If the Jewish threat was supranational, the struggle against it had to become global too, and without compromise. Thus, in an atmosphere suffused with concrete threats and imaginary forebodings, redemptive anti-Semitism seemed, more than ever before, to offer answers to the riddles of the time. And for the anti-Jewish true believers, the ultimate struggle for salvation demanded the unconditional fanaticism of one who could show the way and lead them into action.

V

"Middle-class anti-Semites and young students came. . . . Adolf Hitler spoke." The *Münchner Post* was describing a meeting, in the spring of 1920, of the former DAP (Deutsche Arbeiterpartei, or German Workers' Party), newly renamed NSDAP. "He behaved like a comedian. After every third sentence of lecture, as in a music-hall song, came the 'refrain': the Hebrews are guilty. . . . One thing must be recognized: Herr Hitler himself admitted that his speech was dictated by racial hatred. When the speaker brought up the question of how one should defend oneself against the Jews, calls from the assembly gave the answer: 'Hang them! Kill them!' "[96]

Although Hitler, in the letter (quoted earlier) to Adolf Gemlich, denounced emotional anti-Semitism and insisted on a rational, systematic course in order to achieve total elimination of the Jews, his own style during the first years of his anti-Jewish agitation was very close to the rabble-rousing techniques of other *völkisch* orators, and his arguments did not reach far beyond the usual *völkisch* interpretations of history.[97] "What happened to the city of the easy-going Viennese?" he asked on April 27, in a speech entitled "Politics and Jewry," and in answer exclaimed, "For shame! It's a second Jerusalem!" The police report at this point mentions "stormy applause."[98] None of that, however, amounted to a detailed presentation of Hitler's anti-Jewish credo. A major attempt at this was made for the first time on August 13, 1920, in a three-hour speech in the Hofbräuhaus, a Munich beer hall. The announced title was "Why Are We Anti-Semites?"[99]

At the very outset Hitler reminded his listeners that his party was spearheading a fight against the Jews that was of direct relevance to the workers and their basic problems. There followed a long disquisition on the essence of creative work. In a convoluted way, Hitler argued that work, considered not as imposed necessity but as creative activity, had become the very symbol and essence of the Nordic race, its ultimate form being the construction of the state. This led him back to "the Jew."

Taking the Bible, "which no one can say was written by an anti-Semite," as the basis for his argument, Hitler affirmed that for the Jew work was punishment: The Jew was unable to work creatively and thus unable to build a state. Work for him was but the exploitation of the achievements of others. Starting from this postulate, Hitler then stated the parasitic nature of Jewish existence in history: Throughout millennia, the Jew's subsistence and his racial striving to control the other people of the earth meant the parasitic undermining of the very subsistence of the host peoples, the exploitation of the work of others for the Jew's own racial interests. The absolute character of the racial imperative was unquestionable, and Hitler stated it in absolute terms: "With all that, we must recognize that there is no good or bad Jew; everyone here works according to the imperatives of the race, because the race—or do we prefer to say, the nation?—and all that is linked to it, character and so on, lies, as the Jew himself explains, in the blood, and this blood compels every single individual to act according to these principles. . . . He is a Jew: he is driven only by one single thought: how do I raise my nation to become the dominating nation?"[100]

The National Socialist Party had entered the arena at this crucial moment of the struggle. A new hope had arisen that "finally the day will come when our words will fall silent and action will begin."[101]

As German historian Eberhard Jäckel has emphasized, the broad scope of Hitler's anti-Semitism appeared only in *Mein Kampf*,[102] in which the full force of the apocalyptic dimension of the anti-Jewish struggle found its expression. That may have been an outcome of Hitler's independent evolution; it was probably the result of the ideological input of a man whom Hitler met either in late 1919 or early 1920: the writer, newspaper editor, pamphleteer, drug addict, and alcoholic Dietrich Eckart.

Eckart's ideological influence on Hitler and the practical help he extended to him on several decisive occasions between 1920 and 1923 have often been mentioned. Hitler himself never denied Eckart's impact: "He shone in our eyes like a polar star," he said of him, and added: "At the time, I was intellectually a child at the bottle."[103] *Mein Kampf* was dedicated to Hitler's comrades killed during the 1923 putsch and to Dietrich Eckart (who had died near Berchtesgaden on Christmas Eve 1923).

The notorious "dialogue" between Eckart and Hitler, *Bolschewismus von Moses bis Lenin: Ein Zweigespräch zwischen Adolf Hitler und Mir (Bolshevism from Moses to Lenin: A Dialogue Between Adolf Hitler and Myself)*, published some months after Eckart's death, was written by Dietrich Eckart alone, probably even without Hitler's knowledge.[104] For some historians the *Dialogue* is the expression of Hitler's basic ideological stance with regard to the Jewish issue;[105] for others the text belongs much more to Eckart's rather than to Hitler's way of thinking.[106] Whoever the author of the pamphlet may have been: "Everything we know about Eckart and Hitler lends credence to the document as a representation of the relationship and the ideas they shared."[107]

The themes of the *Dialogue* clearly appear in *Mein Kampf*, wherever Hitler's rhetoric surges to the metahistorical level. What is immediately striking in the *Dialogue*, even in its very title, is that Bolshevism is not identified with the ideology and the political force that came to power in Russia in 1917; Bolshevism is instead the destructive action of the Jew throughout the ages. Indeed, during the early years of Hitler's career as an agitator—and this includes the writing of the text of *Mein Kampf*—political Bolshevism, although always recognized as one of the instruments used by the Jews to achieve world domination, is *not* one of Hitler's central obsessions: It is a major theme only insofar as the Jews from whom it derives are *the* major theme. In other words, the revolutionary period of 1919 is not at center stage in Hitler's propaganda. Thus, to consider Nazism primarily a panic reaction to the threat of Bolshevism, as has been argued by German historian Ernst Nolte, for example, does not correspond to what we know about Hitler's early career.

The *Dialogue* is dominated by the apocalyptic dimension attributed to the Jewish threat. Eckart's pamphlet is certainly one of the most extreme presentations of the Jew as the force of evil in history. At the very end of the text, "he" (that is, Hitler) sums up the ultimate aim of the Jew: "'It is certainly so' he said, 'as you [Eckart] once wrote: "One can understand the Jew only when one knows toward what he aims for in the end. Beyond the domination of the world, toward the *destruction* of the world." ' "[108] This vision of the world ending as a result of the Jew's action reappears almost word for word in *Mein Kampf*: "If, with the help of his Marxist creed, the Jew is victorious over the other peoples of the world," Hitler wrote, "his crown will be the funeral wreath of humanity and this planet will, as it did thousands of years ago, move through [the] ether devoid of human beings."[109]

At the end of the second chapter of *Mein Kampf* comes the notorious statement of faith: "Today I believe that I am acting in accordance with the will of the Almighty Creator: by defending myself against the Jew, I am fighting for the work of the Lord."[110] In Eckart, and in Hitler as he came to state his creed from 1924 on, redemptive anti-Semitism found its ultimate expression.

Some historians have turned Hitler's ideological expostulations into a tight and highly coherent system, a cogent worldview (in its own terms); others have entirely dismissed the significance of the ideological utterances as either a system or as policy guidelines.[111] Here it is argued that Hitler's worldview indicated the goals of his actions, albeit in very general terms, and offered guidelines of sorts for concrete short-term political initiatives. Its anti-Jewish themes, presented in clusters of obsessive ideas and images, had the internal coherence of obsessions, particularly of the paranoid kind. By definition there are no loopholes in such systems. Moreover, although

Hitler's worldview was entirely geared toward political propaganda and political action, it was no less the expression of a fanatical belief. The combination of total belief and a craving for mass mobilization and radical action led naturally to the presentation of the worldview in simple and constantly repeated propositions, whose proof was offered not by means of intellectual constructs but by those of additional apodictic declarations reinforced by a constant stream of violent images and emotionally loaded metaphors. Whether these anti-Jewish statements were original or merely the rehashing of earlier and current anti-Semitic themes (which indeed they were) is basically irrelevant, as their impact stemmed from Hitler's personal tone and from his own individual style of presenting his metapolitical and political beliefs.

Does this mean that Hitler's anti-Jewish obsessions ought to be analyzed in terms of individual pathology? It is a lead that has often been followed;[112] it will not be taken up here. Suffice it to say that any such interpretation usually appears to be highly speculative and often reductive. Moreover, similar anti-Jewish images, similar threats, a similar readiness for violence were shared from the outset by hundreds of thousands of Germans belonging to the extreme right and later to the radical wing of the Nazi Party. If "pathology" there was, it was shared. Rather than an individual structure, we must face the social pathology of sects. It is unusual, however, for a sect to become a modern political party, and it is even more unusual for its leader and his followers to keep to their original fanaticism once they have acceded to power. This, nonetheless, was the unlikely course of things. And this road, which was to lead to domains of unfathomable human behavior, has a well-documented starting point lying in the full light of history: the ranks of a small extremist party in postwar Bavaria, which, after the failure of its 1923 putsch attempt, seemed doomed to oblivion in the German Republic's new atmosphere of increased political stability.

Hitler relentlessly repeated a story of perdition caused by the Jew, and of redemption by a total victory over the Jew. For the future Führer, the Jew's ominous endeavors were an all-encompassing conspiratorial activity extending throughout the span of Western history. The structure of Hitler's tale was not only inherent in its explicit content; it was also the essence of the implicit message the story conveyed. Despite the pretense of a historical analysis, the Jew, in Hitler's description, was dehistoricized and transformed into an abstract principle of evil that confronted a no less metahistorical counterpart just as immutable in its nature and role throughout time—the Aryan race. Whereas Marxism stressed the conflict of changing historical forces, Nazism and particularly Hitler's worldview, considered history as the confrontation of an immutable good and an immutable evil. The outcome could only be envisioned in religious terms: perdition or redemption.

There was another level to Hitler's vision of the Jewish enemy: The Jew was both a superhuman force driving the peoples of the world to perdition and a subhuman cause of infection, disintegration, and death. The first image, that of the superhuman force, raises a question left unanswered both in *Mein Kampf* and in Hitler's speeches: Why did the people of the world offer no resistance, why for centuries had they been driven to ruin by the machinations of the Jew without offering any effective resistance? This question will arise strongly many years later, in connection with Hitler's Reichstag speech of January 30, 1939, when he "prophesied" the extermination of the Jews if they were again to drive the European peoples into a war. How *was* it that the nations of the world were unable to withstand these machinations?

Implicit in this vision is a stupefied, hypnotized mass of peoples completely at the mercy of the Jewish conspiracy. They are the hapless cattle killed by sneering Jewish ritual slaughterers in the final scenes of *The Eternal Jew*, the film whose production was initiated and overseen by Goebbels in 1939–40. But, as Hitler profusely showed in *Mein Kampf*, the image of superhuman control typically gives way to the second one, subhuman threats of contamination, microbial infection, spreading pestilence. These are the swarms of germ-carrying rats that will later appear in one of the most repellent scenes of *The Eternal Jew*. Images of superhuman power and subhuman pestilence are contrary representations, but Hitler attributed both to one and the same being, as if an endlessly changing and endlessly mimetic force had launched a constantly shifting offensive against humanity.

Many of the images, not only in Hitler's vision of the Jew but also in Nazi anti-Semitism generally, seem to converge in such constant transformations. These images are the undistorted echo of past representations of the Jew as endlessly changing and endlessly the same, a living dead, either a ghostly wanderer or a ghostly ghetto inhabitant. Thus the all-pervasive Jewish threat becomes in fact formless and unrepresentable; as such it leads to the most frightening phantasm of all: a threat that looms everywhere, that, although it penetrates everything, is an invisible carrier of death, like poison gas spreading over the battlefields of the Great War.

The last major *written* expression of Hitler's anti-Jewish obsession was the second volume of *Mein Kampf,* published in 1927. Another book by Hitler, completed in 1928, remained in manuscript form:[113] It was politically safer not to disclose the violence of the Führer's views, mainly on international affairs, as he was now donning the garb of a statesman. In his speeches, however, Hitler was less restrained.

In an article of November 5, 1925, headlined HITLER IN BRAUN-SCHWEIG, the *Braunschweigische Landeszeitung* reported a speech delivered by the Nazi leader at a party meeting in the city's concert hall. After mentioning some of the themes of the speech, the story noted that "Hitler dealt with the Jews in well-known form and the usual fashion. One knows what the National Socialists have to say against these citizens, and therefore we may spare ourselves reporting how Hitler held forth on this theme."[114]

The writer of the article could not have put it more concisely or more truthfully. A similar remark appeared in the *Mecklenburger Nachrichten*'s account, on May 5, 1926, of a Hitler speech in Schwerin two days earlier.[115] The hail of insults and threats against the Jews was, if at all, even more massive than in the past. At this time hardly any of Hitler's speeches lacks the kind of anti-Semitic rhetoric established in the early speeches and in *Mein Kampf.* It is as if the failure of the 1923 putsch, as if imprisonment and the temporary disbandment of the Nazi Party, had led to a heightened fury, or as if the needs of political agitation demanded the most aggressive and repetitive slogans that could possibly be mustered. The stock-market Jews and Jewish international capital were brandished side by side with bloodthirsty Jewish revolutionaries; the themes of Jewish race defilement and a Jewish conspiracy to control the world were fed to the delirious party faithful with the same instantaneous effect. In order to hammer home his attacks, Hitler used every rhetorical device, even the rather unusual method of telling well-known Jewish jokes in order to illustrate the perversity of the Jewish soul.[116]

Yet, even in the aftermath of his imprisonment in Landsberg, whenever political expedience dictated caution in the use of gross anti-Jewish outbursts, Hitler knew how to avoid the topic. When, on February 28, 1926, he spoke to the Hamburg National Club of 1919, a conservative-nationalist association whose generally upper-class membership included a number of former high-ranking officers, the Nazi leader simply avoided reference to the Jews.[117] One is reminded of the "detachment" of his later speech to the Association of German Industrialists in Düsseldorf. But what drove Hitler was his anti-Jewish hatred, and it was the calculated restraint that demanded effort. For Hitler the struggle against the Jews was the immutable basis and obsessional core of his understanding of history, politics, and political action.

Sometimes the anti-Jewish stance was rephrased in unexpected terms. Thus, according to a police report, Hitler declared in a speech in Munich on December 18, 1926, that "Christmas was significant precisely for National-Socialists, as Christ had been the greatest precursor in the struggle against the Jewish world enemy. Christ had not been the apostle of peace that the Church had afterward made of him, but rather was the greatest fighting personality that had ever lived. For millennia the teaching of Christ had been fundamental in the fight against the Jew as the enemy of humanity. The task that Christ had started, he [Hitler] would fulfill. National Socialism was nothing but a practical fulfillment of the teachings of Christ."[118]

Hitler's speeches during the decisive year 1932 have not yet been published as this book goes to press, but most of the diatribes of the years 1927–31 are now available:[119] In them anti-Semitic hatred remained prominent. Sometimes, as in Hitler's ferocious polemic against the Bavarian People's Party (Bayerische Volkspartei, or BVP) in the Munich speech of February 29, 1928, not very long before the May national elections, the agitator's venom of the early twenties was back in full force, with the Jews as the central issue because the BVP had rejected anti-Semitism. The themes were the same; the rhetorical devices were the same; the delirious reactions of the crowd were the same: Speaker and audience were thirsting for violence—against the same people, the Jews.[120]

In the 1928 Reichstag elections, the Nazis received only 2.6 percent of the vote (6.1 percent in Bavaria, 10.7 percent in Munich): The breakthrough was yet to come. Anti-Jewish agitation continued. "We see," Hitler exclaimed in his speech of August 31, 1928, "that in Germany, Judaization progresses in literature, the theater, music, and film; that our medical world is Judaized, and the world of our lawyers too; that in our universities ever more Jews come to the fore. I am not astonished when a proletarian says: 'What do I care?' But it is astonishing that in the national bourgeois camp there are people who say: 'This is of no interest to us, we don't understand this anti-Semitism.' *They will understand it when their children toil under the whip of Jewish overseers.*[italics in the original]"[121]

After the stunning success of the NSDAP in the September 1930 elections, and during the almost two and a half years that followed until Hitler acceded to the chancellorship, the Jewish theme indeed became less frequent in his rhetoric, but it did not disappear. And when Hitler did refer to the Jews, as, for example, in a speech on June 25, 1931, the reference carried all the dire predictions of former years. In the first part of that speech, Hitler described how the Jews had destroyed the Germanic leadership in Russia and taken control of the country. In other nations the same process was developing under the cover of democracy. But the finale was more direct and more threatening: "The parties of the middle say: everything is collapsing; we declare: what you see as collapse is the beginning of a new era. There is but one question about this new era: will it come from the German people . . . or will this era sink toward another people? Will the Jew really become master of the world, will he organize its life, will he in the future dominate the nations? This is the great question that will be decided, one way or the other."[122]

For external consumption Hitler sounded far less apocalyptic, far more moderate. In an interview given to the London *Times* in mid-October 1930, he assured the correspondent that he was not to be linked to any pogroms. He merely wanted "Germany for the Germans"; his party did not object to "decent Jews," but if the Jews identified with Bolshevism—and many unfortunately were inclined to do so—he would consider them enemies.[123] Incidentally, in articles published at the same time, Hitler expressed his conviction that recurring reports about the growth of anti-Semitism in the Soviet Union and interpretations of the conflict between Stalin and Trotsky as a struggle between an anti-Semite and a Jew were unfounded and farcical: "Stalin does not have to be circumcised, but nine-tenths of his associates are authentic Hebrews. His actions only continue the complete uprooting of the Russian people with the aim of its total subjugation to the Jewish dictatorship."[124]

Whatever Hitler may have been writing about the Jewish dictatorship in the Soviet Union, in Germany some people were taken in by the apparent ideological change expressed in the *Times* interview. On October 18, 1930, Arthur Mahraun, himself no philo-Semite and the leader of the conservative Jungdeutscher Orden, the youth movement of the newly formed

Deutsche Staatspartei (German State Party), wrote in his organization's periodical: "Adolf Hitler has abandoned anti-Semitism; this much one can now say with certainty. But officially [he has done so] for the moment only vis-à-vis foreign representatives and above all for the consumption of the jobbers in the City and Wall Street. At home, however, National Socialist supporters continue to be taken for a ride with anti-Semitic slogans."[125] Was Mahraun really fooled by Hitler's tactical pronouncements?

Hitler's partial restraint at this time was more than made up for by his subordinates.[126] The prime example was the new Berlin Gauleiter, Joseph Goebbels, and his weekly (later daily), *Der Angriff* (The attack), a paper certainly worthy of its name: it was ruthless and relentless against its main target, the Jews. As the symbol of the Jews' evil machinations and misuse of power, Goebbels chose Dr. Bernhard Weiss, vice president of the Berlin police, whom the Gauleiter dubbed "Isidor." Dozens of anti-Isidor articles appeared from May 1927 (when the police temporarily banned the Nazi Party in Berlin) to the eve of the seizure of power; the articles were given extra punch by Hans Schweitzer's (pen name: "Mjölnir") cartoons. A book of the earliest of these articles by Goebbels, along with the cartoons, was published in 1928 as *Das Buch Isidor* (The Isidor book).[127]

On April 15, 1929, *Der Angriff* turned its attention to a young boy's unexplained death in the vicinity of Bamberg. Goebbels's paper stated that a conclusion could be reached if "one were to ask which existing 'religious community in Germany has already been under suspicion for hundreds of years for containing fanatics who use the blood of Christian children for ritual purposes."[128] A Berlin court dismissed the slander charge that was brought against *Der Angriff* by arguing that Goebbels's paper had not stated that the Jewish community as such encouraged murder and that putting quotation marks around "religious community" meant merely that the author of the article was not certain that the Jews were a religious community.[129] Nazi anti-Jewish propaganda continued without respite throughout the decisive months preceding Hitler's accession to power.[130]

VI

On November 19, 1930, the Hebrew-language theater Habimah presented S. Anski's *The Dybbuk* in the Würzburg municipal theater. A group of Nazis in the crowd tried, without success, to stop the performance. As it was leaving the theater, the predominantly Jewish audience was attacked by the Nazis and several Jews were seriously wounded. When the assailants were taken to court, the judge dismissed the charges, arguing that "the demonstrators did not act from base motives."[131] The Würzburg mayor explained that the police had not intervened because they were certain that the demonstration had "merely" aimed at preventing a show.[132] Although physical assaults of this kind were infrequent during the Weimar years, a pogrom-like anti-Jewish rampage that started in Berlin's Scheunenviertel district on November 5, 1923, went on for several days.[133]

Although there is no straight line between these developments and the events that followed 1933, the trends described here are part of a historically relevant background. Nonetheless, this focus on anti-Semitism should not lead to a skewed perception of the German scene—and particularly of the situation of the Jews in Germany—before 1933. The Jewish influence perceived by the anti-Semites was mythical, but for the great majority of Jews in Germany the Weimar Republic opened the way to social advancement and, indeed, to a greater role in German life. The growth of anti-Semitism was real, but so—for a time at least—was a powerful renaissance of Jewish culture in Germany[134] and, until the onset of the crisis in 1929–30, a wide acceptance of Jews among the liberal and left-wing sectors of German society. On the right, however, anti-Semitism spread unabated, and during the final phase of the republic, it caught on beyond the reaches of the radical, and even the traditional, Right.

No political group shared the rabid anti-Jewish positions of the Nazis, but even during the years of stabilization, between 1924 and 1929, extreme anti-Semitic themes were not uncommon in the political propaganda of the nationalist camp, particularly in that of the German National People's Party (DNVP), whose *völkisch* wing was particularly vehement. At the end of 1922, the most extreme of the anti-Semitic DNVP Reichstag members, Wilhelm Henning, Reinhold Wulle, and Albrecht von Gräfe, left the party to establish their own political organization. But during the debates surrounding this secession, Oskar Hergt, one of the leaders of the DNVP and the former finance minister of Prussia, nonetheless reaffirmed that anti-Semitism remained a fundamental political commitment of the party.[135] For the French journalist Henri Béraud, who himself was to become an extreme anti-Semite in the 1930s, the German right's Jew-hatred seemed completely out of control. "We have no idea in France," Béraud wrote in a report from Berlin in 1926, "of what the anti-Semitism of German reactionaries can be. It is neither an opinion nor a feeling, nor even a physical reaction. It is a passion, a real obsession of addicts which can go as far as crime."[136]

In 1924, the bankruptcy of the brothers Heinrich and Julius Barmat, two Polish Jews who had settled in Germany in 1918, led to a full-scale right-wing anti-Semitic and anti-Republic onslaught. The Barmat brothers were accused of having received loans from the state-sponsored postal savings bank in return for various financial favors to Social Democratic politicians. Given the political ramifications of the affair, the right-wing parties succeeded in setting up an investigation committee that led to the resignation and indictment of several ministers and Reichstag members. But the main target of the right-wing campaign was President Friedrich Ebert, who was accused of having helped the Barmats to obtain a permanent residence permit and even of having dabbled in their food import transactions during the immediate postwar years.[137] There was a similar situation, on a smaller scale, in 1929, with the bankruptcy of the Sklarek brothers.[138] The main casualty this time was the mayor of Berlin, and the political consequence a contribution to the Nazi Party's strong showing in that year's local election.[139]

Political parties soon limited the number of their Jewish Reichstag members—with the exception of the Social Democrats, who retained approximately 10 percent Jewish membership on their Reichstag list to the very end. A telling illustration of the change of mood is to be found in the German Communist Party: In 1924 there were still six Jews among the party's sixty-four Reichstag members; in 1932 not a single one remained.[140] The Communists did not hesitate to use anti-Semitic slogans when such slogans were deemed effective among potential voters.[141]

The most significant political expression of the general climate of opinion was the transformation of the German Democratic Party (DDP), which had often been dubbed the "Jewish Party" because of the prominence of Jews among its founders, the large number of Jews among its voters, and, for a while at least, its espousal of themes identified with the positions of the "Jewish press."[142] In the January 1919 elections, the DDP obtained 18.5 percent of the vote, which made it the most successful of the middle-class liberal parties.[143] That success did not last. Gustav Stresemann's DVP kept attacking the competing DDP as "Jewish," and, as a result, the DDP steadily declined. Within the party itself, personalities associated with the "liberal" right were openly critical of the party's identification with Jewish voters and influence.[144] In 1930 the DDP as such disappeared, to be replaced by the Deutsche Staatspartei (German State Party). This group's leadership became mostly Protestant and some of its components, such as the youth movement Jungdeutscher Orden, did not admit Jews. The DDP's voters had been the pro-Weimar liberal middle classes; the change in party name and policy reflect what were perceived, within these middle-class liberal circles, as electorally useful attitudes regarding the "Jewish problem."

However, neither the "de-Judaization" represented by the Staatspartei nor the hostility of the DVP was of any avail to these parties. Whereas in the elections of 1928 the DDP obtained twenty-five seats and the DVP forty-five, and in those of 1930 the DDP still gained twenty seats and the DVP thirty, in the elections of July 1932, the DDP was reduced to four seats and the DVP to seven.[145] The decline of the liberal parties during the Weimar Republic has been thoroughly analyzed, and the social transformation that underlay it starkly defined.[146] In terms of the changing situation of the Jews of Germany, it meant that their main political basis (apart from the Social Democrats) had simply disappeared.

The "pernicious" influence of Jews on German culture was the most common theme of Weimar anti-Semitism. On this terrain, the conservative German bourgeoisie, the traditional academic world, the majority of opinion in the provinces—in short, all those who "felt German"—came together with the more radical anti-Semites.

The role of Jews in Weimar culture—in modern German culture in general—has been most extensively discussed, and, as we have seen, this theme was not only on the minds of anti-Semites, but often a source of preoccupation for Jews themselves, at least for some of them. In his first book on the subject, the historian Peter Gay showed what role the former "outsider" (mainly the Jew) played in the German culture of the 1920s;[147] later he reversed his position, arguing that, objectively, there was nothing to distinguish Jewish from non-Jewish contributions to German culture and that, as far as cultural modernism in particular is concerned, the Jews were neither more nor less "modern" than their German environment.[148]

Such downplaying of the Jewish dimension may well miss part of the context that provided the anti-Semitic ranters of the twenties with their ammunition.[149] The situation described, for example, in Istvan Deak's study of "Weimar's left-wing intellectuals" seems closer both to reality and to what the general perception was. After surveying the dominant influence of Jews in the press, book publishing, theater, and film, Deak turns to art and literature: "Many of Germany's best composers, musicians, artists, sculptors and architects were Jews. Their participation in literary criticism and in literature was enormous: practically all the great critics and many novelists, poets, dramatists, essayists of Weimar Germany were Jews. A recent American study has shown that thirty-one of the sixty-five leading German 'expressionists' and 'neo-objectivists' were Jews."[150] Deak's presentation in turn demands some nuancing, as, after all, the cultural scene in the twenties was dominated by such figures as Thomas Mann, Gerhart Hauptmann, Bertolt Brecht, Richard Strauss, Walter Gropius; but undoubtedly, in the minds of the middle-class public, be it of the extreme or the moderate right, anything "daring," "modern," or "shocking" was identified with the Jews. Thus, when shortly after (the entirely non-Jewish) Frank Wedekind's death, his "sexually explicit" *Schloss Wetterstein* was staged in Munich (December 1919), the political right did not hesitate to call it Jewish garbage. The police warned that performance of the play would lead to a pogrom,[151] and, sure enough, during the last performance Jews and people who "looked Jewish" in the audience were beaten up.[152] As a police report put it: "One can easily understand that a German who still feels German to some degree and who is not morally and ethically perverted looks with greatest disgust upon the public enjoyment of Wedekind plays."[153] Jewish writers and artists may not have been any more extreme modernists than their non-Jewish colleagues, but modernism as such flourished in a culture in which the Jews played a central role. For those who considered modernism the rejection of all hallowed values and norms, the Jews were the carriers of a massive threat.

More ominous, however, than cultural modernity was left-wing culture in all its aspects. Within months of the end of the war, Jewish revolutionaries were easy targets of the counterrevolution. After Rathenau's murder no Jew (with the exception of the Socialist finance minister Rudolf Hilferding) played any significant role in Weimar politics. On the other hand, left-wing political, social, and cultural criticism and innovation were often "Jewish." "If cultural contributions by Jews were far out of proportion to their numerical strength," Deak writes, "their participation in left-wing intellectual activities was even more disproportionate. Apart from orthodox Communist literature where there were a majority of non-Jews, Jews were responsible for a great part of the leftist literature in Germany. [The periodical] *Die Weltbühne* was in this respect not unique; Jews published, edited, and to a great part wrote the other left-wing intellectual magazines. Jews played a decisive role in the pacifist and feminist movements, and in the campaigns for sexual enlightenment."[154]

Polemics regarding the role of Jews on the cultural scene raged and became more virulent as the Nazi movement grew in strength and as the republic approached its end. One of the most extreme forums of the Right was the Nazi ideologue Alfred Rosenberg's Kampfbund für deutsche Kultur (Fighting League for German Culture), established in 1928; it achieved wide influence by opening its ranks to a variety of antirepublic, anti-Left, anti-Jewish elements—from members of the Bayreuth Circle to conservative Catholics like Othmar Spann, from fanatic anti-Semitic literary specialists like Adolf Bartels to Alfred Heuss, publisher of the *Zeitschrift für Musik*. But sometimes the debates took place in more neutral contexts or were even initiated by Jewish organizations. Thus, in 1930, Max Naumann's Association of National German Jews invited the right-wing literary critic Paul Fechter to lecture on "The Art Scene and the Jewish Question." Fechter did not mince words. He warned his listeners that the "anti-Germanism" of left-wing Jewish intellectuals was a major source of rising anti-Semitism and that the Germans would not tolerate for long the continuation of this state of things. National Jews and national Germans, Fechter suggested, should act in common to oppose such anti-national Jewish intellectual attacks. In a more roundabout way, he hinted at the excessive presence of Jews in German art, literature, and theater. This, too, although unsaid, could be understood as a source of growing anti-Jewish feelings: "I feel obliged to express," declared Fechter, "that a great number of German authors, painters, playwrights go around today with the feeling that it is much more difficult to find a place in German theaters, on the German book market, in the German art business, for things German than for others."[155]

Fechter's lecture was published in the January 1931 issue of Rudolf Pechel's *Deutsche Rundschau*, with the following editorial comment: "We reproduce [the lecture] as it indicates one of the sources of the dangerous growth of anti-Semitism clearly confirmed during the second half of 1930 and as it indicates some ways that still may allow us to counter this danger."[156] A bitter debate followed. It is in this context that the novelist Jakob Wassermann, whose autobiographical essay, "My Way as German and Jew," was possibly the strongest expression of the anguish German Jews felt in the face of the growing tide of anti-Semitism, addressed his question to Rudolf Pechel: "Do the rules of good behavior help against 'Perish, Jew!'?"[157]

One of the more remarkable Jewish contributions to the debate was that of Arthur Prinz, published in the periodical's April 1931 issue under the title "Toward Eliminating the Poison from the Jewish Question." After asking why radical Jewish journalists and literati could provoke such furious anti-Semitic rage in Germany, Prinz ventured an answer that probed deeply into the relations between Germans and Jews: "That sort of journalism and literature would be impossible without that deep and old insufficiency of a healthy state and national feeling in Germany, which threatens to become fatal since the sad outcome of the war and can certainly not be 'compensated for' by the excessive nationalism of the extreme right. The agitation of rootless Jews is poison in a body particularly receptive to it, and precisely this is the main reason for boundless anti-Jewish hatred."[158]

When one turns to the wider reaches of German society as it approached the political turning point of 1933, there is no way of assessing clearly the strength of its anti-Jewish attitudes. For example, the League of Jewish Women (Jüdischer Frauenbund) found its allies in the much larger Federation of German Women's Associations (Bund Deutscher Frauenvereine, or BDF) in their common struggles on feminist issues, but any indication of Jewish identity was not more acceptable to the German women's organization than it was to the surrounding society. In the words of a historian of the league, the attitudes in the BDF "ranged from liberal impatience with Jewish distinctiveness to covert or overt anti-Semitism."[159] As for the nature of this anti-Semitism, one of its most nuanced evaluations remains the most plausible: "More common and widespread than outright hatred or sympathy for the Jews was . . . moderate anti-Semitism, that vague sense of unease about Jews that stopped far short of wanting to harm them but that may have helped to neutralize whatever aversion Germans might otherwise have felt for the Nazis."[160]

In early August 1932 Hitler was negotiating with the consummate schemer and not yet the short-lived last chancellor of the Weimar Republic (November 1932–January 30, 1933) Gen. Kurt von Schleicher, at the time still a close confidant of President Hindenburg, the conditions for his being named to the chancellorship. On the tenth of that month, five SA men forced their way into the home of Konrad Pieczuch, a pro-Communist worker in the small town of Potempa in Upper Silesia, and trampled him to death. "Such brutality once again put a serious obstacle on the path of the Nazi march to power."[161] Hitler had apparently believed that the top position would now be offered to him; what Hindenburg proposed when they finally met was a mere vice-chancellorship. The meeting had been cool, and the official communiqué dismissive of the Nazi leader. Hitler was utterly humiliated and furious. It was exactly then, on August 22, that the court in Beuthen sentenced the five SA men to death. The announcement of the verdict led to tumultuous scenes in the courtroom; outside, Jewish and "socialist" shops were attacked. Hitler reacted with an outburst of rage. He wired the convicted murderers: "My comrades! In view of this incredible criminal verdict I feel myself tied to you in unlimited fidelity. From this moment on, your freedom is our honor, the fight against a government under which such a thing was possible, our duty."[162]

THE JEWS ARE GUILTY! Goebbels thundered in *Der Angriff:* "The Jews are guilty, the punishment is coming. . . . The hour will strike when the state prosecutor will have other tasks to fulfill than to protect the traitors to the people from the anger of the people. Forget it never, comrades! Tell it to yourself a hundred times a day, so that it may follow you in your deepest dreams: the Jews are guilty! And they will not escape the punishment they deserve."[163]

In a moment of sheer frustration, Hitler had abandoned his carefully constructed facade of respectability and given vent to relentless and murderous rage. Nonetheless, during those same weeks of the summer and fall of 1932, Hitler continued to oppose the use of force for toppling the regime and went on negotiating and maneuvering in order to reach his goal.[164] What emerges here with uncanny clarity is a personality in which cold calculation and blind fury coexisted and could find almost simultaneous expression. If a third ingredient—Hitler's ideological fanaticism—is added, an insight into the psychological makeup that led to the Nazi leader's most crucial decisions may be possible, also with regard to the Jews.

Ideological fanaticism and pragmatic calculation constantly interacted in Hitler's decisions. The ideological obsession was unwavering, but tactical considerations were no less compelling. Sometimes, however, the third element, uncontrolled fury, would burst into the open—triggered by some obstacle, some threat, some defeat—sweeping away all practical considerations. Then, fed by the torrent of ideological fanaticism, the murderous fury would explode in an unlimited urge for destruction and death.

CHAPTER 4

The New Ghetto

I

"Cell 6: approximately 5 m. high, window approx. 40 x 70 cm. at a height of 4 meters, which gives the feeling of a cellar. . . . Wooden plank with straw mat and two blankets, a wooden bucket, a jug, a basin, soap, a towel, no mirror, no toothbrush, no comb, no brush, no table, no book from January 12 [1935] until my departure on September 18; no newspaper from January 12 to August 17; no bath and no shower from January 12 to August 10; no leaving of the cell, except for interrogations, from January 12 to July 1. Incarceration in an unlighted cell from April 16 to May 1, then from May 15 to August 27, a total of 119 days."[1]

This was the Würzburg wine merchant Leopold Obermayer writing about the first of his imprisonments in Dachau, in a seventeen-page report, dated October 10, 1935, which he managed to smuggle out to his lawyer. It was seized by the Gestapo and found after the war in their Würzburg files. Obermayer had a doctorate in law (from Frankfurt University); and he was a practicing Jew and a Swiss citizen. October 29, 1934, he had complained to the Würzburg police that his mail was being opened. Two days later, having been ordered to report to headquarters, he was arrested. From then on he became a special case for the local Gestapo chief, Josef Gerum, a Nazi "old fighter" with a bad reputation even among his colleagues. Gerum accused Obermayer of spreading accusations about the new regime. Shortly afterward nude photographs of Obermayer's male lovers were found in his bank safe. Both a Jew and a homosexual: For Gerum this was indeed a rewarding catch.

In his report Obermayer alludes many times to his tormentors' bound-less hatred of the Jews; they assured him that he would never be set free, and tried to drive him to suicide. Why didn't they kill him? Writing about Obermayer's story, Martin Broszat and Elke Fröhlich give no clear expla-nation. It seems, however, that murdering a Swiss citizen, albeit a Jewish one, was not yet done lightly in 1935, all the more so since the Swiss con-sulate in Munich, and later the legation in Berlin, were aware of Obermayer's incarceration; the Ministry of Justice in particular was wor-ried about the possibility of Swiss intervention.[2]

Under interrogation Obermayer was pressed to give details about his lovers; he refused and was beaten up. On May 15, as he was once more being taken to the camp commander's office for interrogation, he asked an SS man named Lang, who had just threatened to shoot him, whether he had any compassion at all. Lang replied: "No, for Jews I have none." Obermayer complained to the commander, SS-Oberführer Deubel, about the way he was being treated. "Thereupon the SS-Truppenführer standing at the window said: 'You are not a human being, you are a beast!' I started to answer: 'Frederick the Great was also one. . . .' Before I could say another word, this Truppenführer hit me in the face: my upper middle tooth was knocked out and I started bleeding from the mouth and nose: 'You Jewish pig, comparing yourself to Frederick the Great!'" Further retribution was immediate: unlighted cell, no straw mat on the wooden plank, arms tied behind the back, manacles left unopened for up to thirty-six hours, so that, Obermayer wrote, he had to defecate and urinate in his trousers.[3]

In mid-September 1935 Obermayer was transferred from Dachau to an ordinary prison in Ochsenfurt, pending court interrogation. In the meantime Obermayer's lawyer, Rosenthal, a Jew, had also been arrested, and it was in his house that Gerum found the incriminating report about the conditions of Obermayer's detention in Dachau. Rosenthal was released and later left Germany: His wife had committed suicide. The court in Ochsenfurt did not keep Obermayer for long. At Gerum's insis-tence the Jewish homosexual was taken back to Dachau on October 12, 1935.[4] Obermayer will reappear in these pages.

At this time Germany and the world were witnessing a dramatic consolida-tion of Hitler's internal and international power. The murder of Ernst Röhm and other SA leaders on the notorious Night of the Long Knives in

June 1934 eliminated even the faintest possibility of an alternative source of power within the party. Immediately following Hindenburg's death, the naming of Hitler as Führer and chancellor on August 2 made him the sole source of legitimacy in Germany. Hitler's popularity reached new heights in 1935: On January 13 an overwhelming majority of the Saar population voted for return of the territory to the Reich. On March 16 general conscription and establishment of the Wehrmacht were announced. No foreign power dared to respond to these massive breaches of the Versailles Treaty; the common front against Germany formed at Stresa by Britain, France, and Italy in April 1935, in order to defend Austria's independence against any German annexation attempt and preserve the status quo in Europe, had crumbled by June, when the British signed a separate naval agreement with Germany. On March 17 of that year, Hitler had been in Munich, and a report for the clandestine Socialist Party vividly captured the overall mood:

"Enthusiasm on 17 March enormous. The whole of Munich was on its feet. People can be forced to sing, but they can't be forced to sing with such enthusiasm. I experienced the days of 1914 and can only say that the declaration of war did not make the same impact on me as the reception of Hitler on 17 March. . . . Trust in Hitler's political talent and honest intentions is getting ever greater, just as generally Hitler has again won extraordinary popularity. He is loved by many."[5]

Between 1933 and 1936 a balance of sorts was kept between the revolutionary-charismatic impulse of Nazism and the authoritarian-conservative tendencies of the pre-1933 German state: "The marriage of an authoritarian system of government with the mass movement of National Socialism seemed to be successful in spite of considerable friction over key points, and also [seemed] to have overcome the shortcomings of the authoritarian system," wrote Martin Broszat.[6] Within this temporary alliance Hitler's role was decisive. For the traditional elites the new "belief in the Führer" became associated with the authority of the monarch. Basic elements of the Imperial state and of the National Socialist regime were linked in the person of the new leader.[7]

Such "belief in the Führer" led quite naturally to an urge for action on the part of state and party agencies according to the general guidelines set by Hitler, without the constant necessity of specific orders from him. The dynamics of this interaction between base and summit was, as British historian Ian Kershaw pointed out, "neatly captured in the sentiments of a routine speech of a Nazi functionary in 1934":

"'Everyone who has the opportunity to observe it knows that the Führer can hardly dictate from above everything which he intends to realize sooner or later. On the contrary, up till now everyone with a post in the new Germany has worked best when he has, so to speak, worked towards the Führer. Very often and in many spheres it has been the case—in previous years as well—that individuals have simply waited for orders and instructions. Unfortunately, the same will be true in the future; but in fact it is the duty of everybody to try to work towards the Führer along the lines he would wish. Anyone who makes mistakes will notice it soon enough. But anyone who really works towards the Führer along his lines and towards his goal will certainly both now and in the future one day have the finest reward in the form of the sudden legal confirmation of his work.'"[8]

Thus the majority of a society barely emerging from years of crisis believed that the new regime offered solutions that, in diverse but related ways, would give answers to the aspirations, resentments, and interests of its various sectors. This belief survived the difficulties of the early phase (such as a still sluggish economy) as a result of a new sense of purpose, of a series of successes on the international scene, and, above all, of unshaken faith in the Führer. As one of its corollaries, however, that very faith brought with it widespread acceptance, passive or not, of the measures against the Jews: Sympathy for the Jews would have meant some distrust of the rightness of Hitler's way, and many Germans had definitely established their individual and collective priorities in this regard. The same is true in relation to the other central myth of the regime, that of the *Volksgemeinschaft*. The national community explicitly excluded the Jews. Belonging to the national community implied acceptance of the exclusions it imposed. In other words, adherence to "positive" tenets of the regime, to mobilizing myths such as the myth of the Führer and that of the *Volksgemeinschaft*, sufficed to undermine explicit dissent against anti-Jewish measures (and other of the regime's persecutions). Yet, as we shall see, despite these general trends, there were nuances in German society's attitudes toward the "outsiders" in its midst.

Hitler's tactical moderation on any issue that could have negative economic consequences shows his conscious alignment with the conservative allies. But when it came to symbolic expressions of anti-Jewish hatred, the Nazi leader could barely be restrained. In April 1935 Martin Bormann, then Rudolf Hess's chief of staff, inquired whether Hitler wished to remove the anti-Jewish placards that were sprouting up all over the Reich.

Fritz Wiedemann, Hitler's adjutant, informed Bormann that the Führer was opposed to their removal.[9] The matter soon resurfaced when Oswald Leewald, president of the German Olympic Committee, complained that these signs were contributing to ongoing anti-Jewish agitation in such major Olympic sites as Garmisch-Partenkirchen. The Olympic Games will be dealt with later on, but with regard to the anti-Jewish notices, Hitler refused at first to act against the initiatives of the regional party chiefs; only when he was told they could cause serious damage to the Winter Olympics did he give the order to remove the offensive signs.[10] Finally a general compromise solution was found. On June 11, 1935, the Ministry of Propaganda ordered that in view of the forthcoming Olympics, signs such as those reading JEWS UNWANTED should quietly be removed from major roads.[11] This may have been asking too much, for a few days before the beginning of the Winter Olympics, Hess's office issued the following decree: "In order to avoid a making a bad impression on foreign visitors, signs with extreme inscriptions should be taken away; signs such as 'Jews Are Unwanted Here' will suffice."[12]

II

On January 1, 1935, a Tübingen Jewish merchant, Hugo Loewenstein, received a medal "in the name of the Führer and Reichskanzler" for his service during World War I.[13] The same distinction was awarded to Ludwig Tannhäuser, a Stuttgart Jewish businessman, as late as August 1, 1935.[14] Yet, nearly a year and a half earlier, on February 28, 1934, Minister of Defense Werner von Blomberg had ordered that the Aryan paragraph be applied to the army.[15] When the Wehrmacht was established, in March 1935, "national" Jews petitioned Hitler for the right to serve in the new armed forces.[16] To no avail: On May 21 military service was officially forbidden to Jews.[17] "Mixed breeds [*Mischlinge*] of the first and second degree" (these categories had already been in use at the Ministry of Defense before the Nuremberg Laws) could, however, be allowed to serve in the armed forces as individual exceptions.[18]

Earlier the army had attempted to help Jewish officers who were being dismissed. On May 16, 1934, a member of the Reichswehr Staff had approached a Chinese diplomat in Berlin with the suggestion that the Chinese Army find positions for some of the younger Jewish Reichswehr officers. Legation Secretary Tan expressed his personal interest in the idea, but was skeptical about its implementation: Nazi Party officials had already been in touch with the Chinese government to dissuade it from hiring German Jewish officers on the grounds that Jews were not representative of the German people, and thus the German Reich saw no value in any activity of theirs abroad.[19]

Goebbels could not lag far behind the military. Less than a month after Blomberg's order, on March 24, 1934, the propaganda minister announced that, as a matter of general principle, all Jews would be excluded from membership in the Reich Chamber of Culture. Preparations started immediately, and in early 1935 the remaining Jewish members of the various specific chambers began to be dismissed.[20] On November 15, 1935, at its annual meeting in Berlin, Goebbels was able to announce—somewhat prematurely, as will be seen—that the Reich Chamber of Culture was now "free of Jews."[21]

The relentlessness of the efforts to segregate the Jews was unmistakable. In ideological terms the most crucial domain was that of physical—that is, biological—separation; much in advance of the Nuremberg legislation, mixed marriages and sexual relations between Germans and Jews became targets of unceasing, often violent party attacks. The party press spearheaded this campaign, and the flow of anti-Jewish abuse spread by a paper such as Streicher's *Der Stürmer* did not remain without effect. On the other hand, however, contrary to the main thrust of party agitation, some groups of the population not only rejected anti-Jewish violence and hesitated to sever their economic ties with Jews, but even at times showed signs of sympathy for the victims. Beyond such reluctance to segregate the Jews completely, the "cleansing" of various areas of German life of any trace of Jewish presence encountered countless other difficulties. Thus, during this early phase of the regime, Jews still remained, in one way or another, in various domains of German life, although as a result of party agitation their situation worsened in the spring and summer of 1935.

The notion of race as such, defined as a set of common physical and mental characteristics transmitted within a group by the force of tradition or even in some biological way, had been used by Jews themselves from Moses Hess to Martin Buber, particularly in Buber's 1911 Prague lectures, published as *Three Speeches on Judaism*. It had not disappeared in postwar Germany. Thus, in a February 1928 speech on the problems facing German Jewry, the director of the Zentralverein, Ludwig Holländer, after asserting that the Jews had been a race since biblical times as a result of their common descent and nonetheless expressing doubts whether the concept of race was applicable to the modern Jew, went on, however, to tell his listeners: "Extraction remains, that is, the racial characteristics are still present, albeit diminished by the centuries; they are present in external as well as mental features."[22] In 1932 a fierce internal Jewish controversy arose around the publication by the Zionist author Gustav Krojanker of a booklet entitled *On the Problem of the New German Nationalism*. According to Krojanker, the Zionist revolt against liberalism, which was in response to a will aroused by the imperatives of the blood, should allow for a deep understanding of the political developments in Germany.[23]

Such rather extreme positions were those of a small minority, but they show the influence of *völkisch* thinking on some German Jews.[24] Here and there some Jewish voices even pleaded for "racial purity of the Jewish stock" and for investigations according to the rules of "racial science" for more ample and precise information regarding "the extent of miscegenation between Jews and Christians [*sic*], thus between members of the Semitic and Aryan race."[25] But these various statements had the connotation neither of racial hierarchy based on biological criteria nor of a struggle between races.

It seems, at the outset at least, that a widespread belief existed in the party that scientific racial criteria for identifying the Jew could be discovered. Thus, in a letter of September 1, 1933, to Baden's minister of the interior (with copies to all relevant authorities in the Reich), Wilhelm Frick made it clear that the identification of the "non-Aryan" was not dependent on parents' or grandparents' religion, but "on descent, on race, on blood." This meant that even if the religious affiliation of parents or grandparents was not Jewish, another criterion could be found.[26] This was the line of thinking that guided the Jena racial anthropologist Hans F. K. Günther in his attempt to identify various external physical characteristics of the Jew, as it did his Leipzig colleague Paul Reche to pursue his yearlong research on racially determined blood types. But even Reche had to admit that "no single blood type was typical among Jews."[27] This failure, however, though soon recognized by most Nazi scientists,[28] did not deter publications specializing in scientific vulgarization from announcing that, on this front as on all others, decisive breakthroughs had been achieved.

In the October 1934 issue of the *Volksgesundheitswacht* (People's health guardian), a Doctor Stähle offered "new research results" concerning "blood and race." He traced some illnesses specifically attributed to Jews (commenting ironically that these were "accumulative diseases"), referring mainly to the work of a Leningrad "scientist" named E. O. Malinoff. This Russian claimed that, with an accuracy of 90 percent, he could distinguish Jewish from Russian blood by chemical means. Stähle conveyed appropriate enthusiasm to his readers: "Think what it might mean if we could identify non-Aryans in the test tube! Then neither deception, nor baptism, nor name change, nor citizenship, and not even nasal surgery could help. . . . One cannot change one's blood!"[29] Stähle was head of the local medical society in Baden-Württemberg.[30]

Despite Stähle's optimism, biological criteria for defining the Jew remained elusive, and it was on the basis of the religious affiliation of parents and grandparents that the Nazis had to launch their crusade for racial purification of the *Volk*.

Almost three years before Hitler's accession to power, the Nazis had unsuccessfully demanded an amendment of the Law for the Protection of the Republic so as to define "betrayal of the race" (*Rassenverrat*) as a crime punishable by imprisonment or even by death. Such an offender would be anyone "who contributes or threatens to contribute to the racial deterioration and dissolution of the German people through interbreeding with persons of Jewish blood or the colored races."[31]

In September 1933 Hanns Kerrl, justice minister of Prussia, and his undersecretary, Roland Freisler, suggested to the party (in a memorandum entitled "National Socialist Criminal Law") that marriages and extramarital sexual relations between "those of German blood" and "members of racially alien communities" be considered "punishable offenses against the honor of the race and endangerment of the race."[32] At the time these proposals were not followed up. After the establishment of the new regime, however, the situation started to change de facto. Officials increasingly referred to the Law for the Restoration of the Professional Civil Service in order to refuse, on the basis of the law's "general national principles," to perform marriage ceremonies between Jews and "those of German blood."[33] The pressure grew to such a point that on July 26, 1935, Frick announced that, since the legal validity of "marriages between Aryans and non-Aryans" would be officially addressed in the near future, such marriages should be "postponed until further notice."[34]

The refusal to perform marriages was an easy matter compared to the other "logical" corollary stemming from the situation: the dissolution of existing mixed marriages. The Civil Code allowed for divorce on the basis of wrongdoing by one of the partners, but it was difficult to equate belonging to a particular race with the notion of wrongdoing.

Paragraph 1333 of the Civil Code did, however, stipulate that a marriage could be challenged if a spouse had been unaware, on contracting the marriage, of "personal qualities" or circumstances that would have precluded the union. But it could only be invoked within six months of the wedding, and racial identity could hardly be defined as a personal quality; finally it is unlikely that partners to a marriage were unaware of such racial identity at the time of their decision. Nevertheless, paragraph 1333 increasingly became the prop of Nazi legal interpretation, on the grounds that "Jewishness" was indeed a personal quality whose significance had become clear only as a result of the new political circumstances. Consequently, the six-month period could be counted from the date when the significance of Jewishness became a major element in public consciousness, that is, from January 30 (Hitler's accession) or even April 7, 1933 (the Civil Service Law's promulgation).[35]

As an increasing number of courts started basing their decisions on the new interpretation of the Civil Service Law, leading Nazi jurists, such as Roland Freisler, had to intervene in order to restore a semblance of order.[36] It was only with the law of July 6, 1938, that "racially" mixed marriages could in fact be legally annulled. The judges, lawyers, and registrars who were intent on the dissolution of mixed marriages were not necessarily members of the party; in their determination to segregate the Jews from society, they went beyond the immediate instructions of the Nazi leadership.

The anti-Jewish zeal of the courts regarding mixed marriages was reinforced by police initiatives and even by mob demonstrations against any form of sexual relations between Jews and Aryans: "Race defilement" was the obsession of the day. Thus on August 19, 1935, a Jewish businessman was arrested on that charge in Stuttgart. As he was brought to the police station, a crowd gathered and demonstrated against the accused. Shortly afterward, according to the city chronicle, a Jewish woman merchant who had had a stall in the market hall since 1923 lost her permit because she allowed her son to have a relationship with a non-Jewish German girl.[37]

Whether the demonstrators assembled in front of the Stuttgart police station were party activists, a mob drummed up by the party, or a random crowd of Germans is hard to say. The agitation against mixed marriages and race defilement reported from all parts of the Reich during the summer of 1935 offers no further clues. Thus a Gestapo report from Pomerania for the month of July 1935 indicates that *Volksgenossen* demonstrated in Stralsund on the 14th "because here various Jews had married Aryan girls," and in Altdamm on the 24th "because here a Jew had committed race defilement with a married Aryan woman."[38]

The party press spared no effort to fan the fury of the *Volksgenossen* against such pollution. Jewish race defilers must be castrated, demanded the *Westdeutscher Beobachter* on February 19, 1935. On April 10 the SS periodical *Das Schwarze Korps* called for dire punishment (up to fifteen years' imprisonment even for the German partner) for sexual relations between Germans and Jews.[39] All aspects of the witch-hunt that was to characterize the period following the passage of the Nuremberg racial laws were already visible.

The presence of Jews in public swimming pools was a major theme, second only to outright race defilement, in the Nazis' pornographic imagination: It expressed a "healthy" Aryan revulsion at the sight of the Jewish body,[40] the fear of possible contamination resulting from sharing the water or mingling in the pool area and, most explicitly, the sexual threat of Jewish nakedness, often alluded to as the impudent behavior of Jewish women and outright sexual harassment of German women by Jewish men. As could be expected, the theme surfaced in Nazi literature. Thus, in Hans Zöberlein's 1937 novel *Der Befehl des Gewissens* (Conscience commands), which takes place during the years immediately after World War I, the Aryan Berta is molested by Jews in an open-air swimming pool in Bavaria: "These Jewish swine are ruining us," she exclaims. "They are polluting our blood. And blood is the best and the only thing we have."[41]

In most German cities the expulsion of Jews from public bathing facilities became a prime party objective. In Dortmund, for example, the party press harped on the danger posed by the presence of Jews in municipal swimming pools until it achieved its goal with the publication of an announcement on July 25, 1935, by the city's mayor: "As a result of various unpleasant occurrences and due to the fact that the immense majority of the members of our German national community feels burdened by the presence of Jews, I have forbidden Jews the use of all public swimming pools, indoor public bathing facilities, and public sun-decks. At all these premises, warning signs will carry the following inscription: ACCESS TO THESE FACILITIES IS FORBIDDEN TO JEWS.[42]

The party press in Stuttgart initiated a similar campaign, the *NS-Kurier* reporting on July 8 that during the preceding week several Jewish women had had to be expelled from the city's swimming pools because of "their impudent behavior." The paper took the opportunity to point out that there were no signs forbidding access to Jews. With the city council divided on the issue, such signs were not finally posted in Stuttgart until after the 1936 Olympic Games.[43]

The process of exclusion seemed to follow a well-established pattern. Sometimes, however, minor hitches occurred. On August 1, 1935, the Bavarian Political Police reported an incident at the Heigenbrücken swimming pool on July 14, when some fifteen or twenty youths chanted for "the removal of the Jews . . . from 'the German baths.'" According to the police report: "A considerable number of other bathers joined in the chanting, so that probably the majority of visitors were demanding the removal of the Jews. In view of the general indignation and the danger of disturbances, the district leader of the NSDAP, Mayor Wohlgemuth of Aschaffenburg, who happened to be in the swimming pool, went to the supervisor of the baths and demanded that he remove the Jews. The supervisor refused on the grounds that he was obliged to follow only the instructions of the baths' administration and moreover, could not easily distinguish the Jews as such. As a result of the supervisor's statement, there was a slight altercation between him and the mayor, which was later settled by the baths' administration. In view of this incident, the Spa Association today placed a notice at the entrance of the baths with the inscription: ENTRY FORBIDDEN TO JEWS."[44]

Among the newspapers spewing a constant stream of anti-Jewish abuse, Streicher's *Der Stürmer* was the most vicious; its ongoing campaign and the wide distribution it achieved by means of public display may have been abhorrent to the educated middle class or even to educated party members, but its appeal among the general population, school youngsters, and the Hitler Youth, possibly because of its pornographic and sadistic streak, seems to have been quite widespread.

On May 1, 1934, *Der Stürmer* published its notorious special issue on Jewish ritual murder. The front-page headline, THE JEWISH MURDER PLOT AGAINST NON-JEWISH HUMANITY IS UNCOVERED, was graced by a half-page drawing of two particularly hideous-looking Jews holding a vessel to collect the blood streaming from the naked bodies of angelic Christian children they have just murdered (one of the Jews is holding a blood-stained knife). In the background stands a cross. The next day the National Representation of German Jews wired Reich Bishop Ludwig Müller: "We feel obliged to draw your attention to the special issue of the *Der Stürmer* of May 1. We have sent the following telegram to the Reich chancellor: '*Der Stürmer* has come out with a special issue which, using incredible insults and horrifying descriptions, accuses Jewry of ritual murder. Before God and humanity, we raise our voice in solemn protest against this unheard-of profanation of our faith.' We are convinced that the deep outrage that we are feeling is shared by every Christian." Neither Hitler nor Reich Bishop Müller replied.[45]

Along with great issues such as ritual murder, *Der Stürmer* also addressed more mundane items (although, in true *Stürmer* fashion, the mundane always led to the broader historical panorama), like one that came up in the summer of 1935. In its August 1935 issue (no. 35), Streicher's paper took up a story previously published by the *Reutlinger Tageblatt* about a Jewish chemist, Dr. R.F., who had been accused of torturing a cat to death. According to *Der Stürmer*, in order to kill the cat, F. had tied it up in a sack, which he then threw onto the concrete in front of his door. "After that, he jumped with both feet on the poor animal, performing a true Negro dance on it. As he could not kill the animal in that way, although it bled through the sack, he took a board and hit the cat with the edge until he killed it." *Der Stürmer* linked the killing of the cat to "the slaughter of 75,000 Persians in the Book of Esther" and to the killing of "millions of non-Jews" in "the most horrible way" in contemporary Russia. "The complacent bourgeois thinks far too little about what would happen in Germany if the Jews came to power once more," *Der Stürmer* concluded.[46] As could be expected, the *Stürmer* story aroused reader reactions. A woman from Munich addressed her letter to the culprit: "The opinion of all my female and male colleagues is that one should not treat you one hairsbreadth better [than the cat] and that you should be kicked and hit until you croak. In the case of such a wretched, disgusting, horrifying, flat-footed, hook-nosed dirty Jew, it would, by God, be no loss. . . . You should croak like a worm."[47]

Streicher's paper did not hesitate to attack the party's faithful conservative allies when any kind of (usually false) information about assistance extended to a Jew reached the paper. Thus on May 20, 1935, Minister of Justice Franz Gürtner himself had to write to Hitler to clear a Stuttgart court of a *Stürmer* accusation that it had helped a Jew named David Isak to change his name to Fritsch (a double scandal, so to speak, given that Fritsch was the name of one of the "great forerunners of the anti-Semitic movement in Germany"). Gürtner went into details: The Isaks were of proven Catholic peasant stock going back more than two hundred years, for which parish records were available. In early 1935 David Isak had asked to change his name to Rudolf Fritsch, as his Jewish-sounding name made for growing difficulties in his work. Despite these easily established facts, *Der Stürmer* had launched a smear campaign against the Ministry of Justice, and Gürtner demanded that Streicher's paper be compelled to recant publicly.[48] A month later Chancellery head Hans Lammers informed the minister that Hitler had agreed to his demand.[49] This incident had far-reaching consequences: the beginning of lengthy administrative debates about Jewish names, name changes, and special names for Jews.

Such complaints against *Der Stürmer* may have convinced Hitler that Streicher's paper could damage the party's reputation. On June 12, 1936, Bormann wrote to the Minister of Justice that, according to the Führer's decision, "*Der Stürmer* is not a mouthpiece of the NSDAP."[50]

The populace appears to have been mainly passive in the face of such ongoing party agitation: Although there was no resistance to it, outright anti-Jewish violence often encountered disapproval. An incident in the spring of 1935 is quite telling. Police interrogation of a man suspected of vandalizing a Jewish cemetery in the Rhineland revealed the following story: The suspect and his friends Gross and Remle had met at a tavern in Hassloch and, after hearing from the local SS leader, Strubel, that "the Jews were to be considered fair game," they set out for the Jewish livestock dealer Heinrich Heene's house. They hurled abuse at Heene and his family while unsuccessfully attempting to break into the courtyard. The people, who by now had gathered in front of Heene's house, gave the three men no aid in their efforts to break down the gate. "When Gross saw . . . that the assembled crowd did not support him," the police report went on, "he yelled at them: 'You call yourselves men, but you're not helping me bring out this pack of Jews.' He then tried with great force to break down the door, kicking against it more wildly than before. The crowd, however, was not in favor of Gross's deed, and one could hear voices growing louder with disapproval—that this was unjust."[51]

Remle, Gross, and the suspect were party members who had taken their cues from the local SS leader and encountered signs of reluctance from a group of townspeople when they moved toward violence. This does not mean that sporadic (and traditional) anti-Jewish violence was unknown in all areas.[52] In one case at least, in Grünzenhausen (Lower Franconia) in the spring of 1934, it led to the deaths of two local Jews.[53] But such occurences were rare.

The peasantry seemed unwilling to forgo the services of the Jew as shopkeeper or cattle dealer:[54] "because of the economic advantages they gained from dealing with Jews who paid cash and sold on credit, they [the peasants] were reluctant to make the move to the Aryan cattle dealers whom the Nazis tried to encourage."[55] On more general grounds, the peasantry often "chose to buy almost solely in Jewish stores," as was reported from Pomerania for the month of June 1935, "because at the Jew's it is cheaper and one has a greater choice [of merchandise]."[56] Probably for the same reasons, a sizable number of *Volksgenossen* still gave preference to Jewish stores and businesses in small towns no less than in large cities. When, according to Victor Klemperer's diary, non-Jews of Falkenstein, in Saxony, were forbidden to patronize local Jewish stores, they traveled to neighboring Auerbach, where they could still patronize the Jewish stores; in turn, non-Jews of Auerbach traveled to Falkenstein for the same purpose. For large-scale purchases, non-Jews from both towns traveled to Plauen, where there was a Jewish department store: "If you happened to know someone you ran into there, neither of you had seen the other. That was the tacit understanding."[57]

What seems to have been most galling to party authorities was the fact that even party members, some in full uniform, were not deterred from doing business with Jews. Thus, in the early summer of 1935, the persistence of such reprehensible behavior was reported from Dortmund, Frankfurt an der Oder, Königsberg, Stettin, and Breslau.[58] In short, while hordes of party activists were beating up Jews, other party members were faithfully buying at Jewish shops. Some party members went even further. According to an SD report, addressed on October 11, 1935, to the party district court in Berlin-Steglitz, party member Hermann Prinz had been seen, six months earlier, in the Bad Polzin area dealing in rugs in partnership with the Jew Max Ksinski; he had even been wearing party insignia while doing this business.[59]

In the summer of 1935, when Jews, as we have seen, were forbidden access to swimming pools and other bathing facilities in numerous German cities, and the very presence of Jews was not allowed in many small towns and villages, a surrealistic situation developed in some of the Baltic seaside resorts, where *Der Stürmer* was widely displayed. It seems that a number of popular guesthouses in these resorts belonged to Jews. In Binz, for instance, a Hungarian Jew owned the most prominent guesthouse, which, according to a Gestapo report, the local population was boycotting, when who should choose to stay there at Whitsuntide but Gauleiter and Reich Governor (Reichsstatthalter) Löper![60] And, adding insult to injury, a month later, in July, it was the Hungarian Jew's guesthouse that was favored by officers and men from the *Köln* on the naval cruiser's visit to Binz.[61] This paradoxical situation lasted for three more years, coming to an end in the spring of 1938, when the director of the Binz office of Baltic Sea resorts announced that "the efforts of recent months have been successful": All the formerly Jewish-owned guesthouses were now in Aryan hands.[62]

The clash between party propaganda against business relations with Jews and the economic advantages brought by such relations was only a reflection of the contradictory nature of the orders from above: on the one hand, no contacts between Jews and *Volksgenossen*; on the other, no interference with Jewish economic activities. This contradiction, which stemmed from two momentarily irreconcilable priorities—the ongoing struggle against the Jews and the need to further Germany's economic recovery—found repeated expression in reports from local authorities. The president of the Kassel administrative district addressed the issue in very direct terms in his monthly report of August 8, 1934: "The Jewish question still plays a significant role. In business life the Jewish presence is still getting stronger. They again have complete control of the cattle market. The attitude of the National Socialist organizations in regard to the Jewish question remains unchanged and is often in conflict with the instructions of the Minister of the Economy, particularly with regard to the treatment of Jewish businesses. I have repeatedly been compelled, together with the State Police, to cancel boycott initiatives as well as other violations by local authorities."[63]

Such contradictions and dilemmas were often particularly visible at the small-town level. On July 2, 1935, a report was sent by Laupheim town officials to the Württemberg Ministry of the Interior: "Under present circumstances, the Jewish question has increasingly become a source of uncertainty for the Laupheim authorities. . . . If the fight against the Jews . . . continues, one has to take into account that the local Jewish businesspeople will emigrate as fast as possible. The municipality of Laupheim will thereby have to expect a further acute loss of income and will have to raise taxes in order to meet its obligations." The author of the report believed that the dying out of the older Jews and the emigration of the younger would cause the Jewish question to resolve itself within thirty years. Meanwhile, he suggested, let the Jews stay as they were, the more so since, apart from a few exceptions, they were a community of well-established families. If Jewish tax revenues were to disappear with no replacement, "the decline of Laupheim into a big village would be unavoidable."[64]

This tension between party initiatives and economic imperatives was illustrated at length in a report devoted entirely to the Jews, sent on April 3, 1935, by the SD "major region Rhine" to SS-Gruppenführer August Heissmeyer in Koblenz. A "quiet boycott" against the Jews is described as having been mainly initiated by the party and its organizations repeatedly asking members in "closed meetings" not to patronize Jewish stores. The report then points to the fact that, "despite more limited possibilities of control in the cities, the boycott is more strictly adhered to there than in rural areas. In Catholic regions in particular, the peasants buy as they did before, mainly from Jews, and this turns in part into an antiboycott movement, which gets its support from the Catholic clergy."

The report continues by describing the growing impact of *Der Stürmer*, "which is sometimes even used as teaching material in schools." But when the paper openly incited its readers to boycott, there was counteraction by state authorities. According to the report, "The Jews conclude from this that the boycott is not wanted by the state. As a result one hears all kinds of complaints about Jewish insolence, which is again coming to the fore."[65]

III

Sometimes genuine sympathy for the plight of the Jews and even offers of help found direct or indirect ways of expression. Thus, in a letter to the *Jüdische Rundschau*, the granddaughter of the poet Hoffmann von Fallersleben, author of the lyrics of the national anthem (the "Deutschlandlied"), offered to put a house on the Baltic shore at the disposal of Jewish children.[66] A different and rather unexpected testimony to both Jewish resilience and Aryan sympathy reached the files of the Göttingen police in early 1935; signed by Reinhard Heydrich, it was a report sent to all Gestapo stations. Its subject: "Performances by Jewish Artists."

"It has recently been observed," Heydrich wrote, "that Jewish artists have attempted in their public appearances to deal, in a veiled way, with government measures as well as with the political and economic situation in Germany and, before an audience of mainly non-Aryans, to exercise by their mimicry and their tone an intentionally destructive criticism, the aim of which is to publicly ridicule the State and the Party." So much for Jewish artists addressing a Jewish public. But there was more: "The State authorities were also provoked by the fact that police intervention resulting from the undesirable cooperation of Aryan with non-Aryan artists has been turned into an occasion for ovations for the non-Aryan artists." And, as if struck by an afterthought, Heydrich added: "The appearance of non-Aryan artists before an Aryan public is fundamentally undesirable, as complications are to be feared." In short the Gestapo was being asked to stop any such shows immediately, although in legal terms some Jews were still exempted from such a prohibition. For Heydrich non-Aryan artists had to limit themselves to a Jewish public. Moreover, in the event that non-Aryan artists were again to allude to the situation in Germany, they must be arrested, "as any interference by non-Aryans in German matters cannot be tolerated."[67]

Repeated orders to party members and civil servants to avoid any further contacts with Jews are indirect proof that such contacts continued into 1935, and not only on economic grounds. On June 7 the mayor of Lörrach in Baden-Württemberg addressed a stern warning to all municipal employees: the Führer had freed Germany from the Jewish danger, and any German "who valued his racial honor" must be grateful to the Führer for this achievement. "If it still happens nonetheless that Germans express their attachment to this foreign race by keeping friendly relations with its members, such behavior shows an absence of sensitivity which must be denounced in the sharpest form."[68]

The undercurrent of sympathy for the persecuted Jews must have been significant enough for Goebbels to address it in a speech that month. Goebbels "attacked those of his countrymen who . . . 'shamelessly,' argued that the Jew, after all, was a human being too." According to Robert Weltsch, who at the time was the editor of the *Jüdische Rundschau*, Goebbels's wrath reveals that a whispering campaign was still going on, indicating some measure of indignation on the part of people whom

Goebbels called bourgeois intellectuals. It was these Germans whom the Gauleiter [of Berlin, Goebbels] wanted to warn."[69]

It may be difficult to prove how effective Goebbels's speech was in intimidating the "bourgeois intellectuals," but it surely had other consequences. In its July 2, 1935, issue, the *Jüdische Rundschau* published an article by Weltsch entitled "The Jew Is Human Too: An Argument Put Forward by Friends of the Jews." It was a subtly ironic comment on the minister's tirade, and it did lead to the banning of the paper.[70] After a few weeks and some negotiating, a letter written in Goebbels's name (but signed "Jahnke") was sent to Weltsch: "The *Jüdische Rundschau* No. 53, dated July 2, 1935, published an article 'The Jew Is Human Too,' which dealt with the part of my speech of 29.6.1935 referring to the Jewish question. My refutation of the bourgeois intellectual view that 'the Jew is human too' was attacked in this article which stated that not only was the Jew human *too*, but had of necessity to be *consciously* human and *consciously* Jewish. Your paper has been banned because of this article. The ban on the paper will be lifted, but in view of the polemic nature of the article I have to reprimand you most severely and expect to have no further cause to object to your publications."[71]

Why would Goebbels have taken the trouble to engage in these maneuvers regarding a periodical written by Jews for Jews? As Weltsch explains it, "One has to keep in mind that the Jewish papers were at that time sold in public. The pretentious main thoroughfare of Berlin's West End, the Kurfürstendamm, was literally plastered with the *Jüdische Rundschau*—all kiosks displayed it every Tuesday and Friday in many copies, as it was one of their best-sellers, especially as foreign papers were banned."[72] This, too, could not last for long. On October 1, 1935, the public display and sale of Jewish newspapers was prohibited.

During these early years of the regime it was difficult entirely to suppress all signs of the Jewish cultural presence in German life. Thus, for example, a 1934 catalog of the S. Fischer publishing house had a front-page picture of its recently deceased Jewish founder and a commemorative speech by the writer Oskar Loerke on the following pages. The catalog also announced volume 2 of Thomas Mann's tetralogy *Joseph and His Brothers* as well as books by the Jewish authors Arthur Schnitzler, Jakob Wassermann, Walther Rathenau, and Alfred Döblin.[73]

The first anniversary of Fritz Haber's death was on January 29, 1935.

Despite the opposition of the Ministry of Education, and the fact that the date was the eve of the second annual national celebration marking Hitler's accession to power, Max Planck decided to hold a memorial meeting under the auspices of the Kaiser Wilhelm Society for the famous Jewish scientist.

A letter sent on January 25 by the Munich party headquarters indicates that the commemoration was also sponsored by the German Society of Physics and the German Society of Chemistry. Headquarters forbade party members to attend, but did not dare, it seems, to rely solely on the argument that Haber was Jewish. The explanation thus included three distinct arguments: "that never before had a German scientist been honored in such a way only a year after his death, and Prof. Dr. Haber, who was a Jew, had been dismissed from his office on October 1, 1933, because of his clearly antagonistic attitude toward the National Socialist state."[74] The special authorizations to attend, which Minister of Education Rust had promised to some of Haber's colleagues, were never granted. Nonetheless the ceremony took place. A wartime coworker of Haber's, a Colonel Köth, spoke, and chemist and future Nobel laureate Otto Hahn delivered the commemorative speech. The hall was filled to capacity with representatives of industry and the spouses of the academics who had been forbidden to attend.[75]

The campaign to cleanse German cultural life of its Jewish presence and spirit had its moments of internal Nazi high drama. In the bitter fight waged between Goebbels and Rosenberg throughout the first months of 1933 for control of culture in the new Reich, Hitler had at first given the preference to Goebbels, mainly by allowing him to establish the Reich Chamber of Culture. Not long afterward, however, an equilibrium of sorts was reestablished by Rosenberg's appointment, in January 1934, as the "Führer's Representative for the Supervision of the General Intellectual and Ideological Education of the NSDAP." The struggle with Goebbels resumed, reaching its climax in "the Strauss case," which lasted for almost a year, from August 1934 to June 1935.

Rosenberg opened hostilities in a letter addressed to Goebbels on August 20, 1934, in which he warned that the behavior of Richard Strauss, the greatest living German composer, president of the Music Chamber of the Reich, and Goebbels's protégé, was threatening to turn into a major public scandal: Strauss had made an agreement that the libretto of his opera Die schweigsame Frau (The silent woman) would be written by "the Jew Stefan Zweig," who, Rosenberg added, "was also the artistic collaborator of a Jewish emigrant theater in Switzerland."[76] At stake was not only ideological purity: Rosenberg was searching for any possible way of undermining Goebbels's dominant position in the domain of cultural politics.

Goebbels, who had just received Hitler's acquiescence to the performance of Strauss's opera in the early summer of 1935 in Dresden, lashed out at the pompous ideologue: "It is untrue that Dr. Richard Strauss has let a Jewish emigrant write the text of his opera. What is true, on the other hand, is that the rewriter of the text, Stefan Zweig, is an Austrian Jew, not to be confused with the emigrant Arnold Zweig. . . . It is also untrue that the author of the text is an artistic collaborator of a Jewish emigrant theater. . . . A cultural scandal could develop as a result of the above-mentioned points only if in foreign countries the matter was treated with the same lack of attention which you display in your letter, which is hereby answered. Heil Hitler!"[77]

The controversy soon became more shrill, Rosenberg in his reply reminding Goebbels of the protection he was granting the Jewish theater director Curt Götz and the difficulties he was thereby causing National Socialist directors. The parting shot was aimed at Goebbels's support of modern, even "Bolshevist," art, particularly the artists belonging to the avant-garde group Der Sturm.[78]

Unfortunately for Goebbels in the spring of 1935 the Gestapo seized a letter from Strauss to Zweig in which the composer explained that he had agreed "to play the role of President of the Music Chamber only to . . . do some good . . . and prevent greater misfortune." As a consequence Strauss was dismissed from his post and replaced by Peter Raabe, a devoted Nazi. Because of Zweig's authorship of the libretto, *Die schweigsame Frau* was banned after a few performances.[79]

The total cleansing of the Reich Music Chamber of its Jewish members took more time, however, than Goebbels had hoped—and announced. Goebbels's diaries repeatedly record his determination to achieve the goal of complete Aryanization. The battle was waged on two fronts: against individuals and against tunes. Most Jewish musicians emigrated during the first three years of Hitler's regime, but to the Nazis' chagrin, it was more difficult to get rid of Jewish tunes—that is, mainly "light" music. "[Arguments] that audiences often asked for such music," writes Michael Kater, "were refuted on the grounds that it was the duty of 'Aryan' musicians to educate their listeners by consistently presenting non-Jewish programs."[80]

Moreover, as far as light music was concerned, intricate commercial relations between Jewish émigré music publishers and partners who were still in Germany enabled a steady flow of undesirable music scores and records into the Reich. Music arrived from Vienna, London, and New York, and it was only in late 1937, when "alien" music was officially prohibited, that Jew hunters could feel more at ease.[81]

Goebbels's Herculean task was bedeviled by the all but insuperable difficulty of identifying the racial origins of all composers and librettists, and by the dilemmas created by well-known pieces tainted with some Jewish connection. Needless to say, Rosenberg's services and related organizations, and even the SD, were unavoidable competitors in that domain. Something of the magnitude of the challenge and of the overall atmosphere can be sensed in an exchange of letters of August 1933 between the Munich branch of Rosenberg's Kampfbund and the Reich Division of German Theaters in Berlin. The Munich people wrote on August 16:

"The Jewish librettists and composers have over the last fifteen years set up a closed circle into which no German author could penetrate, however good the quality of his work. These gentlemen should now be made to see the other side of the coin. A defense action is necessary because in foreign papers hate-filled articles have already been published saying that in Germany things would not work without the Jews.

"In the Deutsches Theater in Munich, the operetta *Sissy* is presently being performed (text by [Ernst] Marischke, music by [Fritz] Kreisler). Kreisler expressed himself in Prague in a most denigrating manner about our Führer. We have expressed a sharp protest against performance of the work, also with the Reich Central Party Office of the NSDAP, to the attention of Party comrade Hess. As we now hear, Director Gruss of the Deutsches Theater is ready to take the work off the program if another work can be found to replace it. We would be grateful to you if you could recommend another such work. It would have to fit a musical-type of performance, as the Deutsches Theater is actually a variety theater and has a concession for musicals only."[82]

By August 23 the answer of the Division of German Theaters was on its way to Munich: The relevant Berlin people already knew about *Sissy* and were probably dealing with it. But other things had to be set straight: In the Munich letter, Franz Lehar and Künnecke had too quickly been cleared of any Jewishness: "Things are not yet clear, as the librettists of both these composers are almost all without exception Jews. I intend shortly to publish a list of all the operettas whose composers and librettists are not Jews."[83] In December of the same year the Division of German Theaters in Berlin received another query from the Munich Kampfbund branch about the Jewishness of Franz Lehar, Robert Stolz, Hans Meisel, Ralph Benatzky, and other composers. Again, the librettists surfaced with regard to Lehar, Stolz, and Benatzky (the other composers were Jews).[84] Hitler, it should be mentioned, was particularly fond of the Lehar operetta *The Merry Widow*. Was the librettist of the most famous of Lehar's pieces by any chance a Jew? And if so, did Hitler know it?

Writing to the Prussian theater council on March 9, 1934, Schlösser cited "the joke about [Meyerbeer's opera] *The Huguenots*: Protestants and Catholics shoot at each other and a Jew makes music about it. Given the unmistakable sensitivity of the wider population about the Jewish question, one has, in my opinion, to bear such an important fact in mind." Schlösser adopted the same attitude toward Offenbach, but mentioned that, owing to contradictory (official) declarations on this issue, a theater in Koblenz had "dug up no fewer than three Offenbach operettas."[85]

All in all, however, the confusion of the new regime's culture masters did not stop the de-Judaization of music in the Reich. Jewish performers such as Artur Schnabel (who had emigrated soon after the Nazis took power), Jascha Heifetz, and Yehudi Menuhin were no longer heard either in concert or on the radio; Jewish conductors had fled, as had the composers Arnold Schoenberg, Kurt Weill, and Franz Schrecker. After some early hesitations, Mendelssohn, Meyerbeer, Offenbach, and Mahler were no longer performed. Mendelssohn's statue, which had stood in front of the Leipzig Gewandhaus, was removed on orders of the city's mayor, Karl Goerdeler (later a key figure in the German resistance against Hitler). But that was far from the end of it: Händel's Old Testament oratorios lost their original titles and were Aryanized so that *Judas Maccabeus* turned into *The Field Marshal: A War Drama* or, alternatively, into *Freedom Oratorio: William of Nassau*, the first version rendered by Hermann Stephani, the second by Johannes Klöcking. Three of the greatest Mozart operas, *Don Giovanni*, *Le Nozze di Figaro*, and *Così fan tutte*, created a special problem: Their librettist, Lorenzo Da Ponte, was of Jewish origin; the first solution was to abandon the original Italian version, but that did not help: The standard German performing version was the work of the Jewish conductor Hermann Levi. There was a last way out: A new translation into purer, nonpolluted German had to be hastily prepared. The new German translations of Da Ponte's libretti to *Figaro* and *Così* were by Siegfried Annheiser, a producer at the theater in Cologne, and by 1938 they had been adopted by seventy-six German opera houses.[86] To cap it all, two major encyclopedias of Jewry and Jews in music, *Judentum in der Musik A-B-C* and *Lexikon der Juden in der Musik*, were to ensure that no mistakes would ever be made in the future. But even encyclopedias did not always suffice: *Judentum in der Musik* was published in 1935; after the Anschluss, however, the new masters of Austria were astonished to discover that there were Jews in "Waltz King" Johann Strauss's extended family, and his birth certificate disappeared from the Vienna archives.[87]

Section II 112 (the Jewish section) of the SD also had its eye on Jewish musicians, dead or alive. On November 27, 1936, it noted the fact that in the hall of the Berlin Philharmonic, the cast for a bust of "the Jew Felix Mendelssohn-Bartholdy" still remained among the casts of famous German composers. As the performance of music by Jewish composers had been forbidden, the note concluded, "the removal of the cast is absolutely necessary."[88] Sometime later the section noticed that a Jewish bass singer, Michael Bohnen, had "recently appeared again in a film." To inform his addressees about Bohnen, the anonymous agent of II 112 quoted the singer's biographical entry in the *Encyclopedia Judaica*.[89]

What would have been the use of cleansing all unseemly Jewish names from the German world of art if the Jews could camouflage their identity by borrowing Aryan names? On July 19, 1935, as a result of the case presented by Gürtner in his complaint against *Der Stürmer*, Frick (who had started to battle against name changes in December 1934) submitted a draft proposal to Hitler that Aryans who bore names commonly considered to be Jewish would be allowed to change them. Generally, Jews would not be allowed to change names unless their name was a source of mockery and insults; in that case another Jewish name could be chosen.[90] On July 31, from Berchtesgaden, Lammers conveyed Hitler's agreement.[91] Frick did not rest with that, and in a communication of August 14, raised with Gürtner the possibility of compelling the descendants of Jews who in the early nineteenth century had chosen princely German names to revert to a Jewish name; this was done, he wrote, on the demand of a Reichstag member, Prince von und zu Loewenstein.[92] It seems that no decision was reached, although Secretary of State Hans Pfundtner at that time ordered the Reich Office for Ancestry Research to prepare lists of German names chosen by Jews since the emancipation.[93] Soon, as will be seen, the strategy was to change: Instead of being forced to abandon their German-sounding names, the Jews would have to take additional—and obviously Jewish—first names.

Hans Hinkel moved to Goebbels's ministry in 1935 to become one of the three supervisors of the Reichskulturkammer. Soon afterward, an unusual title was added to those he already bore: Special Commissioner for the Supervision and Monitoring of the Cultural and Intellectual Activity of All Non-Aryans Living in the Territory of the German Reich."[94] The new title was accurate to the extent that Hinkel, apart from his repeated cleansing forays in the RKK, could now boast of having gently prodded the various regional Jewish Kulturbünde to abandon their relative autonomy and become members of a national association with its seat in Berlin. The decisive meeting, in which the delegates of the Kulturbünde were told in very polite but no uncertain terms that Hinkel considered the establishment of one national organization to be highly desirable, took place in Berlin on April 27 and 28, with Hinkel's participation and in the silent presence of Gestapo representatives.

Hinkel was speaking to the Jewish delegates "in confidence," he said, and any disclosure of the meeting could lead to "unpleasantness"; the decision to form a national organization would really be left to the delegates' "free choice," but the only way of rationally solving a host of technical problems was to establish a single organization. Kurt Singer, who at Hinkel's behest had convened the meeting, was strongly in favor of such unification and seemingly at one with State Secretary Hinkel. He and Singer so briskly managed the meeting that at the end of the first session (the only one Hinkel attended), Singer was able to declare: "I hereby make the official announcement to the State Secretary and to the gentlemen of the State Police that the creation of an umbrella organization of the Jewish Kulturbünde in the Reich was unanimously agreed upon by the delegates present here."[95]

In a 1936 speech Hinkel restated the immediate aim of Nazi cultural policy regarding the Jews: they were entitled to the development of their own cultural heritage in Germany, but only in total isolation from the general culture. Jewish artists "may work unhindered as long as they restrict themselves to the cultivation of Jewish artistic and cultural life and as long as they do not attempt—openly, secretly, or deceitfully—to influence our culture."[96] Heydrich summed up the utility of centralization in slightly different terms: "The establishment of a Reich organization of Jewish Kulturbünde has taken place in order to allow easier control and surveillance of all the Jewish cultural associations."[97] All Jewish cultural groups not belonging to the new national association were prohibited.

IV

From the beginning of 1935, intense anti-Jewish incitement had newly surfaced among party radicals, with discontent and restlessness spreading among the party rank and file and SA members still resenting the murder of their leaders the year before. Lingering economic difficulties, as well as the absence of material and ideological compensations for the great number of party members unable to find positions and emotional rewards either on the local or the national level, were leading to increasing agitation.

A first wave of anti-Jewish incidents started at the end of March 1935; during the following weeks, Goebbels's *Der Angriff* thickened the pogrom-like atmosphere.[98] An announcement by the Ministry of the Interior of forthcoming anti-Jewish legislation and the exclusion of Jews from the new Wehrmacht did not calm the growing unrest.

The first city to witness large-scale anti-Jewish disturbances was Munich, and a carefully drafted police report offers a precise enough description of the sequence of events there. In March and April, Jewish stores were sprayed nightly with acid or smeared with such inscriptions as JEW, STINKING JEW, OUT WITH THE JEWS, and so on. According to the report, the perpetrators knew the police patrol schedule exactly, and could therefore act with complete freedom. In May the smashing of window panes of Jewish shops began. The police report indicates involvement by Hitler Youth groups in one of these early incidents. By mid-May the perpetrators were not only attacking Jewish stores in broad daylight but also assaulting their owners, their customers, and sometimes even their Aryan employees.

On Saturday, May 25, the disturbances took on a new dimension. By midafternoon the attacks had spread to every identifiably Jewish business in the city. According to the police, the perpetrators were "not only members of the Party and its organizations but also comprised various groups of a very questionable nature." In the late afternoon there were clashes outside the central railroad station between police and a crowd of around four hundred people (mainly Austrian Nazis who were training at the SS auxiliary camp at Schlessheim); soon there were other such encounters in other parts of the city. At about six o'clock a crowd tried to attack the Mexican Consulate. Among those arrested there proved to be SS men in civilian clothes. It was not until about nine in the evening that some measure of order was reestablished in the Bavarian capital.[99]

A second major outbreak, one more usually referred to, occurred in mid-July in Berlin, mainly on the Kurfürstendamm, where elegant stores owned by Jews were still relatively active. Jochen Klepper, a deeply religious Protestant writer whose wife was Jewish, wrote in his diary on July 13: "Anti-Semitic excesses on the Kurfürstendamm. . . . The cleansing of Berlin of Jews threateningly announced."[100] A week later Klepper again wrote of what had happened on the Kurfürstendamm: Jewish women had been struck in the face; Jewish men had behaved courageously. "Nobody came to their help, because everyone is afraid of being arrested."[101] On September 7 Klepper, who in 1933 had lost his position with the radio because of his Jewish wife, was fired from the recently Aryanized Ullstein publishing house, where he had found some employment. That day he noticed that the signs forbidding Jews access to the swimming pool were up, and that even the small street in which he took walks with his wife had the same warning on one of its fences.[102]

The exiled German Socialist Party's clandestine reports on the situation in the Reich (the so-called SOPADE [Sozialistische Partei Deutschlands] reports), prepared in Prague, extensively described the spread of anti-Jewish violence throughout Germany during the summer months of 1935. As has been seen, the wrath of Nazi radicals was particularly aroused by Jews who dared to use public swimming pools, by Jewish shops and Jews in marketplaces, and of course by Jewish race defilers. Sometimes the wrong targets were chosen, such as the Gestapo agent from Berlin who on July 13 was mistaken for a Jew in the Kassel swimming pool and beaten up by SA activists.[103] Mostly though, there were no mistakes. Thus, on July 11, for example, approximately one hundred SA men descended on the cattle market in Fulda (as previously mentioned, many cattle dealers were Jews) and indiscriminately attacked both dealers and their customers, causing some to suffer severe injuries. According to the SOPADE report, "The cattle ran aimlessly through the streets and were only gradually brought back together again. The whole of Fulda was in agitation for days on end." The *Jüdisches Familienblatt*, tongue in cheek, said that the Jewish dealers had brought to the market cows that had not been milked for an entire day; this angered the population, causing it to side with the suffering cows and against their Jewish tormentors.[104]

Pressure, violence, and indoctrination were not without their effects. An August 1935 SOPADE report cited an impressive list of new, locally initiated, measures against the Jews: "Bergzabern, Edenkoben, Höheinöd, Breunigweiler, and other places prohibit Jews from moving in and forbid the sale of real estate to them. . . . Bad Tölz, Bad-Reichenhall, Garmisch-Partenkirchen, and the mountain areas of Bavaria do not allow Jews access to their health resorts. . . . In Apolda, Berka, Blankenstein, Sulza, Allstadt, and Weimar, Jews are forbidden to attend cinemas." In Magdeburg, Jews were not allowed to use the libraries; in Erlangen the tramways displayed signs declaring JEWS ARE NOT WELCOME! The report lists dozens of other places and activities forbidden to Jews.[105]

Not all party leaders opposed the spreading of anti-Jewish violence. Gauleiter Grohe of Cologne-Aachen, for example, was in favor of intensifying anti-Jewish actions in order "to raise the rather depressed mood among the lower middle class [*Mittelstand*]."[106] This was not, however, the prevalent position—not because of potential negative reactions among the populace,[107] but mainly because the regime could ill afford to give the impression inside and outside Germany that it was losing control of its own forces by allowing the spread of unbridled violence, particularly in view of the forthcoming Olympic Games. Repeated orders to abstain from unauthorized anti-Jewish actions were issued in Hitler's name by Hess and others, but without complete success.

For Schacht the spread of anti-Jewish violence was particularly unwelcome. In the United States the economic boycott of German goods had flared up again. On May 3 the minister of the economy sent a memorandum to Hitler regarding "the imponderable factors influencing German exports," in which he warned of the economic consequences of the new anti-Jewish campaign. On the face of it at least, Hitler fully agreed with Schacht: At that stage the violence had to stop.[108]

It was in this atmosphere that on August 20, 1935, a conference was called by Schacht at the Ministry of the Economy. Among those present were Minister of the Interior Frick, Justice Minister Gürtner, Prussian Finance Minister Johannes Popitz, Gauleiter and Bavarian Minister of the Interior Adolf Wagner, and representatives of the SD, the Gestapo, and the party's Racial Policy Office.[109]

Frick opened the discussion by describing the additional anti-Jewish legislation, in line with the party program, that was being prepared by the ministry. On the other hand, he took the strongest possible stand against the prevalent unruly anti-Jewish attacks and recommended strong police action.[110]

Wagner concurred. Like Frick he favored further anti-Jewish legal measures, but mentioned that on this matter there were differences of opinion between party and state, as well as among various departments within the state apparatus itself. Not everything had to happen at once; in his opinion further measures had to be taken mainly against full Jews, not against mixed breeds (*Mischlinge*).[111] Yet Wagner insisted that due to demands by a majority of the population for further anti-Jewish measures, new legal steps be taken against the economic activity of Jews.[112] At that stage Wagner's demands went unheeded.

The use of exclusively legal methods was obviously the line adopted at the meeting by the conservative Gürtner: It was dangerous to let the radicals get away with the impression that they were in fact implementing what the government wanted but was unable to do itself because of possible international consequences. "The principle of the Führer-state," argued Gürtner, "had to be imposed against such initiatives."[113]

As could have been expected, Schacht emphasized the damage caused by the anti-Jewish disorders and warned that the developing situation could threaten the economic basis of rearmament. He agreed that the party program had to be implemented, but that the implementation had to take place within a framework of legal instructions alone.[114] Schacht's motives, we have seen, were dictated by short-term economic expediency. The meeting's conclusions were brought to Hitler's attention, and the measures laid out by Frick were further elaborated during late August and early September.[115]

Heydrich, at that time chief of the SD and head of the central office of the Gestapo in Berlin (Gestapa), attended the meeting. In a memorandum sent to all the participants on September 9 he reiterated the points he had made during the conference. In this document Heydrich outlined a series of measures aimed at further segregation of the Jews and, if possible, at the cancellation of their rights as citizens. All Jews in Germany should be subject to alien status. Contrary to what is often stated, however, Heydrich did not indicate that the emigration of all the Jews was to be the central aim of Nazi policy. Only in the last sentence of the memorandum did the SD chief express the hope that the restrictive measures he suggested would direct the Jews toward Zionism and strengthen their incentive to emigrate.[116]

On August 8 both *Der Angriff* and the *Völkischer Beobachter* had published, under the banner headline LAW AND PRINCIPLE IN THE JEWISH QUESTION, an announcement by the chief of the German Police, SS-Obergruppenführer Kurt Daluege, that criminal statistics indicated a pre-eminence of Jews in all areas of crime. Both papers later complained of the lack of attention to this issue in the foreign press; papers abroad that had run the story had interpreted it as a preparation for new anti-Jewish measures, particularly nasty accusations, *Der Angriff* said.[117]

V

On the afternoon of September 15, 1935, the final parade of the annual Nuremberg party congress marched past Hitler and the top leadership of the NSDAP. The Party Congress of Freedom was coming to an end. At 8 P.M. that evening an unusual meeting of the Reichstag opened in the hall of the Nuremberg Cultural Association. It was the first and last time during Hitler's regime that the Reichstag was convened outside Berlin. Nuremberg had last been the site of a German Reichstag (then the assembly of the German Empire's estates) in 1543.[118]

In his speech Hitler briefly addressed the volatile international situation, which had compelled Germany to start rebuilding an army in order to defend its freedom. Ominously, he mentioned Lithuania's control of Memel, a city inhabited by a majority of Germans. The threat posed by international Bolshevism was not forgotten: Hitler warned that any attempt by the Communists to set foot in Germany again would be quickly dealt with. Then he turned to the main topic of his address—the Jews:

The Jews were behind the growing tension among peoples. In New York Harbor, they had insulted the German flag on the passenger ship *Bremen*, and they were again launching an economic boycott against Germany. In Germany itself, their provocative behavior increasingly caused complaints from all sides. Hitler thus set the background. Then he came to his main point: "To prevent this behavior from leading to quite determined defensive action on the part of the outraged population, the extent of which cannot be foreseen, the only alternative would be a legislative solution to the problem. The German Reich Government is guided by the hope of possibly being able to bring about, by means of a singular momentous measure, a framework within which the German *Volk* would be in a position to establish tolerable relations with the Jewish people. However, should this hope prove false and intra-German and international Jewish agitation proceed on its course, a new evaluation of the situation would have to take place."

After asking the Reichstag to adopt the laws that Göring was about to read, Hitler concluded his address with a short comment on each of the three laws: "The first and the second laws repay a debt of gratitude to the Movement, under whose symbol Germany regained its freedom, in that they fulfill a significant item on the program of the National Socialist Party. The third law is an attempt at a legislative solution to a problem which, should it yet again prove insoluble, would have to be assigned by law to the National Socialist Party for a definitive solution. Behind all three laws stands the National Socialist Party, and with it and behind it stands the nation."[119] The threat was unmistakable.

The first law, the Reich Flag Law, proclaimed that henceforth black, red, and white were the national colors and that the swastika flag was the national flag.[120] The second, the Citizenship Law, established the fundamental distinction between "citizens of the Reich," who were entitled to full political and civic rights, and "subjects," who were now deprived of those rights. Only those of German or related blood could be citizens. Thus, from that moment on, in terms of their civic rights, the Jews had in fact a status similar to that of foreigners. The third, the Law for the Defense of German Blood and Honor, forbade marriages and extramarital relations between Jews and citizens of German or kindred blood. Marriages contracted in disregard of the law, even marriages contracted outside Germany, were considered invalid. Jews were not allowed to employ in their households female German citizens under forty-five years of age.[121] Finally, Jews were forbidden to hoist the German flag (an offense against German honor), but were allowed to fly their own colors.

The preamble to the third law revealed all its implications: "Fully aware that the purity of German blood is the condition for the survival of the German *Volk*, and animated by the unwavering will to secure the German nation forever, the Reichstag has unanimously decided upon the following, which is thereby proclaimed."[122] This was immediately followed by paragraph one: "Marriages between Jews and citizens of German and related blood are forbidden." The relation of the preamble to the text of the law reflected the extent of the racial peril represented by the Jew.

According to the September 17 *Völkischer Beobachter*, at a meeting later the same evening with leading party members, "the Führer took the opportunity to underscore the significance of the new laws and to point out that the National Socialist legislation presented the sole means for coming to passable terms with the Jews living in Germany. The Führer particularly stressed that, by virtue of these laws, the Jews in Germany were granted such opportunities in all areas of their own völkisch life as had not hitherto existed in any other land."[123] "In this connection," the report continued, "the Führer renewed the order to the Party that it continue to refrain from taking any independent action against Jews."[124]

In an interview granted on November 27, 1935, to Hugh Baillie, president of the American news agency United Press, Hitler, clearly aiming at the American public, linked the anti-Jewish laws to the danger of Bolshevik agitation.[125]

Taken at face value, the Nuremberg Laws did not mean the end of Jewish life in Germany. "We have absolutely no interest in compelling the Jews to spend their money outside Germany," Goebbels declared at a meeting of propaganda officers held in Nuremberg on the morrow of the congress. "They should spend it here. One should not let them into every public swimming resort, but we should say: We have up there on the Baltic Sea, let's say, one hundred resorts, and into one of them will go the Jews; there they should have their waiters and their business directors and their resort directors and there they can read their Jewish newspapers, of all of which we want to know nothing. It should not be the nicest resort, but maybe the worst of those we have, that we will give them (amusement in the audience)—and in the others, we'll be among ourselves. That I consider right. We cannot push the Jews away, they are here. We do not have any island to which we could transport them. We have to take this into account. . . ."[126]

Two different testimonies from the days following the congress report Hitler's own intentions regarding the future of the Jews. According to Fritz Wiedemann, who was to become his adjutant, the Führer depicted the forthcoming situation to a small circle of Party members: "Out of all the professions, into a ghetto, enclosed in a territory where they can behave as becomes their nature, while the German people look on as one looks at wild animals."[127] From the perspective of 1935, this territorial isolation of the Jews would have had to take place *in* Germany (this is confirmed by the remark about the German people as onlookers). Thus Goebbels was probably repeating what he had heard from Hitler. The second testimony was quite different.

On September 25, 1935, Walter Gross, head of the party's Racial Policy Office, reported to the regional chiefs of his organization the interpretation to the Nuremberg Laws that Hitler gave him, and, mainly, how he saw the next steps of the anti-Jewish policy.

It is worth noting that, once again, after taking a major step in line with his ideological goals, Hitler aimed at defusing its most extreme consequences on a tactical level. In the meeting with Gross, he warned the party not to rush ahead either in extending the scope of the new laws or in terms of direct economic action against the Jews. For Hitler the aim remained the limitation of Jewish influence within Germany and the separation of the Jews from the body of the nation; "more vigorous emigration" from Germany was necessary. Economic measures would be the next stage, but they must not create a situation that would turn the Jews into a public burden; thus carefully calculated steps were needed. As for the *Mischlinge*, Hitler favored their assimilation within a few generations—in order to avoid any weakening of the German potential for war. In the last words of the conversation, however, the pragmatic approach was suddenly gone. According to the Gross protocol, Hitler "declared furthermore, at this point, that in case of a war on all fronts, he would be ready [regarding the Jews] for all the consequences."[128]

CHAPTER 5

The Spirit of Laws

I

A few weeks before the Nuremberg party congress, at the beginning of August 1935, Hitler decided that six Jewish or part-Jewish University of Leipzig professors, hitherto protected by the exception clauses of the Civil Service Law, must retire. On August 26 two officials of the Saxon Ministry of Education arrived for a meeting at the Reich Chancellery; they wanted to know whether, from now on, *all* non-Aryan civil servants were to be retired. Ministerial Councillor Wienstein informed them of the following:

"Basically one should decide case by case, as before. But in each case, however, one should consider that the approach to the treatment of non-Aryans has become stricter. When the Civil Service Law was promulgated, the intention undoubtedly was to give non-Aryans the protection defined in paragraph 3, section 2 of the law, without any restriction. The new development, however, has led to a situation whereby non-Aryans can no longer refer to the above-mentioned instructions in order to claim the right to remain employed. Instead, decisions should, as Ministerial Councillor Wienstein again mentioned, be made "only case by case."[1]

For several months, in fact, Jewish professors still ostensibly protected by the exception clauses had been dismissed. Victor Klemperer had received his dismissal notification in the mail on April 30. Sent via the Saxon Ministry of Education, it was signed by Reichsstatthalter Martin Mutschmann.[2] Within a few months, in the wake of the new Citizenship Law, there were no longer any exceptions, and all remaining Jewish professors were expelled.

*   *   *

Much debate has arisen regarding the origins of the Nuremberg Laws: Were they the result of a haphazard decision or of a general plan aiming at the step-by-step exclusion of the Jews from German society and ultimately from the territory of the Reich? Depending on the view one takes, Hitler's mode of decision making, in both Jewish and other matters, can be interpreted in different ways.

As has been seen, the idea of a new citizenship law had been on Hitler's mind from the outset of his regime. In July 1933 an Advisory Committee for Population and Race Policy at the Ministry of the Interior started work on draft proposals for a law designed to exclude the Jews from full citizenship rights.[3] From the beginning of 1935, the signs pointing to such forthcoming changes multiplied. Allusions to them were made by various German leaders—Frick, Goebbels, and Schacht—during the spring and summer months of that year; the foreign press, particularly the London *Jewish Chronicle* and the *New York Times*, published similar information, and, according to Gestapo reports, German Jewish leaders such as Rabbi Joachim Prinz were openly speaking about a new citizenship law that would turn the Jews into "subjects" (*Staatsangehörige*); their information seemed precise indeed.[4]

Simultaneously, as has also been seen, mixed marriages were encountering increasing obstruction in the courts, to such an extent that, in July, Frick announced the formulation of new laws in this domain as well. In the same month the Justice Ministry submitted a proposal for the interdiction of marriages between Jews and Germans. From then on the issue was the object of ongoing interministerial consultations.[5] Thus, whatever the immediate reason for Hitler's decision may have been, both the issue of citizenship and that of mixed marriages were being discussed in great detail at the civil service level and within the party, and various signs indicated that new legislation was imminent. Incidentally, when Goebbels brought up the topic of "Jewish arrogance" in one of their conversations, Hitler cryptically remarked that "in many things there will soon be changes."[6]

It has been suggested by historians who emphasize the haphazardness of Nazi measures that until September 13, Hitler had been planning to make a major foreign policy statement about the situation in Abyssinia, but that he was dissuaded at the last moment by Foreign Minister Neurath. This hypothesis is not supported by any proof, except for dubious testimony at the Nuremberg Trials by the Interior Ministry's "race specialist,"

Bernhard Lösener. (In the courtroom it was in Lösener's interest to show that there had been no prolonged planning for the 1935 racial laws, for he would necessarily have been involved in such planning.)[7]

In his opening address of September 11 at the Nuremberg party congress, Hitler warned that the struggle against the internal enemies of the nation would not be thwarted by failings of the bureaucracy: The will of the nation—that is, the party— would, if necessary, take over in case of bureaucratic deficiency. It was in these very terms that Hitler ended his September 15 closing speech by addressing the solution of the Jewish problem. Thus it seems that the basic motive for pressing forward with anti-Jewish legislation was to deal with the specific internal political climate already alluded to.

In the precarious balance that existed between the party on the one hand and the state administration and the Reichswehr on the other, Hitler had in 1934 favored the state apparatus by decapitating the SA. Moreover, at the beginning of 1935, when tension arose between the Reichswehr and the SS, Hitler "warned the party against encroachments on the army and called the Reichswehr 'the sole bearer of arms.'"[8] It was time to lean the other way, especially since discontent was growing within the lower party ranks. In short, the Nuremberg laws were to serve notice to all that the role of the party was far from over—quite the contrary. Thus, the mass of party members would be assuaged, individual acts of violence against Jews would be stopped by the establishment of clear "legal" guidelines, and political activism would be channeled toward well-defined goals. The summoning of the Reichstag and the diplomatic corps to the party congress was meant as an homage to the party on the occasion of its most important yearly celebration, irrespective of whether the major declaration was to be on foreign policy, on the German flag, or on the Jewish issue. The preliminary work on the Jewish legislation had been completed, and Hitler could easily switch to preparation of the final decrees at the very last moment.

The conditions under which the drafting of the laws took place are known from another report by Lösener, this one written in 1950, which describes the drafting of the decrees on the last two days of the congress.[9] There was no reason for Lösener to offer a false picture of these two hectic days, except for the suppression of the fact that much preliminary work had been accomplished before then. According to Lösener, on the evening of September 13, he and his Interior Ministry colleague, Franz Albrecht Medicus, were urgently summoned from Berlin to Nuremberg. There,

State Secretaries Pfundtner and Stuckart informed them that Hitler, who considered the flag law an insufficient basis for convening the Reichstag, ordered the preparation of a law dealing with marriage and extramarital relations between Jews and Aryans, and with the employment of Aryan female help in Jewish families. The next day Hitler demanded a citizenship law broad enough to underpin the more specifically racial-biological anti-Jewish legislation. The party and particularly such individuals as Gerhardt Wagner, Lösener wrote, insisted on the most comprehensive definition of the Jew, one that would have equated even "quarter Jews" (*Mischlinge* of the second degree) with full Jews. Hitler himself demanded four versions of the law, ranging from the least (version D) to the most inclusive (version A). On September 15, at 2:30 A.M., he declared himself satisfied with the draft proposals.[10]

Hitler chose version D. But in a typical move that canceled this apparent "moderation" and left the door open for further extensions in the scope of the laws, Hitler crossed out a decisive sentence introduced into the text by Stuckart and Lösener: "These laws are applicable to full Jews only." That sentence was meant to exclude *Mischlinge* from the legislation; now their fate also hung in the balance. Hitler ordered that the Stuckart-Lösener sentence be retained in the official announcement of the laws disseminated by the DNB, the official German news agency.[11] He probably did this to assuage foreign opinion and possibly those sectors of the German population directly or indirectly affected by the laws, but the sentence was to be absent from all further publications of the full text.

There is a plausible reason why, if Hitler was planning to announce the laws at the Nuremberg party congress, he waited until the very last moment to have the final versions drafted: his method was one of sudden blows meant to keep his opponents off balance, to confront them with *faits accomplis* that made forceful reactions almost impossible if a major crisis was to be avoided. Had the anti-Jewish legislation been submitted to him weeks before the congress, technical objections from the state bureaucracy could have hampered the process. Surprise was of the essence.

During the days and weeks following Nuremberg, party radicals close to the Wagner line exerted considerable pressure to reintroduce their demands regarding the status of *Mischlinge* into the supplementary decrees to the two main Nuremberg Laws. Hitler himself was to announce the ruling on "*Mischlinge* of the first degree" at a closed party meeting scheduled for September 29 in Munich. The meeting did take place, but Hitler postponed the announcement of his decision.[12] In fact, the confrontation on the issue of the *Mischlinge* between the party radicals Wagner and Gütt (the latter formally belonged to the Ministry of the Interior) on the one hand, and the Interior Ministry specialists Stuckart and Lösener on the other, lasted from September 22 to November 6, with Hitler's opinion being requested several times by both sides.[13]

Early in the debate both sides agreed that three-quarter Jews (persons with three Jewish grandparents) were to be considered Jews, and that one-quarter Jews (one Jewish grandparent) were *Mischlinge*. The entire confrontation focused on the status of the half Jews (two Jewish grandparents). Whereas the party wanted to include the half Jews in the category of Jews, or at least have a public agency decide who among them was a Jew and who a *Mischling*, the ministry insisted on integrating them into the *Mischlinge* category (together with the one-quarter Jews). The final decision, made by Hitler, was much closer to the demands of the ministry than to those of the party. Half Jews were *Mischlinge*; only as a result of their personal choice (not as the result of the decision of a public agency), either by choosing a Jewish spouse or joining the Jewish religious community, did they become Jews.[14]

The supplementary decrees were finally published on November 14. The first supplementary decree to the Citizenship Law defined as Jewish all persons who had at least three full Jewish grandparents, or who had two Jewish grandparents and were married to a Jewish spouse or belonged to the Jewish religion at the time of the law's publication, or who entered into such commitments at a later date. From November 14 on, the civic rights of Jews were canceled, their voting rights abolished; Jewish civil servants who had kept their positions owing to their veteran or veteran-related status were forced into retirement.[15] On December 21 a second supplementary decree ordered the dismissal of Jewish professors, teachers, physicians, lawyers, and notaries who were state employees and had been granted exemptions.

The various categories of forbidden marriages were spelled out in the first supplementary decree to the Law for the Defense of German Blood and Honor: between a Jew and a *Mischling* with one Jewish grandparent; between a *Mischling* and another, each with one Jewish grandparent; and between a *Mischling* with two Jewish grandparents and a German (the last of these might be waived by a special exemption from the Minister of the Interior or the Deputy Führer).[16] *Mischlinge* of the first degree (two Jewish grandparents) could marry Jews—and thereby become Jews—or marry one another, on the assumption that such couples usually chose to remain childless, as inidicated by the empirical material collected by Hans F. K. Günther.[17] Finally, female citizens of German blood employed in a Jewish household at the time of the law's publication could continue their work only if they had turned forty-five by December 31, 1935.[18]

In a circular addressed to all relevant party agencies on December 2, Hess restated the main instructions of the November 14 supplementary decree to explain the intention behind the marriage regulations that applied to both kinds of *Mischlinge*: "The Jewish *Mischlinge*, that is, the quarter and half Jews, are treated differently in the marriage legislation. The regulations are based on the fact that the mixed race of the German-Jewish *Mischlinge* is undesirable under any circumstances—both in terms of blood and politically—and that it must disappear as soon as possible." According to Hess, the law ensured that "either in the present or in the next generation, the German-Jewish *Mischlinge* would belong either to the Jewish group or to that of the German citizens." By being allowed to marry only full-blooded German spouses, the quarter Jews would become Germans and, as Hess put it, "the hereditary racial potential of a nation of 65 million would not be changed or damaged by the absorption of 100,000 quarter Jews." The Deputy Führer's explanations regarding the half Jews were somewhat more convoluted, as there was no absolute prohibition of their marrying Germans or quarter Jews, if they received the approval of the Deputy Führer. Hess recognized that this aspect of the legislation went against the wishes of the party, declaring laconically that the decision had been taken "for political reasons." The general policy, however, was to compel half Jews to marry only Jews, thus to absorb them into the Jewish group[19]—evidence of Hitler's wish, as stated to Walter Gross, for the dis-appearance of the *Mischlinge*.

To how many people did the Nuremberg Laws apply? According to statistics produced by the Ministry of the Interior on April 3, 1935, living in Germany at the time were some 750,000 *Mischlinge* of the first and sec-ond degree. In this document, signed by Pfundtner and submitted to Hitler by his military adjutant, Col. Friedrich Hossbach, it was not clear how this total had been arrived at. (The ministry, in fact, admitted that there was no precise method for making such an estimate.) Apart from the *Mischlinge*, the document also listed 475,000 full Jews belonging to the Jewish religion and 300,000 full Jews not belonging to it, which made a total of approximately 1.5 million, or 2.3 percent of the population of Germany. A further figure was mentioned, probably at Hitler's demand: Within this total there were 728,000 men, among them about 328,000 of military age.[20]

Even after the proclamation of the laws and the first supplementary decrees in November, Rudolf Hess had the numbers wrong in his circular, giving 300,000 as the overall total of *Mischlinge*.[21] This number was also an exaggeration.

Recent studies have set the number of *Mischlinge* at the time of the decrees at about 200,000.[22] A detailed demographic inquiry conducted by the *CV Zeitung* (the newspaper of the Central Association of German Jews) and published on May 16, 1935, had reached the same result. According to the *CV* inquiry, some 450,000 full Jews (with four Jewish grandparents and belonging to the Jewish religion) were living in Germany at the time. "Non-Jewish non-Aryans"—among them converted full Jews and converted *Mischlinge* with one to three Jewish grandparents—numbered some 250,000. As the author of the inquiry included 50,000 converted full Jews and 2,000 converted three-quarter Jews in his statistics, the numbers according to the Nuremberg degree-categories became the following: full Jews (in racial terms): approximately 502,000 (450,000 plus 50,000 plus 2,000); half Jews: 70,000 to 75,000; quarter Jews: 125,000 to 130,000; total *Mischlinge*: 195,000 to 205,000. (In the *CV* inquiry the half Jews were all converted Jews, and thus, according to the Nuremberg Laws, would not have been counted as Jews but as *Mischlinge* of the first degree.)[23]

II

"In Germany," according to a book published in 1936 by Lösener and Knost, "the Jewish question is simply the race question. How this came about," the authors went on, "need not be described here once again. Here we are dealing only with the solution to this question which has now been decisively set in motion and which represents one of the basic prerequisites for the construction of the new Reich. According to the Führer's will, the Nuremberg Laws are not measures intended to breed and perpetuate race hatred, but measures which mean the beginning of an easing in the relations between the German and the Jewish peoples." Zionism had the right understanding of the issue, the authors asserted, and in general the Jewish people, itself so intent on preserving the purity of its blood over the centuries, should welcome laws intended to defend purity of blood.[24]

The main commentary on the "German racial legislation," published that same year, was coauthored by Secretary of State (in the Ministry of the Interior) Wilhelm Stuckart, and another official from the same ministry, Hans Globke, whose passion for identifying Jews by their names will be encountered later.[25] It reveals starkly some of the most perplexing aspects—even from the Nazi viewpoint—of the Nuremberg Laws. In order to illustrate the absolute validity of religious affiliation as the criterion for identifying the race of the descendants, Stuckart and Globke gave the hypothetical example of a woman, fully German by blood, who had married a Jew and converted to Judaism and then, having been widowed, returned to Christianity and married a man fully German by blood. A grandchild deriving from this second marriage would, according to the law, be considered partly Jewish because of the grandmother's one-time religious affiliation as a Jew. Stuckart and Globke could not but state the following corollary: "Attention has to be given to the fact that . . . [in] terms of racial belonging, a full-blooded German who converted to Judaism is to be considered as German-blooded after that conversion as before it; but in terms of the racial belonging of his grandchildren, he is to be considered a full Jew."[26]

The racial mutation caused by such temporary contact with the Jewish religion is mysterious enough. But the mystery is compounded when it is remembered that in Nazi eugenics or racial anthropology, the impact of environmental factors was considered negligible in comparison with the effect of heredity. Here, however, an ephemeral change in environment mysteriously causes the most lasting biological transformation.[27] But whatever their origins, racial differences could lead to dire consequences in cases of prolonged mixing:

"The addition of foreign blood to one's own brings about damaging changes in the body of the race because the homogeneity, the instinctively certain will of the body, is thereby weakened; in its stead an uncertain, hesitating attitude appears in all decisive life situations, an overestimation of the intellect and a spiritual splitting. A blood mixture does not achieve a uniform fusion of two races foreign to each other but leads in general to a disturbance in the spiritual equilibrium of the receiving part."[28]

Two laws directed against individuals and groups other than Jews followed the September laws. The first of these was the October 18, 1935, Law for the Protection of the Hereditary Health of the German People, which aimed at registering "alien races" or racially "less valuable" groups and imposed the obligation of a marriage license certifying that the partners were (racially) "fit to marry."[29] This law was reinforced by the first supplementary decree to the Law for the Protection of German Blood and Honor of November 14, which also forbade Germans to marry or have sexual relations with persons of "alien blood" other than Jews. Twelve days later a circular from the Ministry of the Interior was more specific: Those referred to were "Gypsies, Negroes, and their bastards."[30]

Proof that one was not of Jewish origin or did not belong to any "less valuable" group became essential for a normal existence in the Third Reich. And the requirements were especially stringent for anyone aspiring to join or to remain in a state of party agency. Even the higher strata of the civil service, the party, and the army could not escape racial investigation. The personal file of Gen. Alfred Jodl, who was soon to become deputy chief of staff of the Supreme Command of the Armed Forces, the Oberkommando der Wehrmacht, contains a detailed family tree in Jodl's handwriting, which, in 1936, proved his impeccable Aryan descent as far back as the mid-eighteenth century.[31]

Exceptions were rarely made. The best-known case was that of the state secretary at the Aviation Ministry, Erhard Milch, a *Mischling* of the second degree who was turned into an Aryan. Incidentally, such rare occurrences rapidly became known, even among the general population. Thus, in December 1937, charges were brought against a Father Wolpert, of Dinkelsbühl, in Bavaria, because he had stated in a religion class that General Milch was of Jewish origin.[32] In every such matter the final decision rested with Hess and often with Hitler himself. Whether Hess consulted Hitler in every instance is hard to know; that he consulted him in highly visible ones is probable. It is unlikely, for example, that Hess decided alone—a few days after the 1938 Kristallnacht pogrom, and after Hitler had told Lammers that he would no longer agree to any exceptions regarding persons of Jewish descent—to issue a "protection letter" for the geopolitician Karl Haushofer's son, Albrecht, a *Mischling* of the second degree according to the Nuremberg Laws.[33] Sometimes Hitler's hypochondriacal worries played a role. It will be remembered that the cancer researcher and "*Mischling* of the first degree" Otto Warburg was transformed into a "*Mischling* of the second degree" on Göring's orders. Something similar occurred in early 1937, when a professor of radiology at the clinic of the Friedrich Wilhelm University in Berlin, Henri Chaoul—who, according to one investigation, was descended from Syrian Maronites and Greek Cypriots, and to another, more plausible one, "was not Aryan in the sense of the Civil Service Law" (in other words, was of Jewish origin)—was shielded from any difficulties on Hitler's explicit demand, and appointed director of a newly established central radiology institute in Berlin.[34]

The investigations probably stopped at the very highest party leadership. Rumors, however, knew no such bounds, and, as is well known, both Hitler and Heydrich, among others, were suspected of hiding non-Aryan ancestors. In both cases the rumors proved unfounded,[35] but under the circumstances the insinuation was certainly meant to be damaging. Sometimes disgruntled party leaders used the accusation of non-Aryan origins against rivals. Thus, in April 1936, Wilhelm Kube, Gauleiter of the Kurmark (part of Prussia), sent an anonymous letter (signed "some Berlin Jews") to the party chancellery stating that the wife of the head of the party tribunal, Walter Buch, and Bormann's mother-in-law were of Jewish origin. An ancestry investigation proved that the accusations were baseless; Kube admitted having written the letter and was temporarily removed by Hitler from all his functions.[36]

The new marriage laws in fact followed the memorandum, drafted in September 1933 by Hans Kerrl and Roland Freisler, that marriages and extramarital sexual relations between "those of German blood" and "members of racially alien communities" be considered "punishable offenses against the honor of the race." During the first three years of the regime, the very strong reactions of a number of Asian and South American countries (including the boycotting of German goods) led, among other reasons, to the shelving of the initiative.[37] There can be no doubt, however, that the early proposals, the third Nuremberg Law, and the marriage laws that followed could be considered the expression of a *general* racial-biological point of view, along with the policies directed against the specific Jewish peril.

A series of exchanges in late 1934 and early 1935 among the Ministry of Foreign Affairs, the Ministry of the Interior, and the Party Racial Policy Office clearly displayed the intertwining and the distinctions between these issues. The Wilhelmstrasse, worried by the impact of the Aryan legislation on the Reich's foreign relations, suggested that the new laws be clearly limited to Jews and that other non-Aryans (such as Japanese and Chinese) be excluded. For Walter Gross, any basic change in the party's attitude to racial questions was impossible, as it lay at the core of the Nazi worldview, but Gross promised that the party would avoid burdening Germany's foreign relations with any inappropriate internal decisions. The replacement of the concept "non-Aryan" by "Jewish" was not yet deemed timely for official use: There was no objection in principle to such a change, but it was feared that the change would be interpreted as "a retreat." In any case exceptions could be made in instances where the Aryan legislation affected non-Aryan, non-Jewish foreigners.[38] Less than two weeks before the opening of the Nuremberg party congress, on August 28, 1935, Hess had expressed the desire that, out of consideration for the Semitic nations, at the rally the term "anti-Semitic" be replaced by "anti-Jewish."[39] For him Lösener and Knost's formula seemed indeed to be of the essence: "In Germany, the Jewish question is simply the race question."

Lösener's report on the final stages preceding the Nuremberg Laws clearly indicates that the September 14–15 discussions centered only on anti-Jewish legislation; this had been the object of party agitation during the preceding months, as it would be that of the discussions that followed (including those involving Hitler's hesitations on September 29 and his decision on November 14). Thus, *the separateness and the compatibility of both the specific anti-Jewish and the general racial and eugenic trends were at the very center of the Nazi system.* The main impetus for the Nuremberg Laws and their application was anti-Jewish; but the third law could without difficulty be extended to cover other racial exclusions, and it logically led to the additional racial legislation of the fall of 1935. The two ideological trends reinforced each other.[40]

III

For the *Mischling* Karl Berthold, the Chemnitz social benefits employee whose story began to be told in chapter 1, the Nuremberg legislation did not solve the problem of his racial purity.[41] On April 18, 1934, the specialist for racial research in the Ministry of the Interior restated his case for Berthold's exclusion from the civil service, arguing that, even if the details about the presumed father, Carl Blumenfeld, were uncertain, Berthold was related to the Blumenfeld family, and his mother had declared that he was the son of a Carl Blumenfeld, a "Jewish artist." His non-Aryan origins could not be doubted.[42]

At this point Berthold's aunt, his mother's sister, briefly entered the scene and testified that his father was an Aryan who, in order to hide his identity, had taken the name Carl Blumenfeld. The main social benefits office in Dresden notified the minister of labor of this new development on June 30. At the end of July, the minister of labor was ready to allow Berthold to remain in public service and merely demanded confirmation by the minister of the interior. The specialist for "ancestry research" at the Ministry of the Interior, was not to be so easily fooled. A detailed report issued on September 14 indicated that the Jew Carl Blumenfeld, whose data had been referred to all along and whose age made it highly improbable that he was the father of Karl Berthold, was, in fact, a distant cousin of the circus artist Carl Blumenfeld, who by now was living in Amsterdam. On November 5 the main office in Dresden forwarded to the minister of labor one more request by Berthold for reexamination of the case, again including the testimony of Berthold's aunt. A few weeks later, as no answer had been received, another petition was addressed to the minister of the interior, this time by Berthold's wife, Frau Ada Berthold. Berthold would be dismissed from his position, she wrote, if a positive answer was not received by March 31, 1936.[43] A new phase of his story was now beginning.

The new laws could in principle introduce a Nazi kind of clarity into some cases where the question of racial belonging had previously received contradictory answers. Thus, an inquiry of October 26, 1934, by the welfare department of the city of Stettin, regarding the treatment of illegitimate children of Jewish fathers and Aryan mothers, had revealed widely different attitudes on the part of welfare departments in various major German cities: In Dortmund such children were considered Aryan and given all the usual assistance, whereas in Königsberg, Breslau, and Nuremberg, the welfare departments considered them "Semitized." The director of the Breslau department volunteered the following comment: "In my view, there is no point in incorporating children of mixed race into the German nation, since, as is well known, they themselves cannot have racially pure children and regulations for the sterilization of racially mixed people do not yet exist. Thus, one should not prevent *Mischlinge* from joining the foreign nation to which they already half belong. In fact, one ought to encourage them to do so, e.g., by letting them attend Jewish kindergartens."[44] The reaction from Nuremberg, Streicher's headquarters, should come as no surprise: "A mother who behaves in such a way," wrote the local welfare director, "is so strongly influenced by Jewish ideas that presumably all attempts to enlighten her will be in vain and the attempt to educate her Jewish child 'according to the principles of National-Socialist leadership' must fail. For the National Socialist *Weltanschauung* which is determined by blood can only be taught to those who have German blood in their veins. In this case, one ought to put into practice Nietzsche's dictum: 'That which is on the point of collapse should be given the final push.'"[45] After passage of the laws, these children must all have become *Mischlinge* of the first degree.

The definition of the two degrees of *Mischlinge* in the Nuremberg Laws, and in the First Supplementary Decree of November 14, 1935, temporarily alleviated their situation both in terms of citizenship rights and with regard to access to professions closed to "full Jews." In principle, at least, young *Michlinge* were accepted in schools and universities like any other Germans. Until 1940 they were allowed to study any subject (except medicine and dentistry).[46] This was merely a reprieve, and, from 1937 on, various new forms of official persecution threatened the professional and economic existence of the *Mischlinge*, not to mention their growing social isolation and eventually the threat to their lives. But sometimes the status of *Mischling* was itself not devoid of ambiguities.

Consider the case of Otto Citron, who in 1937 transferred from the University of Tübingen to that of Bonn. After his departure the Tübingen administration suddenly became suspicious about the student's declaration that he had an Aryan grandmother. If the declaration was false, Citron's status would change, and Tübingen wanted Bonn to start proceedings against the camouflaged Jew. Citron's answer to the charges was impeccable. He had declared *before* the proclamation of the Nuremberg Laws that his half-Jewish grandmother was Aryan, at a time when the only existent distinction was the one that separated the Aryan from the Jew. That is, half Jewishness was a category that did not legally exist before September 1935. After the passage of the Nuremberg Laws, Citron correctly indicated that he was half Jewish according to the other set of grandparents. Since the half-Jewish grandmother on the other side of the family had married an Aryan, Citron again correctly indicated that according to the Nuremberg criteria he was one-eighth Jewish on her side. Thus, he stated, if one added up the two sides of the family, he was five-eighths Jewish. But "according to the supplementary decrees of the Nuremberg Laws," Citron stated, "which I examined with the greatest care before making any written or oral statement, the 5/8 persons are identified with the half-Aryans and the same status is valid for them. The same situation pertains to the 3/8 persons who are considered as one-quarter persons. Thus any attempt at deception or at circumventing the law was entirely foreign to me."

Tübingen University had no choice but to accept Citron's argument and close the proceedings it had started against him.[47]

Citron's case, in fact, was simple enough when compared with some of the potential (or actual) situations described in the form of questions and answers by the Information Bulletin of the Reich Association of Non-Aryan Christians (*Mitteilungsblatt des Reichsverbandes der Nichtarischen Christen*) of March 1936:

"Question: What can be said about the marriage of a half-Aryan with a girl who has one Aryan parent, but whose Aryan mother converted to Judaism so that the girl was raised as a Jew? What can be said, further, about the children of this marriage?

"Answer: The girl, actually half-Aryan, is not a *Mischling*, but is without any doubt regarded as Jewish in the sense of the law because she belonged to the Jewish religious community on the deadline date, i.e., 15th September 1935; subsequent conversion does not alter this status in any way. The husband—a first degree *Mischling*—is likewise regarded as a Jew since he married a statutory Jew. The children of this marriage are in any case regarded as Jews since they have three Jewish grandparents (two by race, one by religion). This would not have been different if the mother had left the Jewish community before the deadline. She herself would have been a *Mischling*, but the children would still have had three Jewish grandparents. In other words, it is quite possible that children who are regarded as Jews may result from a marriage in which both partners are half-Aryan.

"Question: A man has two Jewish grandparents, one Aryan grandmother and a half-Aryan grandfather; the latter was born Jewish and became Christian only later. Is this 62 percent Jewish person a *Mischling* or a Jew?

"Answer: The man is a Jew according to the Nuremberg Laws because of the one grandparent who was of the Jewish religion; this grandparent is assumed to have been a full Jew and this assumption cannot be contested. So this 62 percent Jew has three full Jewish grandparents. On the other hand, if the half-Aryan grandfather had been Christian by birth, he would not then have been a full Jew and would not have counted at all for this calculation; his grandson would have been a *Mischling* of the First Degree."[48]

One of the major hurdles encountered by the legal experts in the interpretation of the Nuremberg Laws was the definition of "intercourse." The basic forms of sexual intercourse were but a starting point, and Stuckart and Globke, for instance, sensed the manifold vistas offered by "acts similar to intercourse such as mutual masturbation." Soon even this extended interpretation of intercourse became insufficient in the eyes of some courts. A district court in Augsburg defined the applicability of the laws in a way that practically eliminated all restrictions on the definition: "Since the law aims at protecting the purity of German blood," the court stated, "the will of the lawmakers must be seen as also making illegal all perverse forms of sexual intercourse between Jews and citizens of German or related kinds of blood. It is furthermore the intention of the relevant law to protect German Honor, in particular the sexual honor of the citizens of German blood."[49]

Litigation on this point reached the Supreme Court, which pronounced its decision on December 9, 1935: "The term 'sexual intercourse' as meant by the Law for the Protection of German Blood does not include every obscene act, but it is also not limited to coition. It includes all forms of natural and unnatural sexual intercourse—that is, coition as well as those sexual activities with the person of the opposite sex which are designed, in the manner in which they are performed, to serve in place of coition to satisfy the sex drive of at least one of the partners."[50]

The Supreme Court encouraged the local courts to understand the intention of the lawmaker beyond the mere letter of the law, thereby opening the floodgate. Couples were found guilty even if no mutual sexual activity had been performed. Masturbation in the presence of the partner, for instance, became punishable behavior: "It would run counter both to healthy popular feeling and to the clear goals of German racial policy if such surrogate acts were to go completely unpunished, thereby creating, with regard to perverse conduct between the sexes, a new stimulus for violating the racial honor of the German people."[51]

The search for ever more precise details about all possible aspects of racial defilement (*Rassenschande*) can be seen not only as one more illustration of Nazi bureaucratic and police thinking, but also as a huge screen for the projection of various "male fantasies."[52] In the Nazi imagination, moreover, Jews were perceived as embodiments of sexual potency and lust, somewhat like blacks for white racists, or witches (and women more generally) in the eyes of the Inquisition or some Puritan elders. Details of the offenses thus became a source of (dangerous) knowledge and of hidden titillation. And, more often than not, the details were graphic indeed. Thus, on January 28, 1937, the district court in Frankfurt sentenced Alfred Rapp, a thirty-four-year-old "full Jew," to two years in prison and the "full-blooded German" Margarete Lehmann to nine months, on the following grounds:

Rapp was an employee in a men's clothing store, and Lehmann was a seamstress there. They were known to be friends and visited each other frequently. According to their testimony, they had had no prior sexual relations. At about 8:30 P.M. on November 1, 1936, Rapp came to Lehmann's apartment, where he also found a Jewish woman named Rosenstock. The three went out for drinks and then went to Rosenstock's apartment. Rosenstock was sent out to buy wine. According to the accused, they then engaged in oral sex. The court report gave some graphic details and added: "This presentation of the facts does not, on its face, appear credible; at least it is incomplete, since common life experience rules out the possibility of a girl having gotten as sexually close to a man without there having been some intimate acts in between, even if—as Lehmann admits—she had two glasses of wine during the preceding two hours. One must add to this that the two accused were also observed in Rosenstock's room by the two witnesses W. and U." The scene as observed by the witnesses follows in the report, again in the greatest detail, as successively confirmed by each of them: "The same observation was made by witness U., whom witness W. . . . then let look through the keyhole. [Then] W. opened the unlocked door and entered the room. Both accused quickly tried to straighten their clothes and hair."[53]

For a Hamburg court the kisses of an impotent man "took the place of normal sexual intercourse" and led to a two-year sentence. Therapeutic massage, needless to say, soon came under suspicion, as in the notorious case of the Jewish merchant Leon Abel. Although the "German-blooded" woman therapist steadfastly denied that Abel had shown any sign of sexual excitation during the one and only massage session, and although, during his trial, Abel himself retracted the confession he had made to the Gestapo, the court sentenced him to two years for "having attained sexual gratification with Miss. M. and thereby 'effecting' the crime of dishonor of the race, whether or not the witness had knowledge of it."[54]

The law regarding female household employees in Jewish families shows that potential situations of race defilement had been taken into consideration. But how could all such potential situations be foreseen? Constant watchfulness was the only possible answer. In November 1937, after asking the minister of the interior to pay attention to the possibilities still existing in the law for the adoption of "full-blooded" German *Volksgenossen* by Jews, Hess brought up a more immediately threatening problem. In those cases in which a German girl grows up in a Jewish family, "some measure should be taken to protect the German side. A way must be found to afford them the same protection . . . as that granted to German female house employees."[55]

In fact, all aspects of everyday life and all professional activities in which the contact between Aryans and Jews could be construed as having some sexual connotation were systematically identified and forbidden. The exclusion of Jews from swimming facilities has already been discussed. In the spring of 1936, most medical faculties prohibited their Jewish students from performing genital examinations of Aryan women (the decision regarding the application of these restrictions was left to the hospital directors responsible for the Jewish gynecology interns).[56]

How far these increasing taboos were welcomed or merely passively accepted by the wider population is hard to surmise. Sometimes, no doubt for economic reasons but also possibly with the intention of expressing a symbolic protest, German women beyond the fertility age were ready to face the corrupting atmosphere of a Jewish family. For instance, on November 14, 1935, the *Frankfurter Zeitung* published the following advertisement: "Cultivated Catholic woman over 45, perfect housekeeper and cook, seeks position in a good Jewish household, also for half-days."[57]

In a study of the almost complete Gestapo files of Würzburg, Robert Gellately has shown that the most important source of information for Gestapo arrests was an influx of informers; the attitude or activity most frequently reported to the Secret Police was "race defilement" or "friendliness to Jews."[58] The Nuremberg Laws offered a kind of vague legal basis informers could use in all possible ways, and during the years following the number of denunciations grew sharply. According to Gellately's analysis of the Würzburg Gestapo files (which deal with a small city, so the numbers should be projected onto the national scale), there were two race defilement denunciations and one denunciation of friendliness to Jews in 1933; one and two respectively in 1934; five and two in 1935; nineteen and twelve in 1936; fourteen and seven in 1937; fourteen and fourteen in 1938; and eight and seventeen in 1939. After the beginning of the war, in September 1939, denunciations decreased, falling to one and one in 1943, and then vanished entirely.[59] By that time, of course, there were no Jews left in Würzburg—nor in Germany.

In the same source, approximately 57 percent of the denunciations came from people who were not party members, and between 30 and 40 percent of the charges were false.[60] Sometimes hotel employees would denounce a couple, neither of whom was Jewish; others were arrested because of information about ties that had ended long before 1933. There were instances of couples whose intimate relations extended back many years now avoiding sexual intercourse, and many cases of women proclaiming readiness to undergo medical examinations to demonstrate that they were virgins.[61]

Goebbels was unhappy with the press reports of race defilement. In March 1936 he asked the press department of the Ministry of Justice to avoid giving undue publicity to *Rassenschande* verdicts against Jews as, in his view, it offered material to anti-German foreign newspapers. Moreover, the releases "were often written so clumsily that the reader did not understand the verdict and rather felt compassion for the accused."[62]

IV

Did public opinion fall further into step with the anti-Jewish policies of the regime after the passage of the Nuremberg Laws? According to Israeli historian David Bankier, a majority of Germans acquiesced in the laws because they accepted the idea of segregating the Jews: "The Potsdam Gestapo fully captured these feelings. The general belief was, it stated, that with the stabilization of the regime the time was ripe to realize this item on the Party's agenda. At the same time, the Gestapo official added, the public hoped that other points of the Nazi program would be acted upon, especially those related to social issues. In Kiel, too, there was approval of the anti-Semitic laws, and people expected the status of the churches to be resolved in an equally satisfactory way."[63]

According to the same analysis, people in various cities and areas of the Reich seemed to have been particularly satisfied with the Law for the Protection of German Blood and Honor, on the assumption that enforcement of the law would put an end to the anti-Jewish terror of the previous months. Tranquility would return, and with it the good name of Germany in the eyes of the world. People believed that under the new laws, the relation to Jewry in Germany was now clearly defined: "Jewry is converted into a national minority and gets through state protection the possibility to develop its own cultural and national life";[64] such was the common opinion reported from Berlin.

For the party radicals, the laws were a clear victory of the party over the state bureaucracy, but many considered the new decrees to be "too mild." The Dortmund Nazis, for instance, regarded the fact that the Jews could still use their own symbols as too much of a concession. Some activists hoped that the Jews would offer new pretexts for action, others simply demanded that the scope of some of the measures be extended: that for example, no German female of any age should be allowed to work in a Jewish (or mixed-marriage) family—or even in the household of a single Jewish woman.[65]

The laws were sharply criticized in opposition circles, mainly among the (now underground) Communists. Some Communist leaflets denounced the Nazis' demagogic use of anti-Semitism and demanded a united opposition front; others demanded the freeing of political prisoners and the cessation of anti-Jewish measures. According to Bankier, however, Communist material at the time, despite its protests against the Nuremberg Laws, continued to reiterate such longtime standard assertions as: "Only poor workers were arrested for race defilement, while rich Jews were not touched by the Nazis," and, "There were no racial principles behind the ban on keeping maids under forty-five years of age; rather, the clause was simply an excuse for firing thousands of women from their jobs."[66]

The churches kept their distance, except for the strongly Catholic district of Aachen and some protests by Evangelical pastors, for instance in Speyer. The Evangelical Church was put to the test when the Prussian Confessing Synod met in Berlin at the end of September 1935: A declaration expressing concern for both baptized and unbaptized Jews was discussed and rejected, but so was too explicit an expression of support for the state. The declaration that was finally agreed on merely reaffirmed the sanctity of baptism, which led Niemöller to express his misgivings about its failure to take any account of the postbaptismal fate of baptized Jews.[67]

To return to the attitudes of the general population, Nazi reports pointed to expressions of anxiety and even protests from Germans employed by Jews—be they German clerks working in Jewish firms or maids employed by Jewish families. But all in all, Bankier leaves little leeway for equivocation and doubts: "To sum up, the vast majority of the population approved of the Nuremberg Laws because they identified with the racialist policy and because a permanent framework of discrimination had been created that would end the reign of terror and set precise limits to anti-Semitic activities."[68]

Although his cases are roughly the same as those later treated by Bankier, the study by another Israeli historian, Otto Dov Kulka, leaves the impression of a more diversified set of reactions. He too mentions Communist opposition as well as Catholic disapproval in some cities such as Aachen and Allenstein, and notes the criticism of some Protestant pastors, particularly in Speyer. He too refers to party activists who find the measures insufficient. In addition he comments on the disapproval that manifested itself among an upper bourgeoisie worried, among other things, about the possibility of economic reprisals in foreign countries. Nevertheless, the overall impression this study gives is that the majority of the population was satisfied with the laws because they clarified the status of Jews in Germany and, it was hoped, would put an end to indiscriminate disorder and violence. A contemporary report from Koblenz seems to reflect the most widespread reactions:

"The Law for the Protection of German Blood and German Honor was mainly received with satisfaction, not least because not only will it psychologically hinder unpleasant individual actions [against the Jews] but even more, it will lead to the desired isolation of Jewry. . . . The question as to how far Jewish blood should be excluded from the German national body is still the object of animated discussions."[69]

The reference to *Mischlinge* is unmistakable. Thus, both studies agree that a majority of Germans were more or less passively satisfied with the laws. In other words, the bulk of the population disliked acts of violence but did not object to the disenfranchisement and segregation of the Jews. It meant, further, that as segregation was now legally established, for a majority of the population the new situation allowed the individual to divest him- or herself of any responsibility for the measures regarding the Jews. The accountability for their fate had been taken over by the state.[70]

There were exceptions, and relations with Jews were maintained, as has already been noted with regard to the period preceding the Nuremberg Laws. On December 3, 1935, the Gestapo sent a general instruction (to all Gestapo stations) indicating that "recently announcements by Jewish organizers and former bandleaders of forthcoming dance events have increased to such a point that it is not always possible for Gestapo stations to control them in an orderly way." And then comes the more interesting piece of news: "It has been repeatedly noticed that Aryans are also allowed to participate in such events."[71]

It seems, incidentally, that the Gestapo was encountering ever greater technical difficulties in controlling Jewish events. The explanation may be simple: The Jews reacted to growing persecution and segregation by intensifying all possible aspects of internal Jewish life, which explains both the number and the diversity of meetings, lectures, dances, and so on; these offered some measure of sanity and dignity, but meant more trouble for the Gestapo. As early as 1934, the State Police complained that many Jewish meetings, particularly those of the Central Association of German Jews, took place in private homes, which made control almost impossible;[72] then, at the end of 1935, Jewish events were allegedly often moved from Saturdays to Sundays and to the Christian holidays, "obviously," according to the Gestapo, "on the assumption that on those days the events would not be controlled. It was difficult to forbid meetings in private homes, but events taking place on Sundays or Christian holidays were, from then on, to be authorized in exceptional cases only."[73] The last straw came in April 1936: the Gestapo stations reported an increasing use of the Hebrew language in public Jewish political meetings. "Orderly control of these meetings," wrote Heydrich, "and the prevention of hostile propaganda have thereby become impossible." The use of Hebrew in public Jewish meetings was therewith forbidden, but the language could continue to be used in closed events, for study purposes, and to prepare for emigration to Palestine.[74] Incidentally, the reports on the use of Hebrew remain somewhat mysterious unless (and this is very unlikely) only meetings of the small minority of East European, Orthodox (though not ultra-Orthodox), and ardent Zionist Jews are being referred to. Any sort of fluency in Hebrew among the immense majority of German Jews was nil.

Among those who may have considered the laws as not being extreme enough there was a hard core of Jew haters who did not belong to the party and were even enemies of National Socialism: Their hatred was such that, in their eyes, even the Nazis were instruments of the Jews. They were not necessarily marginal types. Adolf Schlatter, for example, was a distinguished professor of theology at Tübingen. On November 18, 1935, he published a pamphlet entitled "Will the Jew Be Victorious Over Us?: A Word for Christmas" (*Wird der Jude über uns siegen? Ein Wort für die Weihnacht*). Within a few weeks, some fifty thousand copies had been distributed. "Today," wrote Schlatter, "a rabbi can say with pride: 'Look how the situation in Germany has changed; indeed we are despised, but only because of our race. But until now we were alone in trying to erase from public consciousness the mad message preached at Christmas that Christ has come. Now, however, we have as allies in our fight those who carry the responsibility for the education of the German people, in other words, those to whom the German owes obedience. . . .' One cannot deny to the Jew that in the German sphere the situation has never been as favourable for his world view as it is now."

But there was hope in the closing lines of Schlatter's pamphlet: "It is indeed possible that in the immediate future the Jew will win a powerful victory over us; but his victory will not be final. The Jew did not bring belief in God into the world, and this belief the Jews and the Jew-companions cannot destroy. They cannot destroy it because they cannot cancel the fact that the Christ has come into the world."[75]

Schlatter's antiregime hatred of Jews had its built-in limits in Nazi Germany. On the face of it, the possibilities should have been greater for a member of the SS. Riding as a third-class passenger on the express train from Halle to Karlsruhe on October 22, 1935, SS officer Hermann Florstedt, according to his later testimony, badly needed sleep. As his ticket did not allow him access to a sleeping car, he moved through second class in search of a vacant seat. All the compartments were fully occupied, except for two that, according to Florstedt, were each occupied by a Jew. "I was in uniform," wrote Florstedt in his letter of complaint to the Railways Directorate in Berlin, "and had no desire to spend this long journey in the company of a Jew." Florstedt found the conductor and demanded a place in second class. The conductor led him to the compartments occupied by the Jews; Florstedt protested. "The conductor," wrote Florstedt, "behaved more than strangely. He told me among other things that I had not seen these gentlemen's certificates of baptism and that, moreover, for him, Jews were also passengers."[76]

It seems that in October 1935 an SS uniform did not yet inspire terror. Besides, the conductor's awareness that he was obeying existing administrative rules (an August 1935 decree specifically allowed Jews to use public transportation)[77] must have given him enough self-confidence to answer as he did. The retort that Jews were passengers too can also be associated with the current of opinion (the Jew is human too) Goebbels had attacked in his June 1935 speech.

In Florstedt's complaint to the Railways Directorate, he demanded the name of the conductor, with whom he wanted "to discuss the matter in the *Stürmer*." The letter landed on the desk of Gruppenführer Heissmeyer, head of the SS Main Office, who vindicated the behavior of the railway official and did not take kindly to Florstedt's threat to go public in the *Stürmer*.

Florstedt was soon transferred to the concentration camp administration. Early in the war he was deputy commander of Buchenwald, and in March 1943 he became commandant of the Lublin extermination camp.[78]

V

"Not only are we taking leave of the [Jewish] year, which has come to an end," the *CV Zeitung* announced some two weeks after the proclamation of the Nuremberg Laws, "but also of an epoch in history, which is now drawing to its close."[79] But this apparent understanding that the situation was changing drastically did not lead to any forceful recommendations. Many German Jews still hoped that the crisis could be weathered *in* Germany and that the new laws would create a recognized framework for a segregated but nonetheless manageable Jewish life. The official reaction of the *Reichsvertretung* (which was now obliged to change its name from Reich Representation of German Jews to Representation of Jews in Germany) took at face value Hitler's declaration of the new basis created by the laws for relations between the German people and the Jews living in Germany, and thus demanded the right to free exercise of its activities in the educational and cultural domains. Even at the individual level, many Jews believed that the new situation offered an acceptable basis for the future. According to a study of Gestapo and SD reports on Jewish reactions to the laws, in a significant number of communities "the Jews were relieved precisely because the laws, even if they established a permanent framework of discrimination, ended the reign of arbitrary terror. There was a measure of similarity in the way average Germans and average Jews reacted. The Germans expressed satisfaction while the Jews saw ground for hope. As the author of the report put it: the laws finally defined the relation between Jews and Germans. Jewry becomes a *de facto* national minority, enjoying the possibility of ensuring its own cultural and national life under state protection."[80]

The ultrareligious part of the community even greeted the new situation. On September 19, 1935, *Der Israelit*, the organ of Orthodox German Jewry, after expressing its support for the idea of cultural autonomy and separate education, explicitly welcomed the interdiction of mixed marriages.[81] As for the German Zionists, although they stepped up their activities, they seemed in no particular hurry, the mainstream group Hechalutz wishing to negotiate with the German government about the ways and means of a gradual emigration of the German Jews to Palestine over a period of fifteen to twenty years. Like other sectors of German Jewry, it expressed the hope that, in the meantime, an autonomous Jewish life in Germany would be possible.[82]

The Jews of Germany were, in fact, still confronted with what appeared to be an ambiguous situation. They were well aware of their increasing segregation within German society and of the constant stream of new government decisions designed to make their life in Germany more painful. Some aspects of their daily existence, however, bolstered the illusion that segregation was the Nazis' ultimate aim and that the basic means of economic existence would remain available. For instance, despite the 1933 law on "the overcrowding of German schools" and the constant slurs and attacks against Jewish children, in early 1937, although the majority of Jewish children attended Jewish schools, almost 39 percent of Jewish pupils were still in German schools. In the spring of the following year, the percentage had decreased to 25 percent.[83] As will be seen, many Jewish professionals, benefiting from various exemptions, were still active outside the Jewish community. But it remains difficult to assess accurately the economic situation of the average Jewish family with a retail business or making its living from any of the the various trades.

In 1935 the *Jüdische Rundschau,* which, one would have thought, should have aimed at showing how bad the situation was, quoted statistics published by the *Frankfurter Zeitung* indicating that half the ladies' garment industry was still owned by Jews, the figure rising to 80 percent in Berlin.[84] Whether or not these numbers are exact, the Jews of the Reich still thought they would be able to continue to make a living; they did not, for the most part, foresee any impending material catastrophe.

Yet, even though emigration was slow, as already mentioned, and even though most German Jews still hoped to survive this dire period in Germany, the very idea of leaving the country, previously unthinkable for many, was now accepted by all German-Jewish organizations. Not an immediate emergency flight, but an orderly exodus was contemplated. Overseas (the American continent or Australia, for instance) was higher on the list of concrete possibilities than Palestine, but all German Jewish papers could wholeheartedly have adopted the headline of a *Jüdische*

Rundschau lead article addressed to the League of Nations: "Open the Gates!"[85]

For the many Jews who were considering the possibility of emigration but still hoped to stay in Germany, the gap between public and private behavior was widening: "We must avoid doing anything that will attract attention to us and possibly arouse hostility." Jewish women's organizations warned, "Adhere to the highest standards of taste and decorum in speaking manner and tone, dress and appearance."[86] Jewish pride was to be maintained, but without any public display. Within the enclosed space of the synagogue or the secular Jewish assemblies, this pride and of the pent-up anger against the regime and the surrounding society found occasional expression. Religious texts were chosen for symbolic meaning and obvious allusion. A selection of psalms entitled "Out of the Depths Have I Called Thee," published by Martin Buber in 1936, included verses that could not be misunderstood:

> Be Thou my judge, O God, and plead my cause
> against an ungodly nation;
> O deliver me from the deceitful and unjust man.

A new type of religious commentary, conveyed mainly in sermons— the "New Midrash," as Ernst Simon called it—interwove religious themes with expressions of practical wisdom that were meant to have a soothing, therapeutic effect on the audience.[87]

It seems that occasionally some Jews showed less public humility. William L. Shirer, the American journalist then based in Berlin and soon to become the CBS correspondent there, wrote in his diary on April 21, 1935, while staying at Bad Saarow, the well-known German spa: "Taking the Easter week-end off. The hotel mainly filled with Jews and we are a little surprised to see so many of them still prospering and apparently unafraid. I think they are unduly optimistic."[88]

Self-assertion remained sometimes astonishingly strong among Jews living in even the smallest communities. Thus, in 1936, in Weissenburg, the Jewish cattle dealer Guttmann was accused by the local Nazi peasant leader of stating that he had received official authorization to continue his trade. Although the Jew was arrested, he persisted in asserting his right to do business. The report on the incident concludes with the following words: "Guttmann requests permission to sign the document after the Sabbath is over."[89]

*   *   *

After the proclamation of the Nuremberg Laws, the Zionist leadership in Palestine showed no greater sense of urgency regarding emigration than did the German Jewish community itself. Indeed, the Palestine leadership refused to extend any help to emigrants whose goal was not Eretz Israel. Its list of priorities was increasingly shifting: The economic situation of the Yishuv worsened from 1936 on, while the Arab Revolt of that year increased Britain's resistance to any growth in Jewish immigration to Palestine. Some local Zionist leaders even considered the easier-to-integrate immigrants from Poland by and large preferable to those from Germany, with an exception for German Jews who could transfer substantial amounts of money or property within the framework of the 1933 Haavarah Agreement. Thus, after 1935, the number of immigration certificates demanded for German Jews out of the total number of certificates allocated by the British remained the same as before. This lack of major commitment on the part of the Zionist leadership to encourage Jewish emigration from Germany created a growing tension with some Jewish leaders in the Diaspora.[90]

When a group of Jewish bankers met in London in November 1935 to discuss the financing of emigration from Germany, an open split occurred between Zionists and non-Zionists. The president of the World Zionist Organization, Chaim Weizmann, was particularly bitter about Max Warburg's scheme to negotiate a Haavarah-like agreement with the Nazis to pay for German Jewish emigration to countries other than Palestine.[91] Warburg nonetheless discussed his scheme with representatives of the Ministry of the Economy. The party archives indicate that the Germans made further discussions conditional upon the presentation of a detailed proposal.[92] Nothing came of the project because of the publicity surrounding it and, ultimately, a lack of adequate funding.[93]

VI

"In Bad Gastein. Hitler leads me in animated conversation down an open stairway. We are visible from afar and at the bottom of the stairs a concert is taking place and there is a large crowd of people. I think proudly and happily: now everyone can see that our Führer does not mind being seen with me in public, despite my grandmother Recha."[94] Such was a dream reported by a young girl whom the Nuremberg Laws had just turned into a *Mischling* of the second degree.

Here is a dream of a woman, who had become a *Mischling* of the first degree: "I am on a boat with Hitler. The first thing I tell him is: 'In fact, I am not allowed to be here. I have some Jewish blood.' He looks very nice, not at all as usual: a round pleasant kindly face. I whisper into his ear: 'You [the familiar *Du*] could have become very great if you had acted like Mussolini, without this stupid Jewish business. It is true that among the Jews there are some really bad ones, but not all of them are criminals, that can't honestly be said.' Hitler listens to me quietly, listens to it all in a very friendly way. Then suddenly I am in another room of the ship, where there are a lot of black-clad SS men. They nudge each other, point at me and say to each other with the greatest respect: 'Look there, it's the lady who gave the chief a piece of her mind.'"[95]

The dream world of full Jews was often quite different from that of the *Mischlinge*. A Berlin Jewish lawyer of about sixty dreamed that he was in the Tiergarten: "There are two benches, one painted green, the other yellow, and between the two there is a wastepaper basket. I sit on the wastepaper basket and around my neck fasten a sign like the ones blind beggars wear and also like the ones the authorities hang from the necks of race defilers. It reads: WHEN NECESSARY, I WILL MAKE ROOM FOR THE WASTEPAPER."[96]

Some of the daydreams of well-known Jewish intellectuals living beyond the borders of the Reich were at times no less fantastic than the nighttime fantasies of the trapped victims. "I don't like to make political prophecies," Lion Feuchtwanger wrote to Arnold Zweig on September 20, 1935, "but through the intensive study of history I have reached the, if I may say so, scientific conviction that, in the end, reason must triumph over madness and that we cannot consider an eruption of madness such as the one in Germany as something that can last more than a generation. Superstitious as I am, I hope in silence that this time too the German madness won't last longer than the [1914–1918] war madness did. And we are already at the end of the third year."[97]

Other voices had a very different sound. Carl Gustav Jung tried to delve "deeper" in his search for the characteristics of the Germanic psyche—and for those of the Jewish one as well. Writing in 1934, his evaluation was different: "The Jew, who is something of a nomad, has never yet created a cultural form of his own and as far as we can see never will, since all his instincts and talents require a more or less civilized nation to act as host for their development. . . . The 'Aryan' consciousness has a higher potential than the Jewish; that is both the advantage and the disadvantage of a youthfulness not yet fully weaned from barbarism. In my opinion it has been a grave error in medical psychology up to now to apply Jewish categories—which are not even binding to all Jews—indiscriminately to German and Slavic Christendom. Because of this the most precious secret of the Germanic peoples—their creative and intuitive depth of soul—has been explained as a morass of banal infantilism, while my own warning voice has for decades been suspected of anti-Semitism. This suspicion emanated from Freud. He did not understand the Germanic psyche any more than did his Germanic followers. Has the formidable phenomenon of National Socialism, on which the whole world gazes with astonished eyes, taught them better?"[98]

The "formidable phenomenon of National Socialism" did not, apparently impress Sigmund Freud. On September 29, 1935, he wrote to Arnold Zweig: "We all thought it was the war and not the people, but other nations went through the war as well and nevertheless behaved differently. We did not want to believe it at the time, but it was true what the others said about the Boches."[99]

As for Kurt Tucholsky, possibly the most brilliant anti-nationalist satirist of the Weimar period, now trapped in his Swedish exile, his anger was different from that of Freud, and his despair was total: "I left Judaism in 1911," he wrote to Arnold Zweig on December 15, 1935, but he immediately added: "I know that this is in fact impossible." In many ways Tucholsky's helplessness and rage are turned against the Jews. The unavoidable fate could be faced with courage or with cowardice. For Tucholsky the Jews had always behaved like cowards, now more than ever before. Even the Jews in the medieval ghettos *could* have behaved differently: "But let us leave the medieval Jews—and let us turn to those of today, those of Germany. There you see that the same people who in many domains played first violin *accept* the ghetto—the idea of the ghetto and its realization. . . . They are being locked up; they are crammed into a theater for Jews [*ein Judentheater*—a reference to the activities of the *Kulturbund*] with four yellow badges on their front and back and they have . . . only one ambition: 'Now for once we will show them that we have a better theater.' For every ten German Jews, one has left, nine are staying; but after March 1933, one should have stayed and nine should have gone, ought to, should have. . . . The political emigration has changed nothing; it is business as usual: everything goes on as if nothing had happened. Forever on and on and on—they write the same books, hold the same speeches, make the same gestures. . . ." Tucholsky knew that he and his generation would not see the new freedom: "What is needed . . . is a youthful strength that most emigrants do not have. New men will come, after us. As they are now, things cannot work anymore. The game is up."[100]

Six days later Tucholsky committed suicide.

PART II

The Entrapment

CHAPTER 6

Crusade and Card Index

I

In early 1937, during a meeting on church affairs, Hitler once more gave free rein to his world-historical vision: "The Führer," Goebbels wrote in his diary, "explains Christianity and Christ. He [Christ] also wanted to act against the Jewish world domination. Jewry had him crucified. But Paul falsified his doctrine and undermined ancient Rome. The Jew in Christianity. Marx did the same with the German community spirit, with socialism."[1] On November 30 of the same year, the remarks Goebbels inscribed in his diary were much more ominous: "Long discussion [with Hitler] over the Jewish question. . . . The Jews must get out of Germany, in fact out of the whole of Europe. It will still take some time but it must happen, and it will happen. The Führer is absolutely determined about it."[2] Like his September 1935 declaration to Walter Gross, Hitler's prophecy of 1937 meant the possibility of war: It could be fulfilled only in a situation of war.

On March 7, 1936, the Wehrmacht had marched into the Rhineland, and a new phase in European history had begun. It would unfold under the sign of successive German treaty breaches and aggressions and, in three years, lead to the outbreak of a new conflagration.

The demilitarization of the left bank of the Rhine had been guaranteed by the Versailles and Locarno Treaties. The guarantors of the status quo were Great Britain and Italy, whereas France was the country directly threatened by the German move. Now Italy positioned itself at Germany's side, because the democracies had attempted to impose sanctions on it during the Abyssinian war. In principle, however, France still had the strongest army in Europe. It is now known that a French military reaction would have forced the German units to retreat behind the Rhine—a setback with unforeseeable consequences for the Hitler regime. But although the French government, led by the Radical Socialist Prime Minister Albert Sarrault, threatened to act, it did nothing. As for the British, they did not even threaten; after all, Hitler was merely taking possession of his own "backyard," as the saying went. The French and British policy of appeasement was gaining momentum.

In France the 1936 elections brought the center-left Popular Front to power, and for a large segment of French society the threat of revolution and a Communist takeover became an obsessive nightmare. A few months earlier the Spanish electorate had brought a left-wing government to power. That was a short-lived victory. In July 1936 units of the Spanish army in North Africa, led by Gen. Francisco Franco, rebelled against the new Republican government and crossed over into Spain. The Spanish Civil War—which was to become a murderous struggle of two political mystiques, backed on both sides by a massive supply of foreign weapons and regular troops as well as volunteers—had started. Between the summer of 1936 and the spring of 1939, the battle lines drawn in Spain were the explicit and tacit points of reference for the ideological confrontations of the time.

On the global scene the anti-Comintern pact signed between Germany and Japan on November 25, 1936, and joined by Italy a year later, became, at least symbolically, an expression of the struggle that was to unfold between the anti-Communist regimes and Bolshevism. In the countries of East Central Europe (with the exception of Czechoslovakia) and the Balkans, right-wing governments had come to power. Their ideological commitments included three basic tenets: authoritarianism, extreme nationalism, and extreme anti-Communism. From the Atlantic to the Soviet border, they generally had one more element in common: anti-Semitism. For the European Right, anti-Semitism and anti-Bolshevism were often identical.

The year 1936 also clearly marks the beginning of a new phase on the internal German scene. During the previous period (1933–36), the need to stabilize the regime, to ward off preemptive foreign initiatives, and to sustain economic growth and the return to full employment had demanded relative moderation in some domains. By 1936 full employment had been achieved and the weakness of the anti-German front sized up. Further political radicalization and the mobilization of internal resources were now possible: Himmler was named chief of all German police forces and Göring overlord of a new four-year economic plan, whose secret objective was to prepare the country for war. The impetus for and the timing of both external and internal radicalization also may have been linked to yet unresolved tensions within German society itself, or may have resulted from the fundamental needs of a regime that could only thrive on ever more hectic action and ever more spectacular success.

It was in this atmosphere of accelerated mobilization that the Jewish issue took on a new dimension and fulfilled a new function in Nazi eyes. Now Jewry was again being presented as a worldwide threat, and anti-Jewish action could be used as justification for the confrontation that necessarily was about to come. In the regime's terms, in a time of crisis the Jews had to be expelled, their assets impounded for the benefit of German rearmament, and—as long as some of them remained in German hands—their fate could be used to influence the attitude toward Nazi Germany of world Jewry and of the foreign powers under its control. Most immediately three main lines of action dominated the new phase of the anti-Jewish drive: accelerated Aryanization, increasingly coordinated efforts to compel the Jews to leave Germany, and furious propaganda activity to project on a world scale the theme of Jewish conspiracy and threat.

Accelerated Aryanization resulted in part at least from the new economic situation and the spreading confidence in German business and industrial circles that the risks of Jewish retaliation or its effects no longer had to be taken into account. Economic growth led to gradual coordination of the contradictory measures that, of necessity, had earlier hindered the course of anti-Jewish policy: By 1936 ideology and policy could increasingly progress along a single track. Himmler and Göring's appointments to their new positions created two power bases essential for the effective implementation of the new anti-Jewish drive. And yet, although the framework of the new phase was clearly perceptible, the economic expropriation of the Jews of Germany could not be radically enforced before the beginning of 1938, after the conservative ministers had been expelled from the government in February 1938 and mainly after Schacht had been compelled to leave the Ministry of the Economy in late 1937. During 1938 worse than total expropriation was to follow: Economic harassment and even violence would henceforward be used to force the Jews to flee the Reich or the newly annexed Austria. Within the second phase, 1938 was the fateful turning point.

The anti-Jewish rhetoric expressed in Hitler's speeches and statements from 1936 on took several forms. Foremost, and most massively, was its relation to the general ideological confrontation with Bolshevism. But the world peril as presented by Hitler was not Bolshevism as such, with the Jews acting as its instruments. The Jews were the ultimate threat *behind* Bolshevism: The Bolshevik peril was being manipulated by the Jews.[3] In his 1937 party congress speech, Hitler made sure, as will be seen, that there was no misunderstanding on this point. But Hitler's anti-Jewish harangues were not only ideological (anti-Bolshevik) in a concrete sense; often the Jew was described as the world enemy per se, as the peril that had to be destroyed lest Germany (or Aryan humanity) be exterminated by it. In its most extreme form this apocalyptic vision appeared in the January 1939 speech to the Reichstag, but its main theme was already outlined in the summer of 1936, in the guidelines establishing the Four-Year Plan. The "redemptive" anti-Semitism that had dominated Hitler's early ideological statements now resurfaced. With the conservative agenda crumbling, a new atmosphere of murderous brutality was spreading.

It is at the start of this darkening path that the Nazis achieved one of their greatest propaganda victories: the successful unfolding of the 1936 Olympic Games. Visitors to Germany for the Olympics discovered a Reich that looked powerful, orderly, and content. As the American liberal periodical *The Nation* expressed it on August 1, 1936: "[One] sees no Jewish heads being chopped off, or even roundly cudgeled. . . . The people smile, are polite and sing with gusto in beer gardens. Board and lodging are good, cheap, and abundant, and no one is swindled by grasping hotel and shop proprietors. Everything is terrifyingly clean and the visitor likes it all."[4] Even the president of the United States was deceived. In October of that year, a month before the presidential election, Rabbi Stephen Wise, president of the World Jewish Congress, was invited to meet with Roosevelt at Hyde Park. When the conversation turned to Germany, the president cited two people who had recently "toured" Germany and reported to him that "the synagogues were crowded and apparently there is nothing very wrong in the situation at present." Wise tried to explain to his host the impact of the Olympic Games on Nazi behavior, but left feeling that Roosevelt still regarded accounts of persecution of the Jews as exaggerated.[5]

Signs forbidding access to Jews were removed from Olympic areas and from other sites likely to be visited by tourists, but only very minor ideological concessions were made. The Jewish high-jump finalist Gretel Bergmann, from Stuttgart, was excluded from the German team on a technical pretext; the fencing champion Helene Mayer was included because she was a *Mischling* and thus a German citizen according to the Nuremberg Laws.[6] Only one German full Jew, the hockey player Rudi Ball, was allowed to compete for Germany. But the Winter Games in those days were far less visible than the summer ones.[7]

The negotiations that had preceded the Olympics showed that Hitler's tactical moderation emanated only from the immense propaganda asset they represented for Nazi Germany. When, on August 24, 1935, the Führer received Gen. Charles Sherrill, an American member of the International Olympic Committee, he was still adamant: The Jews were perfectly entitled to their separate life in Germany, but they could not be members of the national team. As for the foreign teams, they were free to include whomever they wanted.[8] Finally, because of the threat of an American boycott of the Olympics, very minor concessions were adopted, as has been seen, which allowed Germany to reap all the expected advantages, the recent passage of the Nuremberg Laws notwithstanding.

The limits of Nazi Olympic goodwill were clearly revealed in the privacy of diaries. On June 20, just before the Olympics opened, Goebbels waxed ecstatic about Max Schmeling's victory over Joe Louis for the world heavyweight boxing championship: "Schmeling fought and won for Germany. The white defeated the black and the white was a German."[9] His entry on the first day of the Olympics was less enthusiastic: "We Germans win a gold medal, the Americans win three, of which two by Negroes. White humanity should be ashamed. But what does that mean down there in that land without culture."[10]

The Winter Games had opened on February 6 in Garmisch-Partenkirchen. The day before, Wilhelm Gustloff, the Nazi Party representative in Switzerland, had been assassinated by the Jewish medical student David Frankfurter. Within a few hours a strict order was issued: Because of the Olympic Games, all anti-Jewish actions were prohibited.[11] And indeed no outbursts of "popular anger" occurred.

*   *   *

Hitler spoke at Gustloff's funeral in Schwerin, on February 12. He recalled the days of defeat when, according to him, Germany had been "delivered a lethal stab at home." During those November days of 1918, the national Germans attempted "to convert [the working masses,] those who, at that time, were the tools of a gruesome supranational power. . . . At every turn we see the same power . . . the hate-filled power of our Jewish foe."[12] A few months later the exiled Jewish writer Emil Ludwig published a pamphlet entitled "Murder in Davos." Goebbels reacted immediately in his diary entry of November 6, 1936: "A nasty, typically Jewish work of incitement to glorify . . . Frankfurter, who shot Gustloff. . . . This Jewish pestilence must be eradicated. Totally. None of it should remain."[13]

The struggle between the new Germany and that gruesome supranational power, the Jewish foe, was now redefined as the total confrontation on the widest international scale with Bolshevism, "the tool of the Jews." At the 1935 party congress the anti-Judeo-Bolshevik declarations had been left to Goebbels and Rosenberg. Soon Himmler joined the fray. In November 1935, at the National Peasants Day (*Reichsbauerntag*) in Goslar, the Reichsführer SS described the threat represented by the Jews in blood-curdling terms: "We know him, the Jew," Himmler exclaimed, "this people composed of the waste products of all the people and nations of this planet on which it has imprinted the features of its Jewish blood, the people whose goal is the domination of the world, whose breath is destruction, whose will is extermination, whose religion is atheism, whose idea is Bolshevism."[14]

Hitler had personally intervened in the new anti-Jewish campaign in his speech at Gustloff's funeral. A no less threatening tone appeared in his secret memorandum of the summer of that same year outlining the goals of the Four-Year Plan. The introductory paragraph addressed the issue of ideology as such: "Politics is the conduct and process of the historical struggle for the life of nations. The aim of these struggles is survival. Idealistic struggles over world views also have their ultimate causes, and draw their deepest motivating power from purposes and aims in life that derive from national sources. But religions and worldviews can give such struggles an especial sharpness and by this means endow them with a great historic effectiveness. They can put their mark on the character of centuries. . . ." In a series of quick associations, this theoretical prologue led to the foreseeable ideological illustration: "Since the beginning of the French

Revolution, the world has been drifting with increasing speed towards a new conflict, whose most extreme solution is named Bolshevism, but whose content and aim is only the removal of those strata of society which gave the leadership to humanity up to the present, and their replacement by international Jewry. . . ."[15]

On Hitler's instructions Goebbels and Rosenberg intensified even further the pitch of their verbal onslaught at the 1936 Congress.[16] For Goebbels "the idea of Bolshevism, that is, the unscrupulous savaging and dissolution of all norms and culture with the diabolical intention of a total destruction of all nations, could only have been born in the brain of Jews. The Bolshevik practice in its terrifying cruelty is imaginable only as perpetrated by the hands of Jews."[17]

In his two programmatic speeches at the congress, Hitler also dealt with the Judeo-Bolshevik danger. In his opening proclamation of September 9, he briefly attacked the worldwide subversive activities of the Jewish revolutionary center in Moscow.[18] But it was in his closing speech, on September 14, that he lashed out at length: "This Bolshevism that the Jewish-Soviet terrorists from Moscow, Lewin, Axelrod, Neumann, Béla Kun, etc., tried to introduce into Germany, we attacked, defeated, and extirpated. . . . And now, because we know and experience daily that the attempt of the Jewish Soviet leaders to interfere in our internal German affairs continues, we are also compelled to consider Bolshevism beyond our borders as our mortal enemy and to recognize no less a danger in its advance."[19]

What Hitler meant was clear enough: The Luftwaffe was now increasingly intervening against the "Bolshevik" forces in Spain. And who was in charge of the Moscow terrorist center that directed subversive activities all over the world? The Jews.

That Hitler rehashed these themes in private conversations is not astonishing; that the approving interlocutor in one such conversation was Munich's Cardinal Faulhaber is somewhat more of a surprise. On November 4, 1936, he met for three hours with Hitler at the Obersalzberg, Hitler's residence in the Bavarian Alps. According to Faulhaber's own notes, Hitler spoke "openly, confidentially, emotionally, at times in a spirited way; he lashed out at Bolshevism and at the Jews: 'How the subhumans, incited by the Jews, created havoc in Spain like beasts,' on this he was well informed. . . . He would not miss the historical moment." The cardinal seemed to agree: "All of this," he noted, "was expressed by Hitler in a moving way in his great speech at the Nuremberg Party rally (Bolshevism could only destroy, was led by the Jews)."[20]

It was at the Party Congress of Labor, in September 1937, that the anti-Judeo-Bolshevik campaign reached its full scope. During the preceding weeks the jockeying for congress preeminence among Hitler's lieutenants had taken a particularly acerbic form. Rosenberg informed Goebbels that, according to Hitler's decision, he (Rosenberg) was to be the first of the two to speak and that, given the time constraints, Goebbels's speech was to be drastically cut down. This must have been a sweet moment for the master of ideology, especially as in the ongoing feud between him and Goebbels, the propaganda minister usually had the upper hand.

On September 11 Goebbels set the tone. In a speech devoted to the situation in Spain, the propaganda minister launched into a hysterical attack against the Jews, whom he held responsible for Bolshevist terror. In his rhetorical fury Goebbels undoubtedly succeeded in outdoing his previous performances. His speech may well be the most vicious public anti-Jewish outpouring of those years. "Who are those responsible for this catastrophe?" Goebbels asked. His answer: "Without fear, we want to point the finger at the Jew as the inspirer, the author, and the beneficiary of this terrible catastrophe: look, this is the enemy of the world, the destroyer of cultures, the parasite among the nations, the son of chaos, the incarnation of evil, the ferment of decomposition, the visible demon of the decay of humanity."[21]

II

On the evening of September 13 Hitler spoke again. All restraint was now gone. For the first time since his accession to the chancellorship, he used the platform of a party congress, with the global attention it commanded, to launch a general historical and political attack on world Jewry as the wire puller behind Bolshevism and the enemy of humanity from the time of early Christianity on. The themes of the 1923 dialogue with Dietrich Eckart were being broadcast to the world.

Never since the fall of the ancient world order, Hitler declared, never since the rise of Christianity, the spread of Islam, and the Reformation had the world been in such turmoil. This was no ordinary war but a fight for the very essence of human culture and civilization. "What others profess not to see because they simply do not want to see it, is something we must unfortunately state as a bitter truth: the world is presently in the midst of an increasing upheaval, whose spiritual and factual preparation and whose leadership undoubtedly proceed from the rulers of Jewish Bolshevism in Moscow. "When I quite intentionally present this problem as Jewish, then you, my Party Comrades, know that this is not an unverified assumption, but a fact proven by irrefutable evidence."[22]

Hitler did not simply leave the concrete aspects of this struggle of world historical significance to his audience's imagination:

"While one part of the 'Jewish fellow citizens' demobilizes democracy via the influence of the press or even infects it with its poison by linking up with revolutionary manifestations in the form of popular fronts, the other part of Jewry has already carried the torch of the Bolshevist revolution into the midst of the bourgeois-democratic world without even having to fear any substantial resistance. The final goal is then the ultimate Bolshevist revolution, i.e., not, for example, consisting of the establishment of a leadership of the proletariat, but of the subjugation of the proletariat to the leadership of its new and alien master. . . . [23]

"In the past year, we have shown in a series of alarming statistical proofs that, in the present Soviet Russia of the proletariat, more than eighty percent of the leading positions are held by Jews. This means that not the proletariat is the dictator, but that very race whose Star of David has finally also become the symbol of the so-called proletarian state."[24]

Hitler usually repeated his main themes in an ever-changing variety of formulas all bearing the same message. The September 13, 1937, speech hammered home the menace represented by Jewish Bolshevism to the "community of Europe's civilized nations."[25] What had been achieved in Germany itself was presented as the example to be followed by all:

"National Socialism has banished the Bolshevist world menace from within Germany. It has ensured that the scum of Jewish *literati* alien to the Volk does not lord it over the proletariat, that is, the German worker. . . . It has, moreover, made our Volk and the Reich immune to Bolshevist contamination."[26] A few months earlier Rudolf Hess had conveyed Hitler's thinking to all party organizations: Germany yearned for relations of friendship and respect with all nations; it was "no enemy of the Slavs, but the implacable and irreconcilable enemy of the Jew and of the Communism he brought to the world."[27]

In private Hitler had expressed puzzlement about the meaning of the events then occurring in the Soviet Union. "Again a show trial in Moscow,"

Goebbels noted in his diary on January 25, 1937. "This time again exclusively against Jews. Radek, etc. The Führer still in doubt whether there isn't after all a hidden anti-Semitic tendency. Maybe Stalin does want to smoke the Jews out. The military is also supposedly strongly anti-Semitic. So, let us keep an eye on things."[28]

Although in the Ministry of Foreign Affairs and in the army efforts were made to maintain a somewhat more realistic assessment of Soviet affairs, the equation of Jewry and Bolshevism remained the fundamental guideline for most party and state agencies. Thus, in 1937 Heydrich circulated a secret memorandum, "The Present Status of Research on the East," which opened with the argument that the importance of the East, mainly of the Soviet Union, for Germany derived from the fact that "this territory had been conquered by Jewish Bolshevism and turned into the main basis of its struggle against National Socialist Germany; all non-Bolshevist forces that are also enemies of National Socialism consider the Soviet Union the most active weapon against National Socialism."[29]

Apart from the axiom that Bolshevism was an instrument of Jewry, Nazi research aimed at proving the link between Jews and Communism in sociopolitical terms. This was shown, for example, in a June 1937 lecture by the head of the Königsberg Institute for the Economy of Eastern Germany, Theodor Oberländer (who was to be a minister in Konrad Adenauer's postwar government), on Polish Jewry: "The east European Jews are, in so far as they are not orthodox but assimilated Jews, the most active carriers of communist ideas. Since Poland alone has 3.5 million Jews, of which over 1.5 million can be regarded as assimilated Jews, and since the Jews live in scarcely credible adverse social conditions in the urban ghetto, so that they are proletarians in the truest sense, they have little to lose but much to gain. They are the ones who are peddling the most militant and succesful propaganda for communism in the countryside."[30]

The link between the Jews and Bolshevism in the Soviet Union could also be proven by erudite, "in-depth" reasoning. "It is not only the numerical importance of the Jews in the higher reaches of the Party and state system or the power exercised by individual Jews that should be simply interpreted as a 'domination' of Bolshevik Russia by the Jews," wrote Peter-Heinz Seraphim, the specialist on East European Jewry at the University of Königsberg. "The question that ultimately needs to be asked is whether there is an ideological linkage and reciprocal influence between Leninist and Stalinist Bolshevism and the Jewish mentality."[31] Published in

1938, Seraphim's massive study, *Das Judentum im osteuropäischen Raum* (Jewry in Eastern Europe), was to become the vade mecum of many Nazi practitioners in the East.

Seraphim started from the postulate that the Jews held a "hegemonical position" within the Bolshevik system.[32] As the argument from sheer numbers and individual influence did not suffice, the question of mental affinity indeed became of central importance. Seraphim had not the least doubt, it seems, about the Jewish features underlying this affinity: "this-worldliness, a materialistic and intellectualistic attitude to the surrounding world, the purposiveness and ruthlessness of the Jewish nature."[33]

Hitler's most threatening anti-Jewish outburst before his Reichstag speech of 1939 was triggered by the apparently minor issue (in Nazi terms) of the identification of Jewish-owned retail stores.

A debate on this issue had been in progress for several years. An April 1935 report prepared by SS main region Rhine tells of the initiative taken by the Frankfurt Nazi trade organization to have its members put up signs marking their own shops German Store, which was one way of solving the problem. According to the report, 80 to 90 percent of the German-owned stores there displayed the sign.[34] This must have been a rather isolated project, as a similar demand put forward by Nazi activists after the passage of the Nuremberg Laws was deemed unmanageable and the marking of Jewish stores advanced as the only possible course of action.

Grass-roots agitation for such marking reached such a pitch that Hitler decided to address the situation at a meeting of district party leaders at the School for Elite Party Youth at (Ordensburg) Vogelsang, on April 29, 1937. Hitler began with a stern warning to party members who wanted to accelerate the anti-Jewish measures in the economic domain. No one should try to dictate the pace of such measures to him, Hitler threatened darkly. He would have a word with "the fellow" who had written in a local party newspaper: "We demand that Jewish shops be marked." Hitler thundered: "What does it mean, 'we demand?' I am asking you from whom does one demand? Who can give the order? Only I! Thus this gentleman, the editor [of that party paper], demands of me, in the name of his readers, that I do this. I would first like to say the following: much before this editor had the least idea about the Jewish question, I had already studied it very thoroughly; second, this problem of a special identification of Jewish businesses has already been considered for two years, three years, and one day will naturally be solved one way or the other." Hitler then inserted a cryptic remark: "And let me add this: the final aim of our policy is obviously clear to us all." Did this mean the total expulsion of the Jews from German territory? Did it hint at other goals? Was it a formula used to cover the uncertainty of the plans? The comments on strategy that followed could accommodate any interpretation: "For me what matters constantly is not to take any step forward that I would have to retract, and not to take any step that could cause us harm. You know that I always move to the most extreme limit of what may be risked, but I never go beyond this limit. One has to have a nose sensitive enough to feel: What can I do more of? What can I not do? [Laughter and applause]"

The finale that followed did not point to any specific measures, but the tone, the words, the images contained a yet unheard ferocity, the intimation of a deadly threat. Without any doubt Hitler was thereby creating an atmosphere in which his listeners could imagine the most radical outcomes:

"I do not immediately want to force an enemy into a fight," Hitler exclaimed. "I do not say 'struggle' because I want to fight, but I do say: 'I want to annihilate you!' And now may cleverness help me to maneuver you into a corner in such a way that you will not manage one single blow; it is then that you get the stab in the heart!"[35] The recording of this secret speech survived the war. By this stage Hitler is shouting at the top of his voice. Then, in an orgiastic spasm, the three last words literally explode: "*Das ist es!*" (That's it!) The audience's applause is frenetic.

After a period of relative rhetorical prudence, the Nazi leaders were returning to the basic themes of the Jewish world conspiracy in their most extreme form. But how were these themes internalized at lower party levels? How were they translated into the language of the party bureaucracy, and of the police bureaucracy in particular?

On October 28, 1935, the Gestapo chief of the Hildesheim district informed the district presidents and mayors under his authority that butchers were complaining about sharp practice on the part of Jewish cattle dealers. The butchers accused the Jews of charging inflated sums for the cattle earmarked for slaughter, thereby driving up the price of meat and sausages: "The suspicion exists," the Gestapo chief wrote, "that these machinations represent a planned attack by Jewry, with the aim of fostering unrest and dissatisfaction among the population."[36] A few days earlier the same Gestapo station had informed its usual addressees that Jewish shoe stores were refusing to buy from Aryan manufacturers. According to the police chief, given the considerable importance of the Jewish shoe business, some Aryan producers were trying to sell their wares to the Jews by declaring that they were not members of the Nazi Party or of any related organization. The Hildesheim Gestapo assumed that the same boycotting was taking place in other parts of the Reich and that it must therefore derive from centrally issued instructions; a report on each local situation was therefore required by November 10.[37]

In each case the existence of a Jewish conspiracy is revealed by the "discovery" of some perfectly mundane event that might be real enough—the price of food did indeed rise in 1935, though this was caused by entirely other factors—or that might be a purely imaginary construct inspired by general economic difficulties. The police turned such random events into elements of a deliberate plot, thus creating a paranoid notion of centrally planned Jewish initiatives aimed at spreading an atmosphere of subversion among the population or intimidating party stalwarts in the business community. The ultimate goal of these "dangerous" Jewish initiatives was obvious: the downfall of the Nazi regime. There is a striking similarity of structure between Hitler's all-encompassing vision of Jewish subversion on a world scale and the dark suspicions of a Gestapo chief in a small German town.

III

In July 1936 a memorandum was submitted to Hitler by the Provisional Directorate of the Confessing Church. It was a forceful document mentioning the concentration camps, the Gestapo's methods, and even the misuse of religious terms and images in worship of the Führer. In an unusually bold departure from previous practice, the memorandum prophesied disaster for Germany if "there were persistence in totalitarian presumption and might contrary to the will of God." The document was leaked and received extraordinary coverage abroad. Such a courageous statement, one could assume, must have given pride of place to the Jewish issue—that is, to the persecution of the Jews. "Yet," in the words of historian Richard Gutteridge, "all that was devoted to this subject was the rather awkward observation that, when in the framework of the National Socialist Weltanschauung a form of anti-Semitism was forced upon the Christian which imposed an obligation of hatred towards the Jews, he had to counter it by the Christian command of love towards one's neighbor.

Here was no disavowal of anti-Semitism as such, including the Christian type, but merely of the militant Nazi version without even an oblique reference to the plight of the Jews themselves. The emphasis was upon the severe conflict of conscience experienced by the devout German Church people."[38] When a declaration of the Confessing Church referring indirectly to the memorandum was read in church by many pastors on August 23, not a single word was directed toward anti-Semitism or hatred of the Jews.[39] A few months later, in March 1937, Pius XI's sharp critique of the Nazi regime, the encyclical *Mit brennender Sorge*, was read from all Catholic pulpits in Germany. Nazi pseudoreligion and the regime's racial theories were strongly condemned in general terms, but no direct reference was made to the fate of the Jews.

For the converted "full Jew" Friedrich Weissler, the memorandum of the Confessing Church was to have fateful consequences. A lawyer by profession, Weissler was employed by the Confessing Church as a legal adviser and was secretly in charge of informing the outside world about its activities. It was probably he who leaked the memorandum to the foreign press. Pretending outrage, the leadership of the Confessing Church asked the Gestapo to find the culprit. Weissler and two Aryan assistants were arrested. Whereas the Aryans were ultimately released, Weissler, for whom the church did not intervene, succumbed in the Sachsenhausen concentration camp on February 19, 1937. Thus a "full Jew" became "the first martyr of the Confessing Church."[40]

Friedrich Meinecke, possibly the most prestigious German historian of his time, had been replaced in 1935 at the editorial helm of the *Historische Zeitschrift*, the leading German historical journal. No doubts could be raised about the ideological orthodoxy of his successor, Karl Alexander von Müller. But from January 1933 on, the *HZ* had not been immune to the new trends, especially since, as has been seen, the academic world found no great difficulty in adapting to the new regime.[41] Contributors were examined as to Jewish origin, and at least one Jewish member of the editorial board, Hedwig Hintze, was ousted.[42]

As could be expected, Müller's initial editorial was a clarion call. The new editor in chief described the fundamental changes the world was undergoing as a mighty context that demanded a renewal of historical insight. Müller's closing words are memorable:

"We are buffeted like few other races by the stormy breath of a great historical epoch. Like few other races, we are granted an insight into the original demonic forces, both stupendous and terrible, that produce such turbulent times. Like few other races, we are filled with the consciousness that in the decisions of the present we shall determine the long-term future of our whole people. Out of what is becoming we seek and relive what has been, and we revive its shades with our blood; out of what is truly past, we recognize and reinforce the power of the living present."[43]

The bombastic hollowness of these lines is in itself revealing. The ideological message of Nazism mobilized an apparently senseless set of images that nonetheless constantly evoked a longing for the sacred, the demonic, the primeval—in short, for the forces of myth. The intellectual and political content of the program was borne by the "stormy breath" of historical events of world-historical significance. Not even the readers of the *Historische Zeitschrift* could be entirely indifferent to the revival of an atmosphere rooted in a German romantic and neoromantic tradition of which many of them partook.

Under the new stewardship the changes went beyond the editorial invocation of "demonic forces." The periodical's remaining Jewish board members, Gerhard Masur, for example, were replaced by Aryans; and, most important, a new permanent section, under the editorial supervision of Wilhelm Grau, was added to deal with the "History of the Jewish Question."[44] In his opening article, "The Jewish Question as the Task of Historical Research," Grau explained that, since hitherto all books dealing with Jewish issues had been reviewed by Jews only, which had naturally led to uncritical praise, his new section would take a somewhat different approach.[45] The first title discussed was a dissertation by a Lithuanian Jew, Abraham Heller (of whom more will be heard), entitled "The Situation of the Jews in Russia from the Revolution of March 1917 to the Present." Grau's immediate contribution to greater objectivity was to add a subtitle that, in his view, conveyed the book's content more accurately: "The Jewish Contribution to Bolshevism."[46]

Young Grau (barely twenty-seven in 1936) had already—in a way— been making a name for himself, having become director of the Jewish Section, the most important research section of the Reich Institute for the History of the New Germany. Inaugurated on October 19, 1935, the institute was headed by Walter Frank, a protégé of Rudolf Hess and a historian of modern German anti-Semitism, mainly of Adolf Stöcker's "Berlin movement." Grau seemed to be a worthy disciple: In 1935 he had already contributed a slim study on Humboldt and the Jews, berating the most famous nineteenth-century German humanist and liberal intellectual for his subservience to Jewish influence. Writing from beyond the borders of the Reich, the Jewish philosopher Herbert Marcuse could afford to be direct: He made mincemeat of Grau's book and showed him as the fool and charlatan that he was. For Walter Frank and his institute, Grau was nonetheless a rising star who would establish a research empire on the Jewish question.[47]

The festive opening, on November 19, 1936, of the Jewish Section took place in Munich, where it was to be located, in the presence of a wide array of national and local celebrities from the party, the government, the army, and the academic world. The Munich chamber orchestra played a Bach suite, and Karl Alexander von Müller, formally Grau's superior, spoke, followed by Walter Frank. According to the summary in the *Deutsche Allgemeine Zeitung*, Frank explained that research on the Jewish question was like "an expedition into an unknown country whose darkness is shrouded in a great silence. Until now, only the Jews had worked on the Jewish problem."[48] Tension soon built up between Frank and the ambitious Grau, and within two years the latter was out, though well on his way to establishing a competing research institute on the Jewish question in Frankfurt, this time under Alfred Rosenberg's aegis.[49]

While Frank and Grau were launching their enterprise, Carl Schmitt was making his own display of anti-Semitic fervor. This luminary of German legal and political theory, whose enthusiastic adherence to National Socialism in 1933 has already been mentioned, apparently deemed it necessary to fortify his newly acquired ideological trustworthiness against the accusations both of exiled intellectuals, such as Waldemar Gurian, and of colleagues who were also members of the SS (such as Otto Köllreuter, Karl August Eckhardt, and Reinhard Höhn), who did not hesitate to allude to his many Jewish friends before 1933 and to his rather sudden political conversion that year.[50]

It was in this atmosphere that Schmitt organized his notorious academic conference, "Judaism in Legal Science," held in Berlin on October 3 and 4, 1936. Schmitt opened and closed the proceedings with two major anti-Jewish speeches. He started his first speech and ended his closing address with Hitler's famous dictum from *Mein Kampf*: "In defending myself against the Jew . . . I am doing the work of the Lord."[51]

In the concrete resolutions Schmitt drafted for the conference, he demanded the establishment of a legal bibliography that would distinguish between Jewish and non-Jewish authors, and the "cleansing" of Jewish authors from the libraries.[52] In the event a Jewish author had to be quoted, he or she was to be identified as such. As Schmitt himself put it on that occasion: "By the very mention of the word 'Jewish,' a healthy exorcism would be effected."[53] Within a few months the implementation of Schmitt's recommendations began.

All this was of little avail to Carl Schmitt himself. *Das Schwarze Korps* attacked him in December 1936, reiterating once again the charges of his prior Jewish contacts. Despite such powerful protectors as Göring and Hans Frank, Schmitt could not withstand the SS pressure: His official/political functions and ambitions were over. His ideological production, however, went on. In his 1938 work on Thomas Hobbes (*Der Leviathan in der Staatslehre des Thomas Hobbes*), Schmitt described the deadly struggle between Leviathan, the great sea powers, and Behemoth, the great land powers; then, turning a Jewish legend about messianic times into an account of bloodshed and cannibalism, he added: "The Jews stand apart and watch as the peoples of the world kill one another; for them this mutual 'slaughter and carnage' (*Schlächten und Schlachten*) is lawful and 'Kosher.' Thus they eat flesh of the slaughtered people and live upon it."[54]

While Schmitt was cleansing legal studies and political science of any remnants of the Jewish spirit, Philipp Lenard, Johannes Stark, and Bruno Thüring, among others, were waging the same purifying campaign in physics.[55] In various ways similar purges were spreading throughout all other domains of intellectual life. Sometimes the thin line between belief and mere compliance was not clear as, for example, in the case of Mathias Göring, Hermann's cousin, who, as director of the Institute for Psychotherapy in Berlin, banished any explicit reference to psychoanalysis and its theories, its Jewish founder, and its mainly Jewish theoreticians and practitioners, while apparently accepting the systematic use of therapeutic methods directly inspired by psychoanalysis.[56]

In some instances the party leadership itself intervened to curtail the initiatives of an ideological orthodoxy that could have significantly negative consequences. Thus, on June 15, 1937, Stark published a full-blown attack in *Das Schwarze Korps* on the famous physicist Werner Heisenberg, then teaching at Leipzig, accusing him of being a "white Jew" and the

"Ossietzky of physics,"* because the young theoretician of quantum physics had adopted various modern theories, in particular Einstein's theory of relativity. At first Heisenberg's protests were of no avail, especially as he had not signed the declaration of support for the new regime circulated by Stark in 1933. However, a highly regarded Göttingen aeronautical engineer, Ludwig Prandtl, intervened with Himmler on Heisenberg's behalf. It took but a few months for Himmler to decide that Heisenberg should be protected from further attacks, on condition that he agreed to restrict himself to purely scientific issues. Orders to that effect were given to Heydrich, and, after the annexation of Austria, Heisenberg was named to the prestigious chair of theoretical physics at the University of Vienna. Heisenberg now acquiesced in all demands without further ado. Thus, although Stark and Lenard represented the most orthodox anti-Semitic line in science, and although Heisenberg had adopted "the Jewish dimension" of physics, Himmler understood the harm that Heisenberg's marginalization or emigration could inflict on Germany's scientific development and decided to shield him.[57] But there were limits to this sort of compromise. Despite receiving the appointment in Vienna, the chair Heisenberg had initially wanted, at Munich, was refused him. Moreover—and this is the main point—Himmler would never have intervened to protect and keep any of the Jewish scientists who were being forced to leave Germany. In Heisenberg's case the basic principle of racial purification had not been infringed.

IV

Heinrich Himmler was appointed head of all German police forces on June 17, 1936, thus becoming Reichsführer SS and chief of the German police.[58] The German police was being withdrawn from the jurisdiction of the state. This decisive reorganization accorded with the new atmosphere of general ideological confrontation, which demanded an effective concentration of the entire surveillance and detention apparatus of the regime. In more concrete terms, it signaled an unmistakable step toward the ever increasing intervention of the party in the state's sphere of competence and a shift of power from the traditional state structure to the party.

---

*Carl von Ossietzky was a left-wing German journalist and passionate pacifist. He was awarded the Nobel Peace Prize in 1935, while imprisoned in the Sachsenhausen concentration camp, where he died in 1938.

On June 26, 1936, Himmler divided the police forces into two separate commands: the Order Police (Ordnungspolizei), under Kurt Daluege, was to comprise all uniformed police units; whereas the Security Police (Sicherheitspolizei, or Sipo), under Heydrich's command, integrated the Criminal Police and the Gestapo into a single organization. Heydrich now had control of both the new Sipo and the Security Service of the SS, the SD. Within the Sipo itself, the new trend was clear from the outset: "Instead of the criminal police reabsorbing the political police and returning them to subordination under the State administration as Himmler's opponents [Frick] had desired, the criminal police assumed more of the extraordinary status of the political police."[59] Although in principle the police forces belonged to the Ministry of the Interior and thus, theoretically, as police chief Himmler was subordinate to Frick, in reality, the Sipo was not submitted to any ordinary administrative or judicial rules; like the Gestapo from its beginnings, its only law was the Führer's will: "It did not need any other legitimation."[60] When he received sole control of Germany's entire repression and terror system, Himmler was thirty-six years old; his right-hand man, Heydrich, was thirty-two.

SS-Untersturmführer Rudolf aus den Ruthen, one of the three young editors of *Das Schwarze Korps*, decided to marry Marga Feldtmann. The future bride's appearance was perfectly Aryan, but her genealogical tree showed an Austrian ancestor named Fried, which, in the Austrian province where he had lived, was most often a Jewish name; Ruthen broke the engagement. In early 1937 he found another prospective wife, Isolina Böving-Burmeister. Born in Mexico of a Cuban mother and a *Volksdeutsch* father, Isolina was a naturalized German. Her appearance did not inspire the investigators' full confidence, and the matter was referred to Himmler. The Reichsführer was soon made aware of a Philadelphia ancestor of Isolina's called Sarah Warner, who might have been Jewish. Finally, there was also a suspicion of some Negro blood on the Cuban mother's side. Himmler first demanded a "full solution" of the problem. When total clarification proved impossible, he at last gave a favorable answer.[61]

Himmler was a stickler for racial purity within his SS. As he explained in a May 22, 1936, speech delivered on the Brockenberg in the Harz Mountains: "Until October 1 of this year, the goal [for the family tree] is set at 1850; by next April 1, it will be set at 1750, until we achieve, within the next three years, for the whole SS and for each new recruit, the goal of

1650." Himmler explained why he did not plan to reach back further in time: Most of the church registers did not exist for the period prior to 1648, the end of the Thirty Years' War.[62] Finally, though, such ideal goals had to be abandoned, and 1800 became the accepted cutoff date for SS members.

The Reichsführer was not above dealing personally with any aspect of ancestry searches. On May 7, 1936, he wrote to Minister of Agriculture Walther Darré, who was also head of the Main Office for Race and Settlement of the SS, to inquire into the ancestry of General Ludendorff's wife, Mathilde von Kemnitz. Himmler strongly suspected her of being of Jewish origin: Otherwise, her troublemaking, as well as her "totally abnormal personal and sexual life would be inexplicable."[63] Two years later Darré was asked by Himmler to deal with the suspected Jewish ancestry of an SS officer on Darré's own staff.[64]

Needless to say, candidates for the SS, or SS members wishing to marry, as we saw in aus den Ruthen's case, made extraordinary efforts to obtain a clean ancestral record, capable of withstanding investigation, for their prospective spouses regarding any Jewish parentage at least as far back as 1800. Thus, to give one more example, on April 27, 1937, SS Master Sgt. (Hauptscharführer) Friedrich Mennecke, a physician who was to become a notorious figure in the euthanasia program, asked for authorization to marry. To his letter he appended forty-one original certificates about his fiancée's ancestors. As the set of necessary documents was not absolutely complete, Mennecke affirmed "with a degree of probability close to certainty that up to 1800 all her ancestors were pure Aryans."[65]

Jewish ancestry was not Himmler's only ideological worry. In March 1938 he wrote a formal letter of protest to Göring about a Luftwaffe court's dismissal of the case against an officer who had sexual relations with a woman identified as Jewish. To Himmler's outrage the case had been dismissed because the officer declared that the woman was not Jewish but a mixed breed of "Negroid" origin.[66]

At lower levels of the SS, racial dogma was set in precise and concrete terms. The educational bulletin (*SS-Leitheft*) of April 22, 1936, posed the question: "Why do we teach about the Jews?" The answer: "In the SS we teach about the Jews because the Jew is the most dangerous enemy of the German people." The explanation insisted on the parasitic aspect of the Jew, who lived off the vital forces of the host people, destroying its racial potential, its thought, its feelings, its morals, its culture. In even more precise terms, the *Leitheft* presented the three symbolic figures of the Jew: "Ahasuerus, the rootless one, who—defiling the race and destroying peoples—driven by unsteady blood, wanders restlessly through the world; Shylock, devoid of soul, who enslaves the peoples economically and as money lender holds them by the throat; Judas Iscariot, the traitor."[67]

The same *Leitheft* had even more lurid details, for those who did not react adequately to Ahasuerus, Shylock, and Judas Iscariot: "The Jew systematically defiles the maidens and women of Aryan peoples. He is equally driven by cold calculation and uninhibited animal lust. The Jew is known to prefer blond women. He knows that the women and maidens whom he has defiled are forever lost to their people. Not because their blood has thereby deteriorated, but because the defiled maiden is spiritually destroyed. She is entrapped by the lust of the Jew and loses all sense of what is noble and pure."[68]

The defiled Aryan maidens could eventually pursue a normal life if their ambitions weren't set too high. But this could not be the case if they aspired to marry an SS officer. Anneliese Hüttemann was interrogated by the SD in August 1935 because of her relation with the Jew Kurt Stern. Both admitted to having had intercourse on several occasions (they were neighbors and had known each other since childhood). What happened to Kurt Stern we can only surmise. For Anneliese Hüttemann, the sin against the blood led to nerve-racking suspense when, nine years later, in May 1944, she was about to marry SS-Obersturmbannführer Arthur Liebehenschel. The 1935 files were brought up by the SD. After a painstaking investigation and endless petitions, Himmler, because a child was expected, assented to the marriage. At this time Liebehenschel was the commandant of Auschwitz.[69]

At first sight there is an apparent contradiction between the ideological importance of the Jewish issue in Nazi Germany in the mid-thirties—and its even greater importance within the SS—and the seemingly subordinate status of the office dealing with Jewish matters within the SD, the SS security service. The SD itself was, in fact, just coming into its own in the years 1935–1936. Elevated as one of the three main offices of the SS at the beginning of 1935, under Heydrich's command from its inception as the party's intelligence arm in August 1931, the SD underwent a major reorganization in January 1936.[70] Three bureaus were established. Amt I, Administration, was headed by Wilhelm Albert, and Amt III, Foreign

Intelligence, was under Heinz Jost. Amt II, Internal Intelligence, under Hermann Behrends and later Franz Albert Six, was subdivided into two main sections: II 1, dealing with ideological evaluations (Erich Ehrlinger and later Six), and II 2, with the evaluation of social conditions/attitudes (Reinhard Höhn and later Otto Ohlendorf). Within II 1, under Dieter Wisliceny, subsection II 11 dealt with ideological opponents; it comprised subsubsections II 111 (Freemasons, also Wisliceny), II 113 (political churches [that is, their political activity], Albert Hartl), and II 112 (Jews). According to Wisliceny, it was only in June 1935 that systematic work regarding the "opponent Jewry" started: Previously surveillance of Jewish organizations had been part of the activities of a section dealing mainly with Freemasonry. Subsection II 112 was successively under the authority of Mildenstein, Kurt Schröder, Wisliceny, and finally, from the end of 1937, Herbert Hagen. It comprised the following "desks": II 1121 (assimilationist Jewry), II 1122 (Orthodox Jewry), and II 1123 (Zionists), the latter headed by Adolf Eichmann.[71]

The Gestapo was organized along roughly the same lines. Its equivalent of the SD's Amt II was Abteilung II, under Heinrich Müller; that of the SD's II 11 was II/1B, under Karl Hasselbacher.[72] The unification of these separate but coordinated lines of command into the Main Office for the Security of the Reich (Reichssicherheitshauptamt, or RSHA), which was to be established under Heydrich's command in September 1939, aimed, in principle at least, at the creation of an entirely integrated system of surveillance, reporting, and arrests.

Heydrich's men were young: In 1936–37 most of the top SD operatives were under thirty. They belonged to the cohort that came of age immediately after World War I. Most of them had been deeply influenced by the war atmosphere, the hardships, and the defeat. They were ruthless, practical, and strongly motivated by the ideological tenets of the extreme-right-wing organizations of the early twenties, in which many of them were active. Intense anti-Semitism (of the rational, not the emotional, kind—according to them) lay at the basis of their worldview.[73]

Although Heydrich's own anti-Jewish initiatives and proposals had been increasingly influential, and while the Gestapo already played a central role in the implementation of anti-Jewish decisions, until 1938 the activities of subsection II 112 of the SD were mainly limited to three domains: gathering information on Jews, Jewish organizations, and on other Jewish activities; drafting policy recommendations; and increasingly active participation in surveillance operations and interrogations of Jews in coordination with the Gestapo. Moreover II 112 unabashedly considered itself the top group of "Jew experts" in Germany and, after March 1936, it systematically organized conferences in which, several times a year, the most updated information was imparted to delegates of other SD sections from the main office and from various parts of Germany. The largest of such conferences, convened on November 1, 1937, brought together sixty-six mostly middle-ranking members of the SD.[74]

One of II 112's pet projects was the compilation of a card index of Jews (Judenkartei), intended to identify every Jew living in the Reich. Franz Six, moreover, ordered II 112 to start compiling another card index of the most important Jews in foreign countries and their mutual connections. As examples Six gave U.S. Supreme Court Justice Felix Frankfurter and the managements of the formerly German banking house Arnhold and the Dutch Unilever Trust.[75]

The Judenkartei was one of the topics on the agenda of the November 1 conference: As SS-Hauptsturmführer Ehrlinger summed up the matter, "for a successful internal struggle against Jewry, a listing in a card index of all Jews and people of Jewish origin living today in Germany is necessary. The aim of this listing is the following: (1) to establish the number of Jews and of people of Jewish origin according to the Nuremberg Laws living today in the Reich; (2) to establish the direct influence of Jewry and eventually the influence it exercises through its connections on the cultural life, the community life, and the material life of the German people."[76]

The general population census of May 1939 was to provide the opportunity for the complete registration of all the Jews in Germany (including half- and quarter-Jews): In each town or village the local police made sure that the census cards of Jews and *Mischlinge* carried the letter "J" as a distinctive mark; copies of all local census registration lists were to be sent to the SD and passed on to II 112.[77] The census took place as planned. The Jews were registered, as planned, and the card files fulfilled their function when the deportations began. (These files were kept in the building that now houses the department of philosophy of the Free University of Berlin.)[78]

A second information-gathering effort was aimed at every Jewish organization in Germany and throughout the world, from the ORT (an organization for vocational training and guidance) to the Agudath Yisrael (ultra-Orthodox Jewry). For the men of II 112 and the SD in general, no detail was too minute, no Jewish organization too insignificant. As the organized enemy they were fighting was nonexistent as such, their own enterprise had to create it *ex nihilo*. Jewish organizations were identified, analyzed, and studied as parts of an ever more complex system; the anti-German activities of that system had to be discovered, its internal workings decoded, its very essence unveiled.

The most astonishing aspect of this system was its concreteness. Very precise—and totally imaginary—Jewish plots were uncovered, names and addresses provided, countermeasures taken. Thus, in his lecture, "World Jewry," at the November 1 conference, Eichmann listed a whole series of sinister Jewish endeavors. An attempt on the life of the Sudeten German Nazi leader Konrad Henlein had been planned at the Paris Asyle de Jour et de Nuit (a shelter for destitute Jews). It had failed only because Henlein had been warned and the murderer's weapon had not functioned. Worse still, Nathan Landsmann, the president of the Paris-based Alliance Israélite Universelle (a Jewish educational organization), was in charge of planning attempts on the Führer's life—and also on Julius Streicher's. To that effect, Landsmann was in touch with a Dutch Jewish organization, the Komitee voor Bizondere Joodsche Belange in Amsterdam, which in turn worked in close cooperation with the Dutch (Jewish) Unilever Trust, including its branches in Germany.[79] This is a mere sample of Eichmann's revelations.

For Heydrich and his men, it was probably inconceivable that connections among Jewish institutions were very loose and of very little importance in Jewish life.[80] As described by him in a pamphlet published at the end of 1935, *Wandlungen unseres Kampfes* (The transformations of our struggle), the network of Jewish organizations acting against the Reich was a deadly threat.[81] It appeared as such on the fictional charts growing apace in the SD offices at 102 Wilhelmstrasse, in Berlin. This was the police face of redemptive anti-Semitism.

In its policy recommendations, II 112 backed any action to accelerate Jewish emigration, including the potentially positive effects of instigated violence.[82] As early as May 1934, an SD memorandum addressed to Heydrich had opened with the unambiguous statement that "the aim of the Jewish policy must be the complete emigration of the Jews." In the context of 1934 the lines that followed were unusual: "The life opportunities of the Jews have to be restricted, not only in economic terms. To them Germany must become a country without a future, in which the old generation may die off with what still remains for it, but in which the young generation should find it impossible to live, so that the incentive to emigrate is constantly in force. Violent mob anti-Semitism must be avoided. One does not fight rats with guns but with poison and gas. . . ."[83] Yet, as has been seen, in September 1935 Heydrich did not set emigration at the center of his policy proposals. It was within the overall shifting of Nazi goals in 1936 that the policy of the SD became an active element in a general drive of all Nazi agencies involved in Jewish matters: For all of them, emigration was the first priority.

Palestine was considered one of the more promising outlets for Jewish emigration, as it had been since 1933. Like the Foreign Ministry and the Rosenberg office (which was mainly in charge of ideological matters, including contacts with foreign Nazi sympathizers), the SD was confronted with the dilemma entailed by the need to encourage Jewish emigration to Palestine on the one hand, and, on the other, the danger that such emigration could lead to the creation of a strategic center for the machinations of world Jewry: a Jewish state. It is in relation to such policy considerations that Heydrich allowed Hagen and Eichmann to visit Palestine in the fall of 1937 and to meet with their Haganah "contact," Feivel Polkes.

For Eichmann at least the mission appears to have raised great expectations: "As during the trip negotiations with Arab princes are foreseen, among other things," the ex-traveling salesman for the Vacuum Oil Company in Upper Austria wrote to the head of II 1, Albert Six, "I will need one light and one dark suit as well as a light overcoat." Eichmann's dreams of Oriental elegance remained unfulfilled; instead both travelers were repeatedly warned about strict secrecy measures: no use of terms like "SS," "SD," "Gestapo"; no sending of postcards to friends in the service, and so on.[84] The mission failed miserably: The British did not allow the two SD men to stay in Palestine more than a day, and their conversations with Polkes—who came to meet them in Cairo—produced no valuable information whatsoever. But the favorable SD view of Palestine as a destination for German Jews did not change. Later on it was with the SD that Zionist emissaries organized the departure of convoys of emigrants to Yugoslav and Romanian ports, from which they attempted to sail for Palestine in defiance of the British blockade.

Finally the SD Jewish subsubsection participated with increasing energy in the surveillance activities of the Gestapo, and in this domain its share of the common work grew throughout 1937. On September 18, for example, the SD main region Rhine submitted a report on a Jewish student named Ilse Hanoch. According to the report, Hanoch ("who supposedly is studying in London") was traveling on the 6:25 train from Trier to Luxembourg, when, "shortly before arriving at the border-control station, Hanoch looked very uncertain and started tearing pieces of paper from her notebook, crumpled them, and threw them into the ashtray." She underwent a thorough search at the border station, but without any result. The SD report assumed, on the basis of Hanoch's travel schedule to and from Germany (as indicated on her passport) and, from the names of various Jewish families that were found on the pieces of paper she had torn and thrown away, that the so-called student was a messenger between Jews who had emigrated and those still living in Germany. Instructions were issued to all border control stations that she "be most thoroughly searched when reentering the country and that she be put under the strictest surveillance during her travels in Germany."[85] It is unknown whether Ilse Hanoch ever returned to Germany.

Strangely enough, however, when no clear instructions were given or when the framework for violence was not preestablished, the anti-Jewish actions of the SS had their built-in limitations, at least in the mid-thirties. Consider the case of SS-Sturmmann (SS Private) Anton Beckmann, of the headquarters staff of the Columbia SS detention center in Berlin. On January 25, 1936, he went into a shop on the Friesenstrasse and bought a pair of suspenders. As he left the shop, a passerby told Beckmann, who was wearing his SS uniform, that he had just been patronizing a Jewish store. He immediately tried to return the suspenders but to no avail: "The Jewess Joel [the store owner] insolently told him that she wouldn't even think of taking back purchased goods, and furthermore, that she had a lot of SS customers, even some high-ranking ones." SS-Obersturmführer Kern, summoned by Beckmann to help him return the suspenders, had no greater success. The commandant of the Columbia detention center sent a report on the matter demanding the arrest of "the Jewess Joel" for spreading false rumors about SS members, adding that "it would be a welcome step in the interest of all National Socialists, if finally, as in other regions, the Jewish shops in Berlin were to be marked."[86]

On receipt of the commandant's letter, the Inspector of concentration camps, Gruppenführer Eicke, had to admit that he felt "powerless in this matter" and transmitted the request to the chief of the SS Personnel Office, Gruppenführer Heissmeyer,[87] who passed it on to the Berlin area SS commander with a comment of his own: "In Berlin, of all places, everyone is in danger of unwittingly buying in Jewish stores, whereas in other cities, Frankfurt, for example, this danger is avoided by the use of a standard sign reading GERMAN ESTABLISHMENT."[88] The evolution of the shop-marking issue has already been noted, but what of the "Jewess Joel"? An absence of orders regarding her and the imminence of the Olympic Games suggest that, despite her "insolence," she might not have been imprisoned.

The Joel incident, as minute as it was, points to an issue that was of central significance for the prewar anti-Jewish Nazi policy. Among the main obstacles faced by the regime in its attempt to eliminate the Jews from Germany was the fact that the victims had been part and parcel of every field of activity in German society. In consequence, if direct violence was not (yet) possible, the system had to elaborate ever new administrative or legal measures in order to undo, stage by stage and step by step, the existing ties between that society and the Jews. And, as we have seen, at each stage, any number of unforeseen exceptions demanded additional administrative solutions. In other words it was not yet easy merely to arrest the "Jewess Joel," who was legally selling her wares and was still protected by the general instructions regarding the economic activity of the Jews: Marking Jewish shops, for example, entailed possible internal and external consequences the regime was not yet ready to face.

V

Although the total number of concentration camp inmates in 1936–37 (about 7,500) was at its lowest point[89], compared to the first two years of the regime and mainly to what happened later, the categories of targeted prisoners were increasing considerably. Apart from political opponents, the inmates were mainly members of religious sects such as Jehovah's Witnesses; homosexuals; and "habitual criminals" or "asocials," a group the Ministry of the Interior defined as follows:

"Persons who through minor, but repeated, infractions of the law demonstrate that they will not adapt themselves to the natural discipline of the National Socialist state, e.g., beggars; tramps (Gypsies); alcoholics;

whores with contagious diseases, particularly sexually transmitted diseases, who evade the measures taken by the health authorities."

A further category of asocials was the "work-shy": "Persons against whom it can be proven that on two occasions they have, without reasonable grounds, turned down jobs offered to them, or who, having taken a job, have given it up after a short while without a valid reason." During the following years, asocials of these various kinds were increasingly picked up by the Gestapo and sent to concentration camps.[90]

The entirely arbitrary nature of the arrests and incarcerations in camps, even by the Third Reich's standards of justice, can be illustrated by two police orders. In September 1935 the Bavarian Political Police demanded that the release date of all prisoners "who had been sentenced by a People's Court be communicated well in advance so that, upon their release, they could immediately be transferred to a concentration camp. In other words, the police were "correcting" the courts' sentences.[91] And on February 23, 1937, Himmler ordered the Criminal Police to rearrest about two thousand habitual criminal offenders and to incarcerate them in concentration camps.[92] These were individuals who had not been sentenced anew; choosing the victims was entirely up to the Criminal Police's judgment—whereby "the overall number of arrests ordered could only encourage the arbitrariness of the choice."[93]

In the thirties the Nazi regime used two different but complementary methods to achieve the complete exclusion of racially dangerous groups from the *Volksgemeinschaft*: segregation and expulsion on the one hand, sterilization on the other. The first method was used in its various aspects against the Jews, Gypsies, and homosexuals; the second method was applied to the carriers of hereditary diseases (physical or mental) and to persons showing dangerous characteristics deemed hereditary, as well as to "racially contaminated individuals" who could not be expelled or put into camps. As for the struggle against the Jew as the world enemy, it took additional and different forms, both on the ideological level and in terms of its all-encompassing nature.

Besides the asocials the main groups designated for segregation and diverse forms of imprisonment in existing camps or newly established camplike areas were the Gypsies and the homosexuals. Like the Jews the Gypsies dwelt in the phantasmic recess of the European mind, and like them they were branded as strangers on European soil. As was seen, the applicability of the Nuremberg Laws to the Gypsies was announced soon after their proclamation. As "carriers of alien blood," the Gypsies were barred from marrying or having sexual contact with members of the German race.[94] But although the decision was applied on the basis of general criteria of appearance and behavior, the task of actually defining the racial nature of "Gypsies" still lay ahead. From 1936 on it became the project of the University of Tübingen's Robert Ritter.

With financing from the state-funded German Research Society (Deutsche Forschungsgemeinschaft, or DFG), the SS, and the Reich Health Ministry, Ritter took upon himself the classification of the thirty thousand Gypsies living in Germany. (Today identified as Sinti and Roma, these ethnic groups were generally called Gypsies [*Zigeuner* in German] long before and during the Third Reich, and most often to this day.) According to the Tübingen specialist, the Gypsies came from northern India and were originally Aryan, but in their migrations they had mingled with lesser races and were now nearly 90 percent racially impure.[95] Ritter's conclusions were to become the basis for the next step on the road to segregation, deportation, and extermination: Himmler's order of December 8, 1938, regarding the measures to be taken against the Sinti and Roma.

The police were not passive while racial laws barring marriage and sexual intercourse between Gypsies and Germans were being promulagated and Ritter and his assistants were researching photographs and measurements. The Sinti and Roma had traditionally been subjected to harassment, mainly in Bavaria; after 1933, however, direct harassment became systematic, with the expulsion from the country of foreign Gypsies, and with others incarcerated as vagrants, habitual criminals, and various other kinds of asocials. Taking the Olympic Games as a pretext, the Berlin police in May 1936 arrested hundreds of Gypsies and transferred whole families, with their wagons, horses, and other belongings to the so-called Marzahn "rest place," next to a garbage dump on one side and a cemetery on the other. Soon the rest place was enclosed with barbed wire. A de facto Gypsy concentration camp had been established in a suburb of Berlin. It was from Marzahn, and from other similar rest places soon set up near other German cities, that a few years later thousands of Sinti and Roma would be sent to the extermination sites in the East.[96]

The Leopold Obermayer case has already given some indication of the system's particular hatred of homosexuals. A measure of liberalization of anti-homosexual laws and regulations had been achieved during the Weimar years, but once the Nazis came to power the prohibitions became harsher, especially after the liquidation of SA leadership in June 1934 (Ernst Röhm and some of the main SA leaders were notorious homosexuals). Homophobia was unusually shrill within the SS. A 1935 article in *Das Schwarze Korps* demanded the death penalty for homosexual activities, and the following year Himmler created a Reich Central Office for Combatting Homosexuality and Abortion.[97] During the Nazi period, some ten to fifteen thousand homosexuals were incarcerated.[98] How many died in the camps is unknown, but according to one Dachau inmate, "the prisoners with the pink triangle did not live very long; they were quickly and systematically exterminated by the SS."[99]

In many ways Obermayer's story remains exemplary.

By mid-October 1935, it will be remembered, Leopold Obermayer was back in Dachau. This time, however, using both the most diverse legal and moral arguments and his status as a Swiss citizen, Obermayer fought back. Most of his letters and petitions were seized by the Dachau censors and passed on to his chief tormentor, the Würzburg Gestapo head Josef Gerum; his defense strategy was undermined by his new lawyer, a run-of-the-mill Nazi; the hopes he set on a decisive intervention by the Swiss authorities never materialized (as Broszat and Fröhlich have put it, the Swiss probably did not consider that the case of a Jewish homosexual was worth an entanglement with Germany).[100] Nonetheless Obermayer's relentless complaints, and the uncertainty of the Bavarian Political Police and the Justice Ministry in Berlin as to just how ready the Swiss would be to turn Obermayer's case into an international scandal, profoundly unsettled Gerum and even some of his superiors in Munich and Berlin.[101] Thus, throughout 1936, the determined resistance of a Jewish homosexual, albeit one benefiting from foreign citizenship, could still induce a measure of uncertainty in the operations of the system. Be that as it may, Obermayer's trial could not be delayed indefinitely. The case was referred to the Würzburg Criminal Court; the trial was scheduled for December 9, 1936. The prosecution intended to concentrate on the accused's homosexual activities, mainly his perversion of German youth (Obermayer himself never denied his homosexuality but steadfastly argued that the relations he had with younger men had never gone beyond the limits set by the law).[102]

In November it dawned on the Würzburg Gestapo and the state prosecution that, given his personality and defiance, Obermayer would be able to use the courtroom to argue that Hitler himself knew of the homosexual relations within the SA leadership and had accepted them until June 30, 1934.[103] Thus the trial had to take place behind closed doors, and whereas Obermayer lost his last chance of embarrassing his persecutors, the propaganda machinery of the party and the Gestapo also lost the opportunity of staging a show trial. (As will be seen, a similar situation was to arise years later with regard to the planned show trial of Herschel Grynszpan, the Jewish youth who in November 1938 shot the German diplomat Ernst vom Rath.)

Little is known about the trial itself, but even the reports in the local Nazi press indicate that Obermayer defended himself astutely. The sentence was a foregone conclusion: life imprisonment. Obermayer was kept in a regular prison until 1942, when he was transferred into the hands of the SS and sent to Mauthausen; there he died on February 22, 1943. After presenting his own version of the events to a de-Nazification court in 1948, Josef Gerum was set free.[104]

Throughout the thirties the sterilization drive inaugurated in July 1933 went steadily forward. When the health argument could not easily be used for racial purposes, other methods were found. Thus the new regime had barely been established when the attention of the authorities was directed to a group probably numbering no more than five to seven hundred, the young offspring of German women and colonial African soldiers serving in the French military occupation of the Rhineland during the early postwar years. In Nazi jargon these were the "Rhineland bastards."[105] Hitler had already described this "black pollution of German blood" in *Mein Kampf* as one more method used by the Jews to undermine the racial fiber of the *Volk*.

As early as April 1933, Göring as Prussian minister of the interior requested the registration of these "bastards," and a few weeks later the ministry ordered that they undergo a racial-anthropological evaluation.[106] In July a study of thirty-eight of these schoolchildren was undertaken by a certain Abel, one of racial anthropologist Eugen Fischer's assistants at the Kaiser Wilhelm Institute. As could be expected, Abel found that his subjects, all of them living in the Rhineland, showed various defects in intellectual ability and behavior. The Prussian ministry reported the findings on March 28, 1934, warning of dire racial consequences if, despite their very small number, these "bastards" were allowed to reproduce. The upshot of the argument was that, since the presence in France of half a million mixed breeds would lead within four or five generations to the bastardization of half the French population, the similar presence of mixed breeds on the German side of the border would lead to local miscegenation and the consequent disappearance of any racial difference between the French and the population of the adjacent western parts of the Reich.[107]

That the matter was not taken lightly is shown by a meeting of the Advisory Committee for Population and Racial Policy of the Ministry of the Interior, which on March 11, 1935, convened representatives of the ministries of the Interior, Health, Justice, Labor, and Foreign affairs as well as eugenicists from the academic world. Walter Gross did not hide the difficulties in handling the problem of what he called the "Negro bastards" (*Negerbastarde*). Their rapid expulsion was impossible; thus, Gross left no doubt about the need for sterilization. But sterilization of a healthy population, if carried out openly, could cause serious internal and external reactions. As the reliability of ordinary practitioners was not to be depended on, Gross saw no other way but to demand the secret intervention of physicians who were also seasoned party members and would understand the imperatives of the higher good of the *Volk*.[108] In the course of 1937, these hundreds of boys and girls were identified, picked up by the Gestapo, and sterilized.[109]

The convolutions of Nazi thinking remain, however, inscrutable. As the party agencies were plotting the sterilization of the "Negro bastards," Bormann sent confidential instructions to all Gauleiters regarding "German colonial Negroes": the fifty or so blacks from the former German colonies living in Germany could not find any employment, according to the Reichsleiter, "because when they found some work their employer encountered hostile reactions and had to dismiss the Negroes." Bormann was ready to have employment authorizations issued to them in order to help them find steady work; any individual action against them was prohibited.[110] The Reichsleiter did not even mention the question of progeny. Were these blacks married to German women? Did they have racially mixed children? Were these children to be sterilized? It seems that none of these questions even crossed the mind of the prime racial fanatic Martin Bormann.

The decision to sterilize carriers of hereditary diseases and the so-called feeble-minded was based on medical examinations and specially devised intelligence tests. The results were submitted to hereditary health courts, whose decisions were in turn forwarded for review to heredity health appellate courts; only their final verdicts were mandatory. Some three-quarters of the total of approximately four hundred thousand individuals sterilized in Nazi Germany underwent the operation before the beginning of the war.[111] But only part of the sterilized population ultimately survived. For mental patients sterilization was often but a first stage: During the late thirties they were the group most at risk in Nazi Germany.

As early as the last years of the republic, patients at mental institutions were increasingly considered to be a burden on the community, "superfluous beings," people whose lives were "unworthy of living." The Nazi regime spared no effort to disseminate the right attitudes toward asylum inmates. Organized tours of mental institutions were meant to demonstrate both the freakish appearance of mental patients and the unnecessary costs that their upkeep entailed. Thus, for example, in 1936 the Munich asylum Eglfing-Haar was toured by members of the SA's Reich Leadership School, by local SS race experts, by instructors of the SS regiment "Julius Schreck," and by several groups from the Labor Front.[112] A crop of propaganda films aimed at indoctrinating the wider public were produced and shown during the same years,[113] and in schools, appropriate exercises in arithmetic demonstrated the financial toll such inmates imposed on the nation's economy.[114]

Whether these educational measures indicated a systematic preparation of public opinion for the extermination of the inmates is unclear, but in this domain— more so than in many others—one can follow the direct impact of ideology on the regime's policies from 1933 on. According to Lammers, Hitler had already mentioned the possibility of euthanasia in 1933, and according to his physician, Karl Brandt, Hitler had discussed the subject with the Reich physicians' leader Wagner in 1935, indicating that such a project would be easier to carry out in wartime.[115] Nonetheless, starting in 1936, mental patients were gradually being concentrated in large state-run institutions, and reliable SS personnel was placed on the staffs of some private institutions. Given this trend, it is not surprising that, in March 1937, *Das Schwarze Korps* had no compunction in praising a father who had killed his handicapped son.[116]

The privately run institutions were well aware of the ominous aspect of these developments. In fact, what is chilling about the documentation of the years 1936–38 is that "the associations established for the care of the handicapped [Protestant religious groups such as the Inner Mission] often . . . denounced those left to their care and thereby helped to bring about their persecution and extermination."[117] Many of the religious institutions that were losing some of their inmates as a result of the regrouping of patients into state institutions did complain—but only about the economic difficulties such transfers were causing them.[118]

The first concrete step toward a euthanasia policy was taken in the fall of 1938. The father of an infant born blind, retarded, and with no arms and legs petitioned Hitler for the right to a "mercy death" for his son. Karl Brandt, was sent to Leipzig, where the Knauer baby was hospitalized, to consult with the doctors in charge and perform the euthanasia, which he did.[119]

As will be seen further on, the planning of euthanasia was accelerated during the first months of 1939. Nonetheless Hitler acted with prudence. He was aware that the killing of mentally ill adults or of infants with grave defects could encounter staunch opposition from the churches, the Catholic Church in particular. This potential obstacle was all the more significant as the largely Catholic population and the ecclesiastical hierarchy of Austria had just given their enthusiastic endorsement to the Anschluss. Thus, in late 1938, Hartl, head of SD desk II 113 (political churches), received (via Heydrich) an order from Viktor Brack, the deputy of Philipp Bouhler (head of the Führer's Chancellery), to prepare an "opinion" about the church's attitude toward euthanasia.[120] Hartl did not feel competent to produce such an evaluation, but he contacted Father Joseph Mayer, professor of moral theology at the Philosophical-Theological Academy in Paderborn, who in 1927 had already written favorably about sterilization of the mentally ill. In the early fall of 1939, Hartl received Mayer's detailed memorandum, which summed up the pros and cons in Catholic pronouncements on the subject. The memorandum has not been found, and we do not know whether the Paderborn cleric expressed his own view on euthanasia, but it seems that even if his conclusion was indeterminate, it left the door open for exceptions.[121]

Through indirect channels Brack's office submitted Mayer's memorandum to Bishop Berning and to the papal nuncio, Monsignor Cesare Orsenigo. On the Protestant side, it was submitted to Pastors Paul Braune and Friedrich von Bodelschwingh. It seems that no opposition was voiced by any of the German clerics—Catholic or Protestant—contacted by Hitler's Chancellery. The pope's delegate, too, remained silent.[122]

CHAPTER 7

Paris, Warsaw, Berlin—and Vienna

I

As the storm gathered over Europe, throughout the Continent the Jews once again became objects of widespread debate, targets of suspicion and sometimes of outright hatred. The general ideological and political cleavages of the mid-thirties were the main source of change, but in countries other than Nazi Germany, a pervasive atmosphere of crisis prepared the ground for a new surge of anti-Jewish extremism.

The first signs of this radicalization had appeared at the beginning of the decade. Growing doubts about the validity of the existing order of things arose as a result of the economic crisis but also because of a more general discontent. By dint of an almost "natural" reaction, the Jews were identified—and not only on the extreme right—with one or another aspect of the apparent social and cultural disintegration, and were held responsible for some of its worst consequences. It was a time when the Catholic writer Georges Bernanos, no fanatic as such, could glorify France's arch-anti-Semite of the late nineteenth century, Louis Drumont, the notorious editor of *La Libre Parole* and author of *La France Juive*, and lash out at the Jewish threat to Christian civilization.

In Bernanos's 1931 book, *La grande peur des bien-pensants* (The great fear of the right-thinking), the values threatened by what he perceived as an ever-increasing Jewish domination were those of Christian civilization and of the nation as a living organic entity. The new capitalist economy was controlled by the concentrated financial power of "*les gros*"—the mythical "two hundred families" that both Right and Left identified with the Jews.[1]

In other words the Jewish threat was, in part at least, that of *modernity*. The Jews were the forerunners, the masters, and the avid preachers of the doctrine of progress. To their French disciples, wrote Bernanos, they were bringing "a new mystique, admirably suited to that of Progress. . . . In this engineers' paradise, naked and smooth like a laboratory, the Jewish imagination is the only one able to produce these monstrous flowers."[2]

La grande peur ends with the darkest forebodings. In its last lines the Jews are left unnamed, but the whole logic of the text links the apocalyptic conclusion to Drumont's lost fight against Jewry. The society being created before the author's eyes was a godless one in which he felt unable to live: "There is no air!" he exclaimed. "But they won't get us . . . they won't get us alive!"[3]

Bernanos's anti-Semitism was passionate without necessarily being racist. It was part and parcel of an antimodernist and antiliberal trend that later would split into opposed camps with regard to Nazi Germany itself. It was the voice of suspicion, of contempt; it could demand exclusion. Such were among others, the anti-Jewish attitudes of a powerful group of European intellectuals steeped in Catholicism, either as believers or as men strongly influenced by their Catholic background: In France, such writers as Thierry Maulnier, Robert Brasillach, Maurice Bardèche, and a whole phalanx of Catholic and nationalist militants of the Action Française (the royalist movement founded by the ultranationalist, anti-Semitic, and anti-Dreyfusard Charles Maurras at the beginning of he century) represented this trend; they either still belonged to Maurras's movement or kept close ties to it. Paradoxically Maurras himself was not a believing Catholic, but he understood the importance of Catholicism for his "integral nationalism." The church banned Action Française in 1926, yet many right-wing believers remained loyal to Maurras's movement. In England such illustrious representatives of Catholicism and Anglo-Catholicism as Hilaire Belloc, G. K. Chesterton, and T. S. Eliot acknowledged their debt to Maurras, yet their anti-Jewish outbursts had a style and a force of their own. Catholic roots were explicitly recognized by Carl Schmitt, and their indirect influence on Heidegger is unquestionable. There was an apocalyptic tone in this militant right-wing Catholicism, a growing urge to engage in the final battle against the forces evoked by Bernanos, forces whose common denominator was usually the Jew.

Simultaneously, however, a growing cultural pessimism—whose political and religious roots were diffuse but that exuded a violent anti-

Semitism of its own—was taking hold of various sectors of the European intellectual scene. Here, too, some of the most prominent French writers of the time took part: Louis-Ferdinand Céline and Paul Léautaud, Pierre Drieu La Rochelle, Maurice Blanchot, Marcel Jouhandeau, Jean Giraudoux, and Paul Morand. But it is not Céline's 1937 *Bagatelles pour un massacre* itself, possibly the most vicious anti-Jewish tirade in modern Western literature (apart from outright Nazi productions), that was most revealing, but André Gide's favorable review of it in the *Nouvelle Revue Française*, under the guise that what Céline wrote in the book was not meant to be taken seriously.[4] And it was not Brasillach's outspoken hatred of the Jews that was most indicative of the prevailing atmosphere, but the fact that Giraudoux, who had just launched a vitriolic attack against Jewish immigrants in *Pleins pouvoirs*, became minister of information during the last year of the Third Republic.[5]

Against the background of this religious-cultural-civilizational crisis and its anti-Jewish corollaries, other, less abstract factors appear as causes of the general exacerbation of anti-Jewish attitudes and anti-Semitic agitation in countries other than Nazi Germany.

The convergence of the worldwide economic crisis and its sequel, decade-long unemployment, with the growing pressure of Jewish immigration into Western countries on the one hand, and economic competition from a large Jewish population in Central and Eastern Europe on the other, may have been the most immediate spur to hostility. But for millions of disgruntled Europeans and Americans the Jews were also believed to be among the beneficiaries of the situation, if not the manipulators of the dark and mysterious forces responsible for the crisis itself. Such constructs had penetrated all levels of society.

In countries such as France, England, and the United States, where some Jews had achieved prominence in journalism, in cultural life, and even in politics, prevailing European pacifism and American isolationism depicted Jewish protests against Nazi Germany as warmongering. The Jews were accused of serving their own interests rather than those of their countries. The French politician Gaston Bergery, a former Radical Socialist who became a collaborator during the German occupation, described in November 1938, in his periodical *La Flèche* (The arrow), how "the Jewish policy" of a war against Nazi Germany was perceived by the wider public: "A war—public opinion senses—less in order to defend France's direct interests than to destroy the Hitler regime in Germany, that is, the death of millions of Frenchmen and [other] Europeans to avenge a few dead Jews and a few hundred thousand unfortunate Jews."[6]

Another immediately apparent factor was—as it had been in the earlier part of the century—the visibility of Jews on the militant left. In both Eastern Europe and France, identification of the Jews with the Marxist peril was partly as phantasmic as it had been in the past, but also partly confirmed by significant left-wing activism by Jews. Such activism arose for the same sociopolitical reasons that had played a decisive role several decades earlier. But in the thirties there were Jews, mainly in Western Europe, who became supporters of the Left in order to find a political expression for their anti-Nazism (at the same time in the Soviet Union, many Jews were falling victim to Stalin's purges). In general terms, however—as had also been the case at the beginning of the century (and has been ever since)—the majority of European Jews identified themselves, and could be identified, with liberalism or social democracy, and, to a lesser extent, with traditional conservatism. At the same time, the crisis of the liberal system and the increasing discontent with democracy led to a growing hostility toward a group that, in addition to its partial identification with the Left, was regarded as the supporter and beneficiary of the liberal spirit both in the economy and in public life.

The spread of anti-Semitism on the European (and American) scene was one of the reasons for the growing difficulties placed in the way of Jewish emigration from Germany, then from Austria and the Sudetenland, and later from the Reich Protectorate of Bohemia-Moravia. Traditional anti-Semitism was also one of the reasons that prompted the Polish government to take measures about the citizenship of nonresident Polish Jews that, as will be seen, gave the Nazis the necessary pretext for expelling thousands of Polish Jews residing in Germany. A few years later, this surge of anti-Jewish hostility was to have much more catastrophic results. The Jews themselves were only partly aware of the increasingly shaky ground on which they stood because, like so many others, they did not perceive the depth of liberal democracy's crisis. The Jews in France believed in the strength of the Third Republic, and the Jews of East Central Europe believed in France. Few imagined that Nazi Germany could become a real threat beyond its own borders.

Eastern Europe's participation in the growing anti-Jewish agitation of the second half of the thirties took place within the context of its own traditions.

The influence of Christian anti-Jewish themes was particularly strong among populations whose majority was still a devout peasantry. Social resentment on the part of budding nationalistic middle classes of the positions acquired by Jews in commerce and the trades, light industry, banking, and the press, as well as in the prototypical middle-class professions of medicine and the law, created another layer of hostility. The latest and possibly strongest ingredient was the fierce anti-Bolshevism of regimes already oriented toward fascism, regimes for which identification of the Jews with Bolshevism was a common slogan—for example in Hungary, where the memory of the Béla Kun government remained vivid. In Poland these diverse elements merged with an exacerbated nationalism that tried to limit the influence of any and all minority groups, be they Ukrainians, Belorussians, Jews, or Germans. By a somewhat different process, the wounded nationalism of the Hungarians and the Slovaks, and the megalomaniacal nationalist fantasies of the Romanian radical right dreaming of a greater Dacia,* led to the same anti-Semitic resentment. "Almost everywhere [in these countries]," writes Ezra Mendelsohn, "the Jewish question became a matter of paramount concern, and anti-Semitism a major political force."[7]

The leaders of the East Central European countries (Miklós Horthy in Hungary, Jósef Beck in Poland after Jósef Pilsudski's death, Ion Antonescu in Romania) were already close to fascism or at least to extreme authoritarianism. All had to contend with ultra-right-wing movements—such as the Endek in Poland, the Iron Guard in Romania, the Hlinka Guard in Slovakia, and the Arrow Cross in Hungary—that sometimes appeared to be allies and sometimes enemies. The right-wing governments, mainly in Romania and Hungary, attempted to take the "wind out of the sails" of the radical right by adopting anti-Semitic policies of their own. Thus, Romania adopted an official anti-Semitic program by the end of 1937, and Hungary in 1938. The results were soon evident. As the Italian journalist Virginio Gayda, a semiofficial representative of the fascist regime, noted at the beginning of 1938, anti-Semitism was the point of "national cohesion" of the political scene in the Danubian states.[8]

Anti-Semitism's deepest roots in Poland were religious. In this profoundly Catholic country, the great majority of whose population still lived on the

---

*Dacia was an ancient kingdom and a Roman province whose borders roughly corresponded to those of the Romanian state of the 1930s.

land or in small towns, the most basic Christian anti-Jewish themes remained a constant presence. In early 1937 Augustus Cardinal Hlond, the primate of Poland, distributed a pastoral letter that, among other things, addressed the Jewish issue. After pointing to the existence of a "Jewish problem" demanding "serious consideration," the head of the Polish Church turned to its various aspects. "It is a fact," Hlond declared, "that the Jews are struggling against the Catholic Church, that they are steeped in free thought, that they are the vanguard of godlessness, of the Bolshevik movement, and of subversive action. It is a fact that Jewish influence on morals is deplorable and that their publishing houses spread pornography. It is true that they are cheaters and carry on usury and white slave traffic. It is true that in the schools the influence of Jewish youth upon the Catholic is in general negative from the religious and moral point of view." But in order to seem equitable, Cardinal Hlond then took a step back: "Not all the Jews are such as described. There are also faithful, righteous, honest, charitable and well-meaning Jews. In many Jewish families there is a wholesome and edifying family spirit. We know some people in the Jewish world who are morally prominent, noble and respectable."

What attitudes did the cardinal therefore recommend to his flock? "I warn you against the moral attitude imported from abroad which is fundamentally and unconditionally anti-Jewish. This attitude is contrary to Catholic ethics. It is permissible to prefer one's people; it is wrong to hate anyone. Not even Jews. In commercial relations it is right to favor one's own people, to avoid Jewish shops and Jewish stalls on the market, but it is wrong to plunder Jewish shops, destroy Jewish goods, break windowpanes, throw bombs at their houses. It is necessary to find protection from the harmful moral influence of the Jews, to keep away from their anti-Christian culture, and in particular to boycott the Jewish press and demoralizing Jewish publications, but it is wrong to attack Jews, to beat, wound or libel them. Even in the Jew we must respect and love the man and neighbor, even though one may not be able to respect the inexpressible tragedy of this people which was the guardian of the Messianic idea and gave birth to our Savior. When God's grace will enlighten the Jew and when he will sincerely join the fold of his and our Messiah, let us welcome him joyfully in the Christian ranks.

"Let us beware of those who endeavor to bring about anti-Jewish excesses. They serve a bad cause. Do you know whose orders they are obeying in so doing? Do you know in whose interest such disorders are fomented? The good cause gains nothing from such inconsiderate acts. And the blood which sometimes flows on such occasions is Polish blood."[9]

This is precisely the translation of Cardinal Hlond's pastoral letter that was sent from Poland to Rabbi Stephen Wise in New York on February 9, 1937. According to the sender, "The statements contained in the first part concerning the moral inferiority and crimes of the Jews have been surpassed in a pronouncement by the prince-bishop of Cracow, Sapieha. But both these pronouncements have been surpassed by the mischief-making public addresses and the recently published book by the prelate Trzeciak, which might compete with the *Protocols of the Elders of Zion*."[10]

Traditional Polish Christian anti-Jewishness was fueled by a particularly difficult demographic and socioeconomic context. When the Polish state was reestablished in the wake of World War I, approximately 10 percent of its population was Jewish (3,113,933 in 1931, i.e., 9.8 percent of the general population). But about 30 percent of the urban population was Jewish (this average was valid for the largest cities such as Warsaw, Cracow, and Lodz, but the Jewish population was more than 40 percent in Grodno and reached 60 percent in Pinsk).[11]

The social stratification of Polish Jewry added to the difficulties created by sheer numbers and urban concentration: the majority, or more than two million, of the Jewish population belonged to the politically crucial petty bourgeoisie.[12] Finally, contrary to the situation in Germany, France, Great Britain, and other Western countries, where the Jews aspired above all to be considered nationals of their respective countries—even though the majority insisted on keeping some form of Jewish identity—in Eastern Europe, particularly in Poland, the self-definition of minorities was often that of a separate "nationality." Thus, in the Polish census of 1921, 73.76 percent of the overall number of Jews by religion also declared themselves to be Jews by nationality, and in the 1931 census, 79.9 percent declared that Yiddish was their mother tongue, while 7.8 percent (an implausibly high number, presumably influenced by Zionism) declared that Hebrew was their first language. That left only a small percentage of Polish Jews who declared Polish to be their mother tongue.[13]

Thus the basically religious anti-Jewish feelings of the Polish population were reinforced by what was perceived as a Jewish hold on a few key professions and on entire sectors of lower-middle class activities, mainly commerce and handicrafts. Moreover, the clear identification of the Jews as an ethnic minority within a state that comprised several other minority groups but aimed, of course, at Polish national supremacy, led the Polish nationalists to consider Jewish religious and national-cultural "separatism" and Jewish dominance in some sectors of the economy to be a compound threat to the new state. Finally the Poles' exacerbated anti-Bolshevism, fed by new fears and an old, deep hatred of Russia, identified Jewish socialists and Bundists with their Communist brethren, thereby inserting the standard equation of anti-Bolshevism with anti-Semitism into a specifically Polish situation. This tendency became more pronounced in the mid–1930s, when the Polish "regime of the colonels" moved to what was in fact a semifascist position, not always very different in its nationalist-anti-Semitic stance from Roman Dmowski's Endek (National Democratic) Party. The Endeks brandished the specters of a Folksfront (that is, a popular front like the one in France; for Poles, the spelling with an "F" signaled Yiddish and thus Jewish origin) and Zydokomuna (in the sense of Jewish communism) to identify the Jews and their political activities.[14] They were for a massive transfer of Jews to Palestine and for a Jewish quota in universities, and their action squads found the smashing of Jewish shops particularly attractive.[15] The trouble was that, despite official declarations, the government and the church were not loath on occasion to encourage similar policies and activities, albeit in an indirect way.[16]

Estimates in the Polish press in 1935 and 1936 that hundreds of Jews died in the pogroms that erupted at the time in no fewer than 150 Polish cities were probably too low.[17] A hidden quota in the universities brought the percentage of Jewish students down from 20.4 percent in 1928–29 to 9.9 percent in 1937–38.[18] What happened in the universities took place more openly in the economic field, with a boycott of Jewish commerce leading to a sharp decrease in the number of Jewish businesses during the years immediately preceding the war.[19] The pauperization of wide sectors of the Jewish population had begun long before the war, but in the post-Pilsudski era, the economic boycott was supported by the government itself. To be sure, anti-Jewish violence was officially condemned, but, as Prime Minister Felician Slawoj-Skladkowski put it in 1936, "at the same time, it is understandable that the country should possess the instinct compelling it to defend its culture, and it is natural that Polish society should seek economic self-sufficiency." The prime minister explained what he meant by self-sufficiency: "economic struggle [against the Jews] by all means—but without force."[20] By 1937–38 Polish professional associations were accepting Gentile members only.

As for the civil service, at the national or at the local level, by then it had entirely ceased employing Jews.[21]

One of the by-products of the "Jewish problem" in Poland was the reemergence in the mid-1930s of an idea that had first been concocted by the German anti-Semite Paul de Lagarde: transfer of part of the Jewish population to the French island colony of Madagascar.[22] In January 1937 the positive attitude of Marius Moutet, the French Socialist colonial minister in Léon Blum's Popular Front government, gave this plan a new lease on life, and soon negotiations between Poland and France regarding practical ways and means for implementing such a population transfer got under way. The Paris government agreed that a three-man Polish investigation commission, two of them Jews, be sent to the island. On their return the Jewish members submitted a report pessimistic about Madagascar's absorptive capacities, but the Polish government adopted the favorable view of the commission's Polish chairman, Mieczyslaw Lepecki. Thus, negotiations with the French continued and, at the beginning of 1938, Warsaw still seemed to be giving serious support to the project.

Whereas at the outset, the European Jewish press was reporting positively on the initiative, and official Nazi comment originating in the Paris and Warsaw embassies appeared only noncommittal, the Nazi press became highly sarcastic once it became clear, at the end of 1937, that the plan had little chance of implementation. "Madagascar could become a 'promised land' for the Jews Poland wants to get rid of," said the *Westdeutscher Beobachter* on December 9, "only if they [the Jews] could lead a life of masters there, without effort of their own, and at the expense of others. It is therefore questionable whether the invitation for an exodus of the Children of Israel to Madagascar will soon free Poland of any great part of these parasites."[23] The plan nonetheless seems to have attracted Heydrich's attention, and on March 5, 1938, a member of his staff sent the following order to Adolf Eichmann:

"Please put together in the near future material for a memorandum which should be prepared for C [Heydrich] in cooperation with II B4 [the Gestapo's Jewish affairs section]. It should be made clear in the memorandum that on its present basis (emigration), the Jewish question cannot be solved (financial difficulties, etc.) and that therefore we must start to look for a solution through foreign policy, as is already being negotiated between Poland and France."[24]

II

There were 90,000 Jews in France at the beginning of the century; in 1935 their number had reached 260,000. On the eve of the war, the Jewish population had risen to approximately 300,000, two-thirds of it in Paris.[25] The most detailed counts of Jews were conducted later by the Vichy government and by the Germans in the occupied zone, in accordance, of course, with their own definition of who was Jewish. The results nonetheless give a more or less precise image of the immediate prewar situation. In mid-1939 approximately half the Jewish population in Paris was French and half was foreign. But even among the French Jews, only half were French-born. In the Paris region 80 percent of the foreign Jews were of East European origin, half of them from Poland.[26] Although there were three million foreigners living in France in the late thirties, of whom only about 5 percent at most were Jews,[27] the Jews were more conspicuous than the others. In the eyes of both the authorities and the population, the foreign Jews were likely to create problems. This was the opinion of many French Jews as well. "As early as 1934," writes Michael Marrus, "R. R. Lambert, editor of the *Univers Israélite* and one of the leading figures of the Franco-Jewish establishment, warned his coreligionists that other Frenchmen were losing their patience: in the current state of affairs, mass emigration [to France] is no longer possible. Foreign Jews should watch their step, should abandon their tendency to cling closely to one another and should accelerate their assimilation into French society."[28]

Actually Lambert was relatively compassionate and did not advocate expelling the refugees; Jacques Helbronner, president of the Consistoire, the central representative body of French Jews, thought otherwise: "France, like every other nation," Helbronner declared as early as June 1933, "has its unemployed, and not all the Jewish refugees from Germany are people worth keeping. . . . If there are 100 to 150 great intellectuals who are worthy of being kept in France since they are scientists or chemists who have secrets our own chemists don't know . . . these we will keep, but the 7, 8, or perhaps 10,000 Jews who will come to France, is it really in our best interests to keep them?"[29] Helbronner continued for years to hold this view; in 1936 he expressed regrets about the liberal French immigration policy of 1933. For him, the Jewish refugees were simply "riff-raff, the rejects of society, the elements who could not possibly have been of any use to their own country."[30] Even after the defeat of France, it should be added, Helbronner, still head of the Consistoire, kept his antipathy toward foreign

Jews. His attitude changed only in the course of 1943. Soon after this change of heart had taken place, the Nazis caught up with him as well: In October of that year he was arrested, deported to Auschwitz with his wife, and murdered.

The Consistoire's position had its effect, and from 1934 on, material help to the refugees almost totally ceased. "Clearly the French Jewish establishment was giving up all efforts to reconcile its competing loyalties and obligations to the refugees and to France. In this struggle, French interests . . . dominated. The refugees were quite simply abandoned."[31]

The first official measures against foreigners (expulsion of those whose papers were not in order) were taken during the first half of the thirties, mainly in 1934 under the premiership of Pierre-Etienne Flandin.[32] After a brief improvement under the Blum governments, anti-immigrant measures became ever more draconian, culminating in the highly restrictive law of November 1938, which facilitated the immediate expulsion of aliens and made their assigned residence in some remote corner of the country a matter of simple administrative decision. Stripping naturalized foreigners of their newly acquired French nationality also became possible, and a number of professional groups that considered recently arrived Jews to be dangerously competitive began to lobby for their exclusion from various domains such as medicine and the law.[33]

However, there was more to the rapid rise of French anti-Semitism in the mid-thirties than the problems of Jewish immigration.[34] As the economic crisis was worsening, in late 1933 the Stavisky Affair, a scandal involving a series of shady financial deals in which the central role was played by a Russian Jew named Serge-Alexandre Stavisky and in whose mysterious ramifications major French political figures were implicated, came to a head. In the early days of 1934, Stavisky's body was discovered near Chamonix in the French Alps. The Radical Socialist government of Camille Chautemps was brought down and replaced by the ephemeral premiership of Édouard Daladier, also a Radical Socialist, and the entire array of extreme right-wing organizations, from the Action Française of Maurras and Daudet to the Croix de Feu, the war veterans' organization headed by François de La Rocque, was in an uproar. A putsch attempt was quelled in Paris on February 6, 1934: Eighteen rightists were killed by the police on the Place de la Concorde and the rue Royale as they tried to storm the Chambre des Députés. The republic survived the crisis, but the internal rift that had divided French society since the Revolution and dominated the political life of the country from the time of the restoration to that of the Dreyfus affair was wide open again.

A turning point came with the confrontations that preceded and followed the 1936 elections, with the overwhelming victory of the Popular Front led by Léon Blum. When, on June 6, the new government was sworn in, Xavier Vallat, the future Vichy delegate general for Jewish Affairs, turned to Blum at the rostrum of the Chambre des Députés: "Your accession to power, Mr. Prime Minister, is an undeniably historic occasion. For the first time, this ancient Gallo-Roman country will be governed by a Jew. I dare say aloud what the country thinks: that it would be preferable to put at the head of this country a man whose roots belong to its soil rather than a subtle Talmudist."[35]

Much of what Blum did during his two brief tenures as prime minister of the Popular Front government seemed to play into the hands of the Right. Admirable as his social achievements—the forty-hour work week and the six-week paid annual vacation—were, they appeared manifestly to contradict his urge to speed up rearmament in the face of the Nazi menace. In any event, if it was somewhat incongruous to see traditional pacifists turn into the military guardians of France, it was certainly much worse to watch the shift of right-wing nationalists toward outright appeasement of Nazi Germany, among other reasons out of hatred for the enemy within. "Better Hitler than Blum" was just one of the slogans; worse were to come.

As in Germany in previous decades, notwithstanding the visibility of some Jewish left-wing activism, the majority of the Jews in France were in fact anything but politically supportive of the Left. The Consistoire was an essentially conservative body that did not hesitate to welcome the presence of right-wing organizations, such as La Rocque's Croix de Feu, at its commemorative occasions; it openly backed, at least until 1935, a Jewish patriotic and ultraconservative movement, Édouard Bloch's Union Patriotique des Français Israélites.[36] Even among the immigrants from Eastern Europe, support for the Left was not pervasive. In the 1935 Paris municipal elections and in the decisive 1936 elections for the legislature, official immigrant bodies were readier to give their support to right-wing than to Communist candidates.[37]

Blum himself often seemed impervious to the role played by anti-Semitism in the mobilization of right-wing opinion against his leadership. Or possibly his awareness was of the detached and fatalistic kind that characterized Rathenau's acceptance of the hatred directed against him in the months preceding his assassination. In February 1937 Blum himself was slightly wounded by right-wing demonstrators as his car passed the funeral cortege of the Action Française historian Jacques Bainville.[38] Blum's imperviousness made it easy for the extreme right to point to the number of Jewish ministers in his cabinets.[39] And personal slander was in any case in no need of well-established facts; repeated unfounded attacks, for instance, drove the Jewish Socialist minister of the interior and mayor of Lille, Roger Salengro, to suicide.

Anti-Semitism did not play a central role in the programs or the propaganda of the French parties closest to fascism, at least during the thirties. Although anti-Jewish slogans were part of the repertory of Marcel Déat's Solidarité Française, Jacques Doriot's Parti Populaire Français became anti-Semitic only after 1938 in order to attract voters from among the notoriously anti-Semitic French settlers in North Africa.[40] But anti-Jewish themes were the major staple of a host of right-wing periodicals that carried the message to hundreds of thousands of French homes: *L'Action Française*, *Je suis partout*, and *Gringoire* were merely the most widely read among them. On April 15, 1938, *Je suis partout* published the first of its special issues on "the Jews." The articles carried such titles as "The Jews and Germany," "Austria and the Jews," "The Jews and Anti-Semitism," "The Jews and the Revolution," "When Israel Is King: The Jewish Terror in Hungary," and so on. Brasillach's lead article demanded that the Jews in France be put under alien status.[41] The continuing stream of anti-Semitic articles reached such proportions that, in April 1939, a law was passed to prohibit press attacks "against a group of persons belonging by their origin to a given race or religion, when these attacks aim at inciting hatred among citizens or inhabitants." The perceived need for such a law was a sign of the times. Another such sign, also in April 1939, was that the newly elected pope, Pius XII, repealed the ban on *Action Française*. Neither the ban nor its repeal had anything to do with anti-Semitism, but nonetheless, as of 1939 Maurras's doctrine of anti-Jewish hatred was no longer beyond the official Catholic pale.

Nazi Germany encouraged the spread of anti-Semitism all over Europe and beyond. Sometimes these initiatives were indirect: In France the France-Allemagne Committee, organized by Joachim von Ribbentrop's Foreign Policy Office and guided by the future Nazi ambassador to occupied France, Otto Abetz, carefully supported various cultural activities, most of which carried a subtle pro-Nazi ideological slant.[42] On the other hand, the function of Nazi organizations, such as the Stuttgart-based press agency Weltdienst, was worldwide anti-Jewish propaganda.[43] Yet it was not the Nazi-like and sometimes Nazi-financed groups of French, Belgian, Polish, and Romanian Jew-haters who were of significance during the immediate prewar period. The really ominous aspect in these countries was the exacerbation of homegrown varieties of anti-Semitism; Nazism's contribution was that of an indirect influence. At this time the upsurge of anti-Jewish passion, with or without Nazi incitement, had some immediate impact both on attitudes toward local Jewish communities and on immigration policies toward Jews trying to flee from Germany, Austria, and the Czech Protectorate. In more general terms, it prepared the ground for active collaboration by some, and passive acquiescence by many more, in the sealing of the fate of European Jewry only three or four years hence.

III

On September 29, 1936, the state secretary in the German Ministry of the Interior, Wilhelm Stuckart, convened a conference of high officials from his own agency, from the Ministry of the Economy, and from the Office of the Deputy Führer in order to prepare recommendations for a meeting of ministers regarding the further steps to be taken in regard to the jews at this post-Nuremberg stage. As the Office of the Deputy Führer represented the party line, the Ministry of the Interior (though headed by the Nazi Wilhelm Frick) often represented middle-of-the-road positions between the party and the conservative state bureaucracy, and the Ministry of the Economy (still headed by Schacht), was decidedly conservative, it is remarkable that, at this conference, the highest officials of the three agencies were entirely in agreement.

All those present recognized that the fundamental aim now was the "complete emigration" of the Jews and that all other measures had to be taken with this aim in mind. After restating this postulate, Stuckart added a sentence that was soon to find its dramatic implementation: "Ultimately one would have to consider carrying out compulsory emigration."[44]

Most of the discussion was concentrated on dilemmas that were to bedevil German choices until the fall of 1938: First, what measure of social and economic activity should be left to Jews in the Reich so as to prevent their becoming a burden to the state and yet not diminish their incentive to emigrate? Second, toward which countries was Jewish emigration to be channelled without it leading to the creation of new centers of anti-German activity? The participants agreed that all emigration options should be left open, but that German means should be used only to help the emigration to Palestine. In answer to the question whether the press was not slowing down Jewish emigration to Palestine by reporting the Arab anti-Jewish unrest there, Ministerial Director Walther Sommer (from the Deputy Führer's Office) indicated that "one could not reproach other nations for defending themselves against the Jews." No measures regarding the press reports were to be taken.[45] And no decision was made regarding the problem of the identification of Jewish businesses.[46]

The September 1936 conference was the first high-level policy-planning meeting devoted to the regime's future anti-Jewish measures in which the priority of total emigration (compulsory emigration: that is, expulsion if need be) was clearly formulated. Before the passage of the Nuremberg Laws, segregation had been the main goal, and it was only in September 1935 that Hitler, in his declaration to Walter Gross, mentioned "more vigorous emigration" of the Jews from Germany as one of his new objectives. Thus, some time at the end of 1935 or in 1936, Hitler's still tentative formulations became a firm guideline for all related state and party agencies. The move to new objectives tallied, as has been seen, with the new radicalization in both the internal and the external domains.

Simultaneously the "cleansing" process was relentlessly going forward: The major initiatives stemmed from Hitler, yet, when other initiatives were submitted to him by cabinet ministers or high party leaders, his approval was far from being automatic.

On April 1, 1933, some 8,000 to 9,000 Jewish physicians were practicing in Germany. By the end of 1934, approximately 2,200 had either emigrated or abandoned their profession, but despite a steady decline during 1935, at the beginning of 1936, 5,000 Jewish physicians (among them 2,800 in the Public Health Service) were still working in the Reich. The official listing of the country's physicians for 1937 identified Jewish physicians as Jews according to the Nuremberg criteria; by then their total was about 4,200, approximately half the number of those listed in 1933,[47] but in Nazi eyes still too many by far.

On December 13, 1935, the minister of the interior submitted the draft of a law regulating the medical profession. According to the protocol of the cabinet meeting (which gave no details of the draft), Frick drew the ministers'attention to the fact that articles 3 and 5 "settled the Aryan issue for the physicians." The proposal was accepted.[48] It seems, however, that for an unspecified reason the final drafting of the law was postponed for more than a year.

On June 14, 1937, Wagner met with Hitler in the presence of Bormann: "As I submitted to the Führer that it was necessary to free the medical profession of the Jews," Wagner wrote, "the Führer declared that he considered such cleansing exceptionally necessary and urgent. Nor did he consider it right that Jewish physicians should be allowed to continue to practice [in numbers] corresponding to the percentage of the Jewish population. In any case, these doctors had also to be excluded in case of war. The Führer considered the cleansing of the medical profession more important than for example that of the civil service, as the task of the physician was in his opinion one of leadership or should be such. The Führer demanded that we inform State Secretary Lammers of his order to prepare the legal basis for the exclusion of the Jewish physicians still practicing (cancellation of licenses)."[49]

Two months later Lammers informed State Secretary Pfundtner that the issue of Jewish physicians was on the agenda for a meeting, scheduled for September 1, of state secretaries with Hitler.[50] Within a year the professional fate of the remaining Jewish physicians in Germany would be sealed.

Interior Minister Wilhelm Frick, a party stalwart if ever there was one, nevertheless seemed to have underestimated the stepped-up pace of radicalization. It appears, from a November 25, 1936, Education Ministry memorandum, that at the beginning of the year, Frick had decided that there was no legal basis for the dismissal of Aryan civil servants with Jewish wives. In the memorandum's words, "[Frick's] position has not received the approval of the Führer and Reich Chancellor." The corollary was simple: Frick's initiative was invalid.[51]

A few months later Frick made up for his initial lack of creative legalism. On April 19, 1937, he issued the following ordinance: "My memorandum of December 7, 1936, which forbids the raising of the national colors over the house of a German living in a German-Jewish mixed marriage, also applies to civil servants. As a situation in which a civil servant cannot raise the national flag at home is not tenable in the long run, civil servants married to a Jewish wife are usually to be pensioned off."[52] Some exceptions were allowed, but the legal basis for dismissing civil servants with Jewish spouses had been found.

Generally, however, Frick could boast of outright success. On July 21, 1937, he solved another major problem: safety measures to be taken regarding the presence of Jews in health resorts and related establishments. Jews were to be housed only in Jewish-owned hotels and guesthouses, on condition that no German female employees under forty-five worked on the premises. The general facilities (for bathing, drinking spa waters, and the like) were to be accessible to Jews, but there was to be as much separation from the other guests as possible. As for facilities with no immediate health function (gardens, sports grounds), these could be prohibited to Jews.[53]

But as in previous years, Hitler hesitated when a measure could create unnecessary political complications. Thus, on November 17, 1936, he ordered further postponement of a law on Jewish schooling,[54] a draft of which had been submitted to him by the minister of education. It seems that at the time Hitler was still wary of implementing the segregation of Jewish pupils on racial lines, as it would have entailed the transfer of Jewish children of Christian faith into Jewish schools and added further tension to relations with the Catholic Church.[55]

At times the cleansing measures turned into a totally surrealistic imbroglio. The issue of doctoral degrees for Jewish students was one such instance.[56] The problem was apparently raised at the end of 1935 and discussed by the minister of the interior: Any restrictions on the right to obtain a doctoral degree were not to apply to foreign Jewish students; for German Jews the issue remained unresolved. At the beginning of 1936, it was brought up again by the notorious Wilhelm Grau, who was about to become head of the Jewish Section in Walter Frank's Institute for the History of the New Germany. On February 10, 1936, Grau wrote to the secretary of state for Education that he had been asked to evaluate a dissertation on the history of the Jews of Ulm in the Middle Ages, submitted by a Jew at the faculty of philosophy of Berlin University. "Whereas in the above-mentioned case," wrote Grau, "the dissertation is already inadequate from a scientific viewpoint, a general question also arises, namely whether Jews should be allowed to obtain a doctorate at all in a German university on such historical subjects. As our university professors unfortunately have little knowledge and even less instinct regarding the Jewish question, the most incredible things happen in this area." Grau continued with a story mentioned in the discussion of his first contribution to the *Historische Zeitschrift*; "Last October, an Orthodox Jew called Heller obtained his doctorate at the University of Berlin with a dissertation on Jews in Soviet Russia, in which he attempted to deny entirely the Jewish contribution to Bolshevism by using a method that should raise extreme indignation in the National Socialist racial state. Heller simply does not consider those Jews he finds unpleasant, such as Trotsky and company, to be Jews but anti-Jewish 'internationalists.' With reference to this, I merely want to raise the question of the right of Jews to obtain a doctorate."[57]

The discussion on this topic, which developed throughout 1936 and the early months of 1937, involved the Ministry of Education, the deans of the philosophy faculties at both Berlin and Leipzig Universities, the rectors of these universities, the Reichstatthalter of Saxony, and the Office of the Deputy Führer. The Ministry of Education's attitude was to adhere to the law regarding Jewish attendance at German universities: As long as Jewish students were allowed to study in German universities, their right to acquire a doctoral degree could not be canceled. The best way of handling the situation was to appeal to the national feelings of the professors and prevail upon them not to accept Jews as doctoral students.[58] But some deans (particularly the dean of the philosophical faculty at Leipzig) declared that, as party members, they could no longer bear the thought of signing doctoral degrees for Jews.

On February 29, 1936, the philosophy dean at Berlin University emphasized the negative consequences that stemmed from the rejection of the dissertations of all four Jewish doctoral candidates (Schlesinger, Adler, Dicker, and Heller) in his faculty. Since in each instance the dissertation topics had been suggested by "Aryan members of the faculty," rejection of the theses also affected the professors concerned. The dean cited one of them, Professor Holtzmann, sponsor of "the Jew Dicker's" rejected thesis on the Jews of Ulm: "Filled with anger, Holtzmann declared that he had had enough, and that he would no longer direct the doctoral work of any Jew."[59]

On October 15, 1936, Bormann intervened. For him, appealing to "the national consciousness of the professors" was not the right way to handle the matter. "In particular," Bormann wrote to Frick, "I would not want the implementation of basic racial tenets that derive from the worldview of National Socialism to be dependent upon the goodwill of university professors." Bormann did not hesitate: A law prohibiting the award of doctoral degrees to Jewish students was necessary, and it was to be aimed at the professors, not the students. As for foreign reactions, Bormann thought that the impact of the law would be beneficial; in justifying this claim he used an argument whose significance extended well beyond the issue at hand: "Furthermore, I believe that the decree will fall on favorable ground, particularly in racially alien countries, which feel slighted by our racial policy, as thereby Jewry will once more be consciously set apart from other foreign races." There was no objection to granting the doctoral degree to Jewish students who had already fulfilled all the necessary requirements.[60]

A decree reflecting Bormann's view was drafted by the minister of education on April 15, 1937: The universities were ordered not to allow Jewish students of German citizenship to sit for doctoral exams. Exemptions were granted to *Mischlinge* under various conditions, and the rights of foreign Jews remained as before.[61]

The matter seemed settled. But only a few days later, on April 21, a telegram from Dean Weinhandel of the Kiel University philosophy faculty arrived at the Ministry of Education requesting "a decision whether reservations exist against acceptance of anthropology doctoral dissertation when candidate has Jewish or not purely Aryan wife."[62]

The purification process also duly progressed at the local level. Thus, the Munich city fathers, who had excluded the Jews from public swimming pools in 1935, took a further bold step in 1937. Now the Jews were to be forbidden access to municipal baths and showers. But as the matter was weighty, Bormann's authorization was requested. It was refused,[63] although it is not clear what Bormann's reasons were.

Slowed down in one area, the Munich authorities pushed ahead in another. Since 1933 the city streets that bore Jewish names had gradually been renamed. At the end of 1936, however, Mayor Karl Fiehler and the Construction Commission discovered that eleven Jewish street names still remained. During 1937, therefore, with assistance from the municipal archive, the names that were undoubtedly Jewish were changed. But as an archive official put it, there was always the possibility that "as a result of more thorough research, one or more street names might be identified as being Jew-related.[64]

In Frankfurt the problems created by Jewish street names were worse. It seems that the first person to raise the issue publicly was a woman party member, who on December 17, 1933, wrote an open letter to the *Frankfurter Volksblatt*: "Please do me the great favor of seeing whether you could not use your influence to change the name of our street, which is that of the Jew Jakob Schiff. Our street is mainly inhabited by people who are National Socialist–minded, and when flags are flown, the swastika flutters from every house. The 'Jakob Schiff' always gives one a stab to the heart."[65] The letter was sent to the municipal chancellery, which forwarded it to the city commission for street names. In March 1934 the commission advised the mayor of all the donations made by the Jewish-American financier Jacob Schiff to various Frankfurt institutions, including the university, and therefore suggested rejecting the proposed name change, especially since, given the importance of the Jacob Schiff private banking house in the United States, such a change would be widely reported and could lead to a demand for restitution of the monies that had been given to the city.[66]

The letter in the *Volksblatt* had, however, triggered a number of similar initiatives, and on February 3, 1935, after a lengthy correspondence, the city commission for street names requested the mayor's agreement to the following proposal: The names of fourteen streets or squares were to be changed immediately, starting with Börne Square, which was to become Dominicans' Square. When Nazi propaganda "discovered" that Schiff had heavily financed the Bolsheviks, Jakob-Schiff-Strasse became Mumm-Strasse (in honor of a former Frankfurt mayor).[67] Twelve more streets were to be renamed in 1936, and twenty-nine others whose renaming had been suggested were to keep their names, either because their real meaning could be explained away (Mathilden-Strasse, Sophien-Strasse, Luisen-Strasse, and Luisen-Platz, all in fact named after women of the Rothschild family, would now be regarded as merely named for generic women) or because no sufficient or valid reason could be found for the change. In the case of Jakoby-Strasse, for instance, the name's possibly Aryan origins had still to be researched; as for Iselin-Strasse, "Isaac Iselin was not a Jew (the biblical first name was common among Calvinists from Basel)."[68]

In Stuttgart the exclusion of Jews from public swimming pools was postponed until after the Olympic Games; anti-Jewish initiatives did not, however, lag behind those in other German cities. Quite the contrary. The local party leaders were infuriated by the fact that, at least until 1937, the Jewish population of the city was growing rather than declining. Jews from the small towns and villages of surrounding Württemberg were fleeing to the city in the hope of finding both the protection of anonymity and the support of a larger community. Thus, whereas during the first seven months of 1936, 582 Jews left Stuttgart, 592 moved in. It was only at the end of 1937 that the four-thousand-strong Jewish population started to decline.[69]

The city council decided to take Jewish matters in hand. After asking for advice from, of all places, Streicher's Nuremberg, the council decided at its September 21, 1936, meeting that old people's homes, nursery schools, and (finally) swimming pools belonging to the city were forbidden to Jews; in hospitals Jews were to be separated from other patients; city employees were forbidden to patronize Jewish shops and consult Jewish physicians; Jewish businessmen were forbidden to attend markets and fairs; and the city canceled all its own real estate and other business transactions with Jews.[70]

Paradoxically these initiatives led to a clash with the state administration of Württemberg, when the latter demanded that a Stuttgart Jewish developer be exempted from the building limitations. The city council complained to the Württemberg Ministry of the Interior, and Stuttgart mayor Karl Strölin mentioned the incident as an example of the differences that could arise between city and state authorities regarding the implementation of anti-Jewish policies.[71]

Such confrontations, mainly between regional bureaucracies and local party members, were actually not unusual. In Offenburg, in the Breisgau district, one started on March 19, 1937, with a complaint sent by a Jewish attorney, Hugo Schleicher, to the Offenburg district office in the name of the local Jewish community and of the Jews of Gengenbach, an Offenburg suburb. A grocer there, a certain Engesser, had refused to sell groceries and milk to a Jewish customer named Ferdinand Blum. The reason, it soon appeared, was that the mayor of Gengenbach, who also chaired the finance committee of the local hospital, had informed Engesser that he would not be allowed to sell his wares to the hospital if he continued to sell goods to Jews. As all grocers in Gengenbach were allowed to sell to the hospital, the mayor's tactics would quickly achieve a result that Schleicher clearly defined in his letter: "The final consequence of this measure will be that the Jewish population of Gengenbach will no longer be provided with food and milk."[72]

The Offenburg district office forwarded the complaint to Gengenbach's mayor and asked for an answer. On April 2 the mayor wrote back "concerning the complaint of the Jew H. Schleicher": "The facts presented in the complaint are correct. At the crow-black Engesser's ["crow-black" meant that Engesser was a devout Catholic], the customers, apart from the Jews, are the blackest types of Gengenbach, so that his store has become a meeting place for all the obscurantists of our time.* I confronted Engesser with the option of giving up either his deliveries to the hospital or his Jewish customers. He immediately declared that he was ready to give up his Jewish customers. Whether the Jews here get food or whether they croak is one and the same to me; they can leave for more fertile regions where milk and honey already flowed in Abraham's time. In no way shall I permit deliveries to an institution under my authority to be made by Jew lackeys; neither will I allow myself to be held responsible because of a Jew's complaint, and as a National Socialist I reject the demand for explanations and answers. I ask that the Jew be given the appropriate answer."[73]

The district office soon answered. On April 5 the mayor's letter was sent back to him because of its "entirely irrelevant and incredible tone, totally inappropriate and unacceptable in addressing superior authority." This was the message throughout: "When superior authority demands a report, it is the duty of your office to present it in a factual and relevant way. I am now expecting such a concretely formulated report, which will also state whether and how the provision of milk will be assured in Gengenbach to the Blum family."[74]

IV

For Jews and Germans alike, the fundamental criterion for measuring the success of the anti-Jewish segregation policies was the level of Jewish economic presence in Germany. Some local occurrences seemed, on occasion, to point to unexpected resilience. Thus, on February 2, 1937, the Stuttgart *NS-Kurier* published a lengthy article on a particular instance of "wretchedness and lack of character." The wife of the director of a city enterprise (whose name was withheld) had been seen buying laundry soap in the Jewish department store Schocken.[75] Still worse, on March 20 that same year, the *NS-Kurier* must have deeply angered its readers when it reported that the Munich Jewish-owned fashion house Rothschild had presented its designs at the Marquardt Hotel, and that "some German women, rich and accordingly devoid of convictions," had accepted the Jewish invitation to attend.[76]

Sometimes silence was a safer option for the local party press. No

---

*This tirade was in keeping with the party's vituperative anti-Catholic campaign of the late thirties: Its main ideologue was Alfred Rosenberg, but soon Martin Bormann was to become its principal driving force.

Munich newspaper published anything about the four-hour visit paid in 1936 by Göring, accompanied by his adjutant, Prince Philipp von Hessen, to Otto Bernheimer's carpet and tapestry store. Although Bernheimer's was well known as a Jewish-owned business, Göring paid 36,000 Reichsmarks for two rare carpets, which were duly sent to their lofty destination in Berlin.[77]

Indeed, Göring was no exception, nor were the Stuttgart society ladies. Gestapo reports from various parts of the Reich indicate that at the end of 1935 and in 1936, many Germans were still not hesitating to do business with Jews. Despite the party's growing concern, the cattle trade in rural areas remained largely Jewish; according to a Gestapo report on the month of November 1935 "the Jews almost totally control the cattle trade [in Hesse]. They have transferred their activities to the late evening hours or to night time. Sometimes it even happens that Volksgenossen put themselves at the Jews' disposal as hidden representatives, i.e., under their own name but for the benefit of the Jews, and do business for them in the large markets of cattle for slaughter in Frankfurt, Wiesbaden, and Koblenz."[78] Almost one year later, a report from the Franconian district of Hipoltstein sounded the alarm: "The peasants' business relations with Jews have assumed such a dimension that the political leadership felt prompted to intervene energetically."[79]

In the cities the annual late-winter sales at Jewish stores were big occasions. Thus in February 1936, the Munich police directorate reported that the sale at the Jewish-owned textile house Sally Eichengrün had drawn "large crowds." At times as many as three hundred eager female customers stood in line on the street outside the store.[80] And various SD reports indicate that even in 1937 economic relations between Germans and Jews still remained active in several domains, with, for example, members of the aristocracy, of the officer corps, and of the high bourgeoisie still keeping their assets in Jewish banks.[81]

It is difficult to assess what was paid—as an average percentage of value—to the tens of thousands of Jewish owners of small businesses during this early phase of Aryanization. As noted in chapter 1, recent research indicates that the considerable scope of Aryanization at the medium- and small- business level was not indicative of the situation at the higher levels of the economy: There the competition was more limited, and the attitude toward extortion still negative, because the enterprises involved had higher international visibility. The Nazis decided, therefore, to avoid any head-on clash.[82]

Dozens of Jews remained on boards of directors and in other important managerial positions at companies such as Mannesmann, IG Farben, Gesellschaft für Elektrische Unternehmungen, and so on. The Dresdner Bank, for instance, "still had 100 to 150 Jewish employees in Berlin in 1936, and five directors retained their posts until the period 1938 to 1940."[83]

When Aryanization did take place at the big-business level, there are indications in some very significant instances that fair prices were being offered to the owners until the end of 1937, when the situation was to change drastically. Self-interest was obviously part of the motivation for this kind of seeming restraint and fairness. The economic recovery remained uncertain. Some of the largest German firms, eager to avoid additional taxation of their new profits or to escape the effects of eventual devaluation, used the costly acquisition of tested yet depreciable enterprises to improve their accountable benefits. In any case, this is both the Nazi and the business press interpreted the acquisition by Henkel of the Jewish-owned Norddeutsche Hefeindustrie above par, and a similar operation by Unilever's main German subsidiary.[84] In general, however, the overall economic situation of the Jews in Germany was steadily worsening.

A remarkable contemporary summary appeared in December 1935 in the Austrian *Reichspost*: "The Jewish merchants in small- and medium-size [German] provincial towns have, for some time now, been fighting a difficult battle. In these towns, the weapon of the boycott can be utilized far better than in a place like Berlin, for example. The consequence is that there is now a massive sell-off of Jewish retail shops. . . . There are reports . . . from certain areas . . . that an average of 40 to 50 percent of all Jewish businesses have already been transferred to Aryan ownership. Along with this, there are many small towns in which the last residues of Jewish business activity have already been liquidated. This is also the reason for the fact that various small congregations are offering their synagogues for sale. Only recently, a farmer in Franconia was able to purchase such a building for the price of 700 marks—for the purpose of storing grain."[85]

In villages and small cities, harassment was often the easiest way to compel Jews to sell their businesses at a fraction of their value and move away or emigrate. In the larger cities and for larger businesses, credit restrictions and other boycott measures devised by Aryan firms led to the same result. Those Jews who clung to their economic activity were increasingly confined to the rapidly shrinking Jewish market. Excluded from their occupations, Jewish professionals became peddlers, either selling wares out of their homes or traveling from place to place—a reversal of the historic course of Jewish social mobility. Barkai has noted that, since peddling had to be registered, the state and party authorities were sometimes under the misapprehension that Jewish economic activity was growing. After the Nuremberg Laws forbade Jews to employ female Aryans under forty-five in their homes, young Jewish girls moved into the newly vacant positions, again reversing a trend that modern Jewish women had been supporting, and fighting for, for decades.[86]

This overall evolution is unquestionable; yet it demands to be nuanced if we are to rely on SD reports. Thus, the annual report for the year 1937 of the SD's Jewish section gives the impression that attitudes toward the Jews among some sectors of the population remained mixed, and were fed not only by economic but also by religious and possibly some political motives:

"The year covered by the report has shown that large parts of the population, and even of the Party community, do not bother anymore even about the most basic demand, namely not to buy from the Jew. This kind of sabotage is particularly strong in strictly Catholic areas and among the supporters of the Confessing Church, who partly from ideological motives—the solution of the Jewish question by way of baptism and the inclusion of the Jews in the Christian community—but also partly in order to strengthen the opposition to National Socialism, try to hamper the work of the Reich with regard to Jewry. The best proof of the success of this oppositional activity is the fact that, in contrast to other parts of the Reich, in mid and lower Franconia as well as in Swabia a move of the Jewish population is taking place from the cities to the rural areas, where the Jews, under the moral protection of the Church, are less directly affected by the measures taken by the Reich. A similar trend can be noticed in the Catholic areas of the Prussian province of Hesse-Nassau and in Hesse."[87]

Although the SD report only described the situation in some areas, and although—since the contrary trend is generally documented—the movement of Jews from the cities to the countryside must have been very limited, anti-Semitism was apparently not becoming an *active* force within the overall population. The words "do not bother anymore" even indicate a growing indifference, on this subject, to party propaganda. Yet, as before, during these two years some religious constraints and economic self-interest seem to have been the main motivations for such "lax" attitudes toward the Jews. But the forthcoming disappearance of almost all Jewish economic activity, coupled with more violent official pressure, would soon make themselves felt.

Once Hitler had taken concrete steps to launch the Reich on the course of a major military confrontation, the fate of the conservatives was sealed. At the end of 1937, Schacht would be on his way out, replaced by the Nazi Walther Funk. At the beginning of 1938, other conservative ministers, including Foreign Minister Neurath and Defense Minister Blomberg, would follow. At the same time, the army chief of staff, Gen. Werner von Fritsch, left in disgrace on trumped-up charges of homosexuality. Hitler himself became the commander of the armed forces, which henceforward were led de facto by a new Supreme Command of the Wehrmacht (Oberkommando der Wehrmacht, or OKW), under Gen. Wilhelm Keitel. The ever weaker and ever more ambiguous protection offered by the conservatives against radicalization of the regime's anti-Jewish policies had therewith disappeared.

In the directive establishing the Four-Year Plan, Hitler demanded passage of a law that "would make the whole of Jewry responsible for all damage some individual members of this gang of criminals caused the German economy and thereby the German people."[88] In order to punish the Jews for Gustloff's death (Gustloff, it will be remembered, was the Nazi representative in Switzerland who was murdered by a Jewish student in early February 1936), the decree concerning the collective fine Hitler wanted to impose on the Jews of Germany was to be ready by the end of the assassin's trial in Switzerland. The deadline was missed because discussions between the Ministries of Finance and the Interior on technicalities regarding the fine continued throughout 1937 and the first half of 1938. But the postponement really resulted from Göring's hesitations about the potential effects of such a decree on the Reich's foreign currency and raw materials situation.[89] It would be Göring, however, who finally announced the imposition of a collective fine on the Jews of Germany after the Kristallnacht pogrom that followed Ernst vom Rath's assassination.

The waning of conservative influence, particularly with regard to the economic situation of the Jews, became palpable at various levels, as well as in the tone of the exchanges between party grandees and the Ministry of the Economy. In the fall of 1936, the Chemnitz clothing manufacturer Königsfeld became a target of growing harassment from the local party organizations. As the owner of the firm was a Mischling of the first degree married to a German woman and therefore still entitled to the status of full-fledged German citizen (*Reichsbürger*), and as, according to a Ministry of the Economy memorandum, no Jewish influence could be perceived in that enterprise, the party authorities in Saxony were requested to put an end to their campaign against the Königsfeld company. On December 6 Reichsstatthalter Mutschmann responded to this request in a letter to Councillor Hoppe at the ministry. Mutschmann was "astounded" by Hoppe's stance regarding the "non-Aryan" Königsfeld enterprise: "Such a position is contrary to the National Socialist worldview and is, in my opinion, a sabotage of the Führer's orders. I request you therefore not to change any aspect of the existing situation; otherwise I would be compelled to take countermeasures that might be quite unpleasant. In time I shall present your position to the Führer very clearly. In any case, I am not willing to transmit your instructions to the officials who are under my orders; quite the contrary, I am of the opinion that you have proven by your attitude that you are totally in the wrong job."[90]

Party activists now took it upon themselves to publish in the press the names of *Volksgenossen* who patronized Jewish stores; for good measure, the culprits' addresses were added. Bormann had to react. In an order of October 23, 1937, he took issue with these initiatives by pointing out a well-known circumstance: Many shoppers were not aware that a particular store was Jewish, and thus found themselves exposed in the press for a totally unintentional misdeed. Names should therefore be carefully checked before publication, and party members who were in an area unfamiliar to them should avoid buying in Jewish stores by inquiring beforehand about the proprietors' identity.[91]

By 1936 it was clear that the Haavarah Agreement had brought Germany no economic or political advantages, but, quite the contrary, that channeling Jewish emigration toward Palestine could foster the creation of an independent Jewish state. Such a state could become a center of agitation against Nazi Germany or, worse still, could enhance and coordinate world Jewry's power. The issue seemed to become particularly urgent from the end of 1936 and into 1937, when Britain's Peel Commission recommended the division of Palestine into separate Jewish and Arab-

Palestinian states, with other areas to remain under British control. What should Germany's diplomatic stance be? By April 1937 Ernst von Weizsäcker, head of the political division of the Wilhelmstrasse and future secretary of state, had adopted a position, consistently promoted by the Foreign Ministry's Department Germany against the creation of a Jewish state; in concrete terms, however, the policy remained one of non-interference, which meant, among other things, no active support for the Arab national movement.[92]

The Wilhelmstrasse's anti-Zionist position became more adamant, at least on the level of principle, when, in June 1937, Foreign Minister Neurath himself took a stand: "The formation of a Jewish State or of a Jewish political entity under British Mandate is not in Germany's interest," Neurath cabled to his diplomatic representatives in London, Jerusalem, and Baghdad, "given the fact that such a state in Palestine would not absorb all the Jews of the world but would give them a new power position, under the cover of international law, something comparable to what the Vatican represents for political Catholicism or Moscow for the Comintern. That is why it is in the interest of Germany to contribute to the strengthening of the Arab world in order to offset, if need be, the increased power of world Jewry. Clearly, one cannot expect the direct intervention of Germany in order to influence the evolution of the Palestinian problem. However, it would be good if the interested foreign governments were not left uninformed of our position."[93]

Neurath's position and the general trend of thought prevailing at the Foreign Ministry encouraged opponents of the Haavarah through the year 1937, although it was becoming clear that the recommendations of the Peel Commission were leading nowhere, mainly as a result of violent Arab opposition. But no one dared to take any concrete measures against the agreement, as Hitler had not yet expressed his viewpoint. His decision, announced at the end of January 1938, clearly implied maintenance of the Haavarah: Further Jewish emigration by all possible means. The bureaucracy was left with only one choice: Comply. And so it did.[94]

A few days before Hitler's decision, a somewhat less weighty matter was resolved in court: A Jewish businessman was sentenced for selling swastika flags and other national emblems. The court argued that, just as the law forbade Jews to display the national colors because they had no possible "inner relation" to the symbols of the movement or were even hostile to them, so trading in these symbols by Jews—an even more demeaning action—represented an offense against the honor of the movement and of the German people.[95]

V

On November 5, 1937, Hitler convened a wide array of military, economic, and foreign affairs experts to inform them of his strategic plans for the next four to five years. In the near future Hitler envisioned taking action against Czechoslovakia and against Austria (in that order), given the Western democracies' glaring weakness of purpose. In fact Austria came first, due to an unforeseen set of circumstances cleverly exploited by Hitler.

In the German-Austrian treaty of 1936, the Austrian Chancellor Kurt von Schuschnigg had promised to include some Nazi ministers in his cabinet. As, in the Nazis' eyes, Schuschnigg was going neither far nor fast enough in acceding to their requirements, Hitler summoned him to Berchtesgaden in February 1938. Under threat of military action, Schuschnigg accepted the German dictator's demands. Yet, once back in Vienna, he tried to outwit Hitler by announcing a plebiscite on Austrian independence. Hitler responded by threatening an immediate invasion of Austria if the plebiscite was not canceled. Berlin's further demands—including Schuschnigg's resignation and his replacement by an Austrian Nazi, Arthur Seyss-Inquart—were all accepted. Nonetheless Hitler's course was now set: On March 12, 1938, the Wehrmacht crossed the Austrian border; the next day Austria was annexed to the Reich. On March 15 Hitler spoke from the balcony of the Hofburg to hundreds of thousands of ecstatic Viennese assembled on the Heldenplatz. His closing words could hardly have been surpassed: "As Führer and Chancellor of the German nation and Reich, I now report to history that my homeland has joined the German Reich."[96]

On March 16, as the Gestapo was coming to arrest him, the Jewish playwright and historian of culture, Egon Friedell, jumped to his death from the window of his Vienna apartment. Five Jews had committed suicide in Vienna in January 1938, four in February. In the second half of March, seventy-nine Viennese Jews killed themselves.[97]

In Austrian author Thomas Bernhard's last play, *Heldenplatz*, the Jewish professor Robert Schuster, originally from Vienna, returns from Oxford to the Austrian capital sometime in the 1980s. For himself and his wife, he discovers, the past remains hauntingly present:

My brother Josef may speak of luck
that he managed such a spontaneous departure
I always admired those who committed suicide
I never believed that my brother would be capable
of doing it. . . .

Later he alludes to his wife:

For months, she again hears the really frightening way
in which the masses were shouting on the Heldenplatz
You know: on March fifteenth Hitler arrives
at the Heldenplatz . . .[98]

CHAPTER 8

An Austrian Model?

I

On June 4, 1938, Sigmund Freud, aged eighty-two, was allowed to depart from Vienna, the city that had been his home since he was four years old. His apartment had twice been searched by the Gestapo, and his daughter Anna summoned for interrogation. Finally, after the Nazis had impounded part of his possessions and imposed the emigration tax, they demanded his signature on a declaration that he had not been ill treated. Freud dutifully signed, and added: "I can most highly recommend the Gestapo to everyone." The Gestapo men were too dull witted to perceive even such heavy-handed sarcasm, but the risk of such a comment was considerable—and one may wonder "whether there was something at work in Freud making him want to stay, and die, in Vienna."[1]

As a result of the Anschluss, an additional 190,000 Jews had fallen into Nazi hands.[2] The persecution in Austria, particularly in Vienna, out-paced that in the Reich. Public humiliation was more blatant and sadistic; expropriation better organized; forced emigration more rapid. The Austrians—their country renamed Ostmark and placed under the authority of Gauleiter Josef Bürckel, who received the title Reich Commissary for the Reunification of Austria with the Reich—seemed more avid for anti-Jewish action than the citizens of what now became the Old Reich (Altreich). Violence had already started before the Wehrmacht crossed the border; despite official efforts to curb its most chaotic and moblike aspects, it lasted for several weeks. The populace relished the public shows of degradation; countless crooks from all walks of life, either wearing party uniforms or merely displaying improvised swastika armbands, applied threats and extortion on the grandest scale: Money, jewelry, furniture, cars, apartments, and businesses were grabbed from their terrified Jewish owners.

In Austria in the early 1930s, the Jewish issue had become an even more potent tool for right-wing rabble-rousing than had been the case in Germany during the last years of the republic.[3] When the Nazi campaign against Engelbert Dollfuss reached its climax in early 1934, it harped unceasingly on the domination of the chancellor by the Jews.[4] The incitement intensified after Dollfuss's assassination, on July 25, and during the entire chancellorship of his successor, Kurt von Schuschnigg, which ended with the German invasion of March 1938. According to police sources, anti-Semitism was of "decisive importance 'for the success of Nazi propaganda' during the Schuschnigg years. 'The most dangerous breach in the Austrian line of defense [against Nazism] was caused by anti-Semitism,' wrote the ultraconservative Prince Ernst Rüdiger Starhemberg, the commander of the Heimwehr and head of the Patriotic Front, in his postwar memoirs. 'Everywhere people sniffed Jewish influence and although there was not a single Jew in any leadership position in the whole Patriotic Front, the Viennese were telling each other. . . of the Judaization of this organization, that after all the Nazis were right and that one should clean out the Jews."[5]

The wild aggression following the Anschluss quickly reached such proportions that by March 17 Heydrich was informing Bürckel that he would order the Gestapo to arrest "those National Socialists who in the last few days allowed themselves to launch large-scale assaults in a totally undisciplined way [against Jews]."[6] In the overall chaos, such threats had no immediate effect, nor did the fact that the violence was officially attributed to the Communists change the situation. It was only on April 29, when Bürckel announced that the leaders of SA units whose men took part in the excesses would lose their rank and could be dismissed from the SA and the party, that the violence started to ebb.[7]

In the meantime the official share of the takeover of Jewish property was rapidly growing. On March 28 Göring had issued orders "to take quiet measures for the appropriate redirecting of the Jewish economy in Austria."[8] By mid-May a Property Transfer Office (Vermögensverkehrsstelle) with nearly five hundred employees was actively promoting the Aryanization of Jewish economic assets.[9] Within a few months, 83 percent of the handicrafts, 26 percent of the industry, 82 percent of the economic services, and 50 percent of the individual businesses owned by Jews were taken over in Vienna alone; of the eighty-six Jewish-owned banks in Austria's capital, only eight remained after this first sweep.[10] The funds made available by the confiscations and expropriations were used in part to compensate the losses suffered by "Nazi fighters" (*NS-Kämpfer*) in "Jewish-Socialist Vienna" and to give some support to the pauperized Jewish population that was unable to emigrate.[11] The compensation idea actually offered a wide array of possibilities. On July 18 the Office of the Führer's Deputy sent to Bürckel a draft of the Law for the Compensation of Damages Caused to the German Reich by Jews. The law had not yet been announced, the letter indicated, "as it is not yet clear how the compensation fund should be set up after implementation of the measures planned against the Jews by Göring."[12]

Some measures were not in need of any law. A few days after the Anschluss, SA men took the board chairman of the Kreditanstalt, Austria's leading bank, Franz Rothenberg, for a car ride and threw him out of the moving vehicle, killing him. Isidor Pollack, the director general of the chemical works Pulverfabrik, received a visit from the SA in April 1938 and was so badly beaten up during the "search" of his home that he died shortly afterward. The Deutsche Bank confiscated the Rothschild-controlled Kreditanstalt, while Pulverfabrik, its subsidiary, was taken over by I. G. Farben.[13]

The overall Aryanization process continued to unfold with extraordinary speed. By mid-August 1939 Walter Rafelsberger, the head of the Property Transfer Office, could announce to Himmler that within less than a year and a half his agency "had practically completed the task of de-Judaizing the Ostmark economy." All Jewish-owned businesses had disappeared from Vienna. Of the 33,000 Jewish enterprises that had existed in the Austrian capital at the time of the Anschluss, some 7,000 had already been liquidated before the setting up of the Transfer Office in May 1938. "Of the other 26,000, approximately 5,000 were Aryanized and the remaining 21,000 liquidated in an orderly way."[14]

Simultaneously Jewish dwellings began to be confiscated throughout the country, particularly in Vienna. By the end of 1938, out of a total of approximately 70,000 apartments owned by Jews, about 44,000 had been Aryanized. After the beginning of the war, the rate of occupancy in the remaining Jewish apartments was approximately five to six families per apartment. Often there were neither plumbing nor cooking facilities, and only one telephone was available in every building.[15]

Herbert Hagen arrived in Vienna on March 12 with the first units of the Wehrmacht; a few days later Adolf Eichmann, who had just been promoted to second lieutenant in the SS (SS Untersturmführer), joined him. On the basis of lists that had been prepared by the SD, employees of Jewish organizations were arrested and documents impounded.[16] After this first sweep, some measure of "normalization," allowing for the implementation of more far-reaching plans, took place. Eichmann was appointed adviser on Jewish affairs to the inspector of the Security Police and SD, Franz Stahlecker. In a letter dated May 8, he informed Hagen about his new activities: "I hope that I will shortly be in possession of the Jewish yearbooks of the neighboring states [probably Czechoslovakia and Hungary], which I will then send to you. I consider them an important aid. All Jewish organizations in Austria have been ordered to make out weekly reports. These will go to the appropriate experts in II 112 in each case, and to the various desks. The reports are to be divided into a report on the situation and a report on activities. They are due each week on Monday in Vienna and on Thursday in the provinces. I hope to be able to send you the first reports by tomorrow. The first issue of the Zionist *Rundschau* is to appear next Friday. I have had the [printer's copy] sent to me and am now on the boring job of censorship. You will get the paper, too, of course. In time this will become 'my paper' up to a point. In any case, I have got these gentlemen on the go, you may believe me. They are already working very busily. I demanded an emigration figure of 20,000 Jews without means for the period from April 1, 1938, to May 1, 1939, from the Jewish community and the Zionist organization for Austria, and they promised to me that they would keep to this."[17]

The idea of establishing a Central Office for Jewish Emigration (Zentralstelle für Jüdische Auswanderung) apparently came from the new head of the Jewish community, Josef Löwenherz. The community services assisting the would-be emigrants had been overwhelmed by the tens of thousands of requests for departure authorizations; a lack of coordination among the various German agencies involved in the emigration process turned obtaining these documents into a lengthy, cumbersome, and grueling ordeal. Löwenherz approached Eichmann, who transmitted the suggestion to Bürckel.[18] Berlin gave its agreement, and on August 20, 1938, the central office was established under the formal responsibility of Stahlecker and the de facto responsibility of Eichmann himself.[19] The procedure inaugurated in the former Rothschild palace, at 20–22 Prinz Eugen Strasse, used, according to Eichmann, the "conveyor belt" method: "You put the first documents followed by the other papers in at one end and out comes the passport at the other."[20] One more principle was implemented: Through levies imposed on the richer members of the Jewish community, the necessary sums were confiscated to finance the emigration of the poorer Jews. Heydrich later explained the method: "We worked it this way: Through the Jewish community, we extracted a certain amount of money from the rich Jews who wanted to emigrate. . . . The problem was not to make the rich Jews leave but to get rid of the Jewish mob."[21]

Aside from hastening legal emigration by all available means, the new masters of Austria started to push Jews over the borders, mainly those with Czechoslovakia, Hungary, and Witzerland. What had been a sporadic Nazi initiative in some individual cases until March 1938 became a systematic policy after the Anschluss. According to Göring and Heydrich, some five thousand Austrian Jews were expelled in that way betwen March and November 1938.[22] And even tighter control was imposed on those Jews who had not left. Sometime in October 1938, Himmler gave the order to concentrate all Jews from the Austrian provinces in Vienna. According to an internal memo of the SD's Jewish section, Eichmann discussed the transfer of an estimated 10,000 Jews still living outside the capital with Odilo Globocnik, the Gauleiter of Lower Danube, and himself set out on October 26 to tour the Austrian provinces in order to inform the SD chiefs in each region "that with the help of the Gestapo stations, they advise the Jews either to leave the country by 15/12/1938 or to move to Vienna by 31/10/38 [probably an error for 31/12/38]."[23] Within six months of the Anschluss, 45,000 Austrian Jews had emigrated, and by May 1939, approximately 100,000, or more than 50 percent, had left.[24] The Jewish exodus from Austria had an unexpected side benefit for the Nazis. Each emigrant had to attach three passport photos to the forms. The Vienna SD drew the attention of the party's Racial Policy Office to such an outstanding collection; Walter Gross's office responded immediately: It was "exceptionally interested" in this immense inventory of Jewish faces.[25]

The Germans had some other plans as well. In October 1938 Rafelsberger suggested the setting up of three concentration camps for 10,000 Jews each in areas empty of population, mainly in sandy regions and in marshes. The Jews would build their own camps; costs would be kept to about ten million marks, and the camps would provide work for approximately 10,000 unemployed Jews. It seems that one of the technical problems was to find enough barbed wire.[26] Nothing came of this idea—for a short while at least. Another idea—not directly related to anti-Jewish policies, and deadlier in the immediate future—was, however, quickly implemented.

"Mauthausen," writes its most recent historian, "sits amid lovely rolling hills whose fields cover the Austrian landscape like the bedspread of a giant. The town nuzzles peacefully along the north bank of the Danube, whose swift current is quickened by the nearby confluence of the Ems, a major Alpine waterway. . . . Mauthausen lies just 14 miles downriver from Linz, the provincial capital of the province of Upper Austria; 90 miles to the east the spire of St. Stephen's Cathedral, the landmark of Vienna, rises to meet the sky. . . . Of all the area's treasures, however, the most significant to our story are the great yawning pits of granite."[27]

A few days after the Anschluss, in March 1938, Himmler, accompanied by Oswald Pohl, chief of the administrative office of the SS-Hauptamt, made a first inspection of the quarries. The intention was clear: excavation of the granite would bring considerable financial benefits to an SS-operated enterprise, the German Earth and Stone Works Corporation (DEST), which was about to be established in April; a concentration camp on location would provide the necessary work force. The final decision must have been taken quickly as, according to a report in the London *Times* of March 30, "Gauleiter Eingruber, of Upper Austria, speaking at Gmunden, announced that for its achievements in the National Socialist cause his province was to have the special distinction of having within its bounds a concentration camp for the traitors of all Austria. This, according to the *Völkischer Beobachter*, aroused such enthusiasm in the audience that the Gauleiter could not continue his speech for some time."[28]

A second visit took place at the end of May; this time it included Theodor Eicke, the inspector of concentration camps, and Herbert Karl of the SS construction division.[29] The first 300 inmates, Austrian and German criminals from Dachau, arrived on August 8, 1938. By September 1939 Mauthausen held 2,995 inmates, among them 958 criminals, 1,087 Gypsies (mainly from the Austrian province of Burgenland), and 739 German political prisoners:[30] "The first Jewish inmate was a Viennese-born man arrested as a homosexual, who was registered at Mauthausen in September 1939 and recorded as having died in March 1940. During 1940 an additional 90 Jews arrived; all but 10 of them were listed as dead by the year's end."[31]

According to Götz Aly and Susanne Heim, it was in Austria that the Nazis inaugurated their "rational" economically motivated policy regarding the Jewish question, which from then on dictated all their initiatives in this domain, from the "model" established in Vienna to the "Final Solution." The Viennese model (*Modell Wien*) was basically characterized by a drastic restructuring of the economy as a result of the liquidation of virtually all the unproductive Jewish businesses on the basis of a thorough assessment of their profitability prepared by the Reich Board for Economic Management (Reichskuratorium für Wirtschaftlichkeit);[32] by a systematic effort to get rid of the newly created Jewish proletariat by way of accelerated emigration whereby, as we saw, wealthy Jews contributed to the emigration fund for the destitute part of the Jewish population; by establishing labor camps (the three camps planned by Walter Rafelsberger), where the upkeep of the Jews would be maintained at a minimum and financed by the labor of the inmates themselves.[33] In essence those in charge of the Jewish question in annexed Austria were supposedly motivated by economic logic and not by any Nazi anti-Semitic ideology. The argument seems bolstered by the fact that not only was the entire Aryanization process in Austria master-minded by Göring's Four-Year Plan administration and its technocrats, but the same technocrats (such as Rafelsberger) also planned the solution of the problem of impoverished Jewish masses by way of forced-labor concentration camps that appeared to be early models of the future ghettos and eventually of the future extermination camps.

In fact, as has been seen, the liquidation of Jewish economic life in Nazi Germany had started at an accelerated pace in 1936, and by late 1937, with the elimination of all conservative influence, the enforced Aryanization drive had become the main thrust of the anti-Jewish policies, mainly in order to compel the Jews to emigrate. Thus what happened in Austria after the Anschluss was simply the better organized part of a general policy adopted throughout the Reich. The link between economic expropriation and expulsion of the Jews from Germany and German-controlled territories did continue to characterize *that stage* of Nazi policies until the outbreak of the war. Then, after an interim period of almost two years, another "logic" appeared, one hardly dependent on economic rationality.

II

After the Anschluss the Jewish refugee problem became a major international issue. By convening a conference of thirty-two countries in the French resort town of Evian from July 6 to 14, 1938, President Roosevelt publicly demonstrated his hope of finding a solution to it. Roosevelt's initiative was surprising, because "he chose to intrude into a situation in which he was virtually powerless to act, bound as he was by a highly restrictive immigration law."[34] Indeed, the outcome of Evian was decided before it even convened: The invitation to the conference clearly stated that "no country would be expected to receive a greater number of emigrants than is permitted by its existing legislation."[35]

The conference and its main theme, the fate of the Jews, found a wide and diverse echo in the world press. "There can be little prospect," the London *Daily Telegraph* said on July 7, "that room will be found within any reasonable time."[36] According to the *Gazette de Lausanne* of July 11: "Some think that they [the Jews] have got too strong a position for such a small minority. Hence the opposition to them, which in certain places has turned into a general attack." "Wasn't it said before the first World War that one-tenth of the world's gold belonged to the Jews?" queried the *Libre Belgique* on July 7.[37]

Not all of the press was so hostile. "It is an outrage to the Christian conscience especially," said the London *Spectator* on July 29, "that the modern world with all its immense wealth and resources cannot get these exiles a home and food and drink and a secure status."[38] For the future postwar French Foreign Minister Georges Bidault, writing in the left-wing Catholic paper *L'Aube* on July 7, "One thing is clearly understood: the enlightened nations must not let the refugees be driven to despair."[39] The mainstream French Catholic newspaper *La Croix* urged compassion: "We cannot stand aside," it pleaded on July 14, "in view of the suffering of human beings and fail to respond to their cry for help. . . . We cannot be partners to a solution of the Jewish question by means of their extinction, by means of the complete extermination of a whole people."[40] But no doors opened at Evian, and no hope was offered to the refugees. An Intergovernmental Committee for Refugees was established under the chairmanship of the American George Rublee. Rublee's activities, which ultimately achieved no result, will be discussed further on.

Nazi sarcasm had a field day. For the SD Evian's net result was "to show the whole world that the Jewish problem was in no way provoked only by Germany, but was a question of the most immediate world political significance. Despite the general rejection by the Evian states of the way in which the Jewish question has been dealt with in Germany, no country, America not excepted, declared itself ready to accept unconditionally any number of Jews. It was remarkable that the Australian delegate even mentioned that Jewish emigration would endanger his own race."[41] There was no fundamental difference between the German assessment and the biting summary of Evian by the *Newsweek* correspondent there: "Chairman Myron C. Taylor, former U.S. Steel head, opened the proceedings: 'The time has come when governments . . . must act and act promptly.' Most governments represented acted promptly by slamming their doors against Jewish refugees."[42] The *Völkischer Beobachter* headlined triumphantly: "Nobody wants them."[43]

For Hitler too, this was an opportunity not to be missed. He chose to insert his comments into the closing speech of the party rally on September 12. Its main theme, the Sudeten crisis, riveted the attention of the world; never since 1918 had the danger of war seemed closer, but the Jews could not be left unmentioned: "They complain in these democracies about the unfathomable cruelty that Germany—and now also Italy—uses in trying to get rid of their Jews. In general, all these great democratic empires have only a few people per square kilometer, whereas Germany, for decades past, has admitted hundreds and hundreds of thousands of these Jews, without even batting an eye.

"But now, as the complaints have at last become too strong and as the nation is not willing any more to let itself be sucked dry by these parasites, cries of pain arise all over. But it does not mean that these democratic countries have now become ready to replace their hypocritical remarks with acts of help; on the contrary, they affirm with complete coolness that over there, evidently, there is no room! Thus, they expect that Germany with its 140 inhabitants per square kilometer will go on keeping its Jews without any problem, whereas the democratic world empires with only a few people per square kilometer can in no way take such a burden upon themselves. In short, no help, but preaching, certainly!"[44]

*   *   *

The Evian debacle acquires its full significance from its wider context. The growing strength of Nazi Germany impelled some of the countries that had aligned themselves with Hitler's general policies to take steps that, whether demanded by Germany or not, were meant to be demonstrations of political and ideological solidarity with the Reich. The most notorious among such initiatives were the Italian racial laws, approved by the Fascist Grand Council on October 6, 1938, and taking effect on November 17.

In Italy the Jewish community numbered barely more than fifty thousand and was fully integrated into the general society. Anti-Semitism had become rare with the waning of the church's influence, and even the army—and the Fascist Party—included prominent Jewish members. Finally Mussolini himself had not, in the past, expressed much regard for Nazi racial ideology. Devised on the Nuremberg pattern, the new anti-Jewish laws caused widespread consternation among Italian Jews and many non-Jews alike.[45]

The October laws had been preceded, in mid-July, by the Racial Manifesto, a declaration setting forth Mussolini's concoction of racial anti-Semitism and intended as the theoretical foundation of the forthcoming legislation. Hitler could not but graciously acknowledge so much goodwill. He duly did so on September 6, in the first of his speeches to the Nuremberg party rally: "I think that I must at this point announce, on my own behalf and on that of all of you, our deep and heartfelt happiness in the fact that another European world power has, through its own experiences, by its own decision and along its own paths arrived at the same conception as ourselves and with a resolution worthy of admiration has drawn from this conception the most far-reaching consequences."[46] The first anti-Jewish law introduced in Hungary, in May 1938, was greeted with less fanfare than Mussolini's decision, but it pointed to the same basic evidence: The shadow of Hitler's anti-Jewish policy was lengthening over ever larger parts of Europe.[47]

While the Jews were becoming targets of legal discrimination in a growing number of European countries, and while international efforts to solve the problem of Jewish refugees came to naught, an unusual step was being taken in complete secrecy. In the early summer of 1938, Pope Pius XI, who over the years had become an increasingly staunch critic of the Nazi regime, requested the American Jesuit John LaFarge to prepare the text of an encyclical against Nazi racism and Nazi anti-Semitism in particular. LaFarge had probably been chosen because of his continuous antiracist activities in the United States and his book *Interracial Justice*, which Pius XI had read.[48]

With the help of two other Jesuit priests, the French Gustave Desbuquois and the German Gustav Gundlach, LaFarge completed the draft of *Humani Generis Unitas* (The unity of humankind) by the autumn of 1938 and delivered it to the general of the Jesuit order in Rome, the Pole Wladimir Ledochowski, for submission to the pope.[49] In the meantime Pius XI had yet again criticized racism on several other occasions. On September 6, 1938, speaking in private to a group of Belgian pilgrims, he went further. With great emotion, apparently in tears, the pope, after commenting on the sacrifice of Abraham, declared: "It is impossible for Christians to participate in anti-Semitism. We recognize that everyone has the right to self-defense and may take the necessary means for protecting legitimate interests. But anti-Semitism is inadmissible. Spiritually, we are all Semites."[50]

In this declaration, made in private and thus not mentioned in the press, the pope's condemnation of anti-Semitism remained on theological grounds: He did not criticize the ongoing persecution of the Jews, and he included a reference to the right of self-defense (against undue Jewish influence). Nonetheless his statement was clear: Christians could not condone anti-Semitism of the Nazi kind (or for that matter, as it was shaping up in Italy at the very same time).

The message of the encyclical was similar: a condemnation of racism in general and the condemnation of anti-Semitism on theological grounds, from the viewpoint of Christian revelation and the teachings of the church regarding the Jews.[51] Even so, the encyclical would have been the first solemn denunciation by the supreme Catholic authority of the anti-Semitic attitudes, teachings, and persecutions in Germany, in Fascist Italy, and in the entire Christian world.

Ledochowski was first and foremost a fanatical anti-Communist who moreover hoped that some political arrangement with Nazi Germany remained possible. He procrastinated. The draft of *Humani Generis Unitas* was sent by him for further comment to the editor in chief of the notoriously anti-Semitic organ of the Roman Jesuits, *Civiltà Cattolica*.[52] It was only after LaFarge had written directly to the Pope that, a few days before his death, Pius XI received the text. The pontiff died on February 9, 1939.

His successor, Pius XII, was probably informed of the project and probably took the decision to shelve *Humani Generis Unitas*.[53]

III

Even in 1938, small islands of purely symbolic opposition to the anti-Jewish measures still existed inside Germany. Four years earlier, the Reich Ministry of Education had ordered the German Association for Art History to expel its Jewish members. The association did not comply but merely reshuffled its board of directors. Internal ministry memoranda indicate that Education Minister Rust repeated his demand in 1935, again apparently to no avail. In March 1938 State Secretary Zschintsch sent a reminder to his chief: All funds for the association were to be eliminated, and, if the order was not obeyed, it would no longer be allowed to call itself "German." "The Minister must be interested," Zschintsch concluded, "to have the association finally comply with the principles of the National Socialist world-view."[54] We do not know what the association then decided to do; in any case its Jewish members were certainly not retained after the November 1938 pogrom.

There were some other—equally unexpected—signs of independence. Such was to be the case at the 1938 Salzburg Festival. After the Anschluss, Arturo Toscanini, who had refused to conduct at Bayreuth in 1933, turned Salzburg down as well.

Salzburg was emblematic in more ways than one. From the very outset, in 1920, when Hugo von Hofmannsthal and Max Reinhardt had organized the first festival around a production of Hofmannsthal's *Jedermann* (*Everyman;* based on the medieval mystery play of the same name), the Austrian anti-Semitic press had raved against the Jewish cultural invasion and the exploitation by three Jews (the third was the actor Alexander Moissi) of Christianity's loftiest heritage.[55] Hofmannsthal's *Jedermann* nonetheless opened the festival year in year out (except for performances of his *Welttheater* in 1922 and 1924). In 1938 *Jedermann* was of course removed from the repertory.[56] The Jewish invasion had been stemmed.

Wilhelm Furtwängler agreed to take Toscanini's place at Salzburg. Throughout his career in Nazi Germany, Furtwängler showed himself to be a political opportunist who had moments of courage. In Salzburg he agreed to conduct Wagner's *Meistersinger* on condition that the Jew Walter Grossmann be kept as the understudy in the role of Hans Sachs. As it happened, on opening night Karl Kammann, the scheduled Hans Sachs, fell ill, and Walter Grossmann sang: "A glittering crowd headed by Joseph

Goebbels and his entourage sat dutifully enthralled through the Führer's favorite opera, while Grossmann brought Nuremberg's most German hero to life."[57] But neither the actions of the art historians' association nor Walter Grossmann's performance could stem the ever growing tide—and impact—of Nazi anti-Jewish propaganda.

"The Eternal Jew" (*Der ewige Jude*), the largest anti-Jewish exhibition of the prewar years, opened on November 8, 1937, in Munich's Deutsches Museum. Streicher and Goebbels gave speeches. On the same evening the director of the Bavarian State Theater organized a cultural event in the Residenz Theater, which, according to the *Deutsche Allgemeine Zeitung*, expressed "the basic themes of the exhibition." The first part of the program offered a staged rendition of excerpts from Luther's notorious pamphlet *Wider die Juden und ihre Lügen* (Against the Jews and their lies); the second part presented readings from other anti-Jewish texts, and the third, the Shylock scenes from Shakespeare's *The Merchant of Venice*.[58]

A SOPADE report written a few weeks after the opening stressed that the exhibition "did not remain without effect on the visitors." In the first hall the viewer was faced with large models of Jewish body parts: "Jewish eyes . . . , the Jewish nose, the Jewish mouth, the lips," and so on. Huge photographs of various "racially typical" Jewish faces and mannerisms followed—Trotsky gesticulating, Charlie Chaplin, and so on—"all of it displayed in the most repulsive way." Material (extracts from the Book of Esther, for instance), and caricatures, slogans, and descriptions of "Jews in politics," "Jews in culture," "Jews in business"—and accounts of Jewish goals and methods in these various domains—filled room after room. According to the report, "Jews in film" was particularly effective: An unbearably kitschy commercial production was shown in that section; at the end Alfred Rosenberg appeared on the screen and declared: "You are horrified by this film. Yes, it is particularly bad, but it is precisely the one we wanted to show you."

The author of the SOPADE report admits that he was deeply impressed on leaving the exhibition; so was his companion. She asked questions about what they had seen: "I couldn't tell her the truth," he admits. "I did not have sufficient knowledge for that."[59] Some SA units were so inspired by the exhibition that they started a boycott action of their own as an "educational follow-up" to what they had learned at the Deutsches Museum.[60]

An exhibition such as *The Eternal Jew* was merely the most extreme expression of the ongoing effort to assemble any kind of damning material about the Jews. Diverse forms of this endeavor were encountered during the first years of the regime. Now, at the end of 1937 and throughout 1938, the search went on with renewed inventiveness. On February 24, 1938, the minister of justice informed all prosecutors that it was no longer necessary to forward a copy of every indictment against a Jew to the ministry's press division, as it had already acquired a sufficient perspective on the criminality of Jews. The kinds of criminal acts by Jews that still had to be included were "cases that raised new legal points; those in which the perpetrator had demonstrated a particularly evil intention or had used particularly objectionable methods; those in which the crime had been perpetrated on an especially large scale or had caused particularly great damage or aroused uncommon interest among the public; finally, cases of racial defilement in which the perpetrator was a recurrent offender or had abused a position of power."[61] Such instances of Jews in Germany abusing their positions of power in order to commit *Rassenschande* must have been rather rare in the year of grace 1938. . . .

In March 1938 the issue of Jewish *Mischlinge* and persons related to Jews still in government employment came to the fore. The order for an investigation seems to have originated with Hitler himself, since it was a member of the Führer's Chancellery, Hans Hefelmann, who on March 28, 1938, asked the SD, and specifically section II 112, to collect all the relevant documentation. The II 112 officials pointed out that the forthcoming population census would give an exact account of this particular group and that, in any case, such files as existed were most probably to be found in the higher reaches of each ministry, as any promotion had to take into consideration the candidate's partly Jewish origin or Jewish family connections.[62]

By the beginning of 1938 all German Jews had had to turn in their passports (new ones were issued only to those Jews who were about to emigrate).[63] But another identification document was soon decided upon. In July 1938 the Ministry of the Interior decreed that before the end of the year all Jews had to apply to the police for an identity card, which was to be carried at all times and shown on demand.[64] On August 17 another decree, prepared by Hans Globke, announced that from January 1, 1939, Jews who did not bear the first names indicated on an appended list were to add the first name Israel or Sara to their names.[65] The appended list of men's names started with Abel, Abieser, Abimelech, Abner, Absalom, Ahab, Ahasja,

Ahaser,[66] and so on; the list of women's names was of the same ilk. (Had these lists been compiled under other circumstances, they could stand as an appropriate illustration of the mind-set of bureaucratic half-wits.)

Some of the names on Globke's lists were entirely fictitious and others were grotesque choices manifestly resulting from a compounded intention of identification and degradation. A surprising inclusion among the typically Jewish names was that of Isidor. As has been pointedly remarked, "Saint Isidor of Seville, the anti-Jewish church father, and Saint Isidor of Madrid, the patron saint of so many village churches in Southern Germany, would have been astonished."[67] But it may well be that Globke was merely following the current custom: In Germany at the time, Isidor was a name borne mainly by Jews.[68]

A few months after the Anschluss, Streicher demanded from Himmler that his researchers be granted access to the Rothschild archives in Vienna in order to collect material for a "monumental historical work about Jews, Jewish crimes and Jewish laws in Germany from past to present." Himmler agreed but insisted on the presence of an SD representative during the perusal of the documents.[69] The Rothschild archives exercised a widespread fascination. Rosenberg planned an official exhibition at the September 1938 party congress, whose theme was to be "Europe's Fate in the East." His office turned to SS-Hauptsturmführer Hartl of the Vienna Gestapo, who had impounded the Rothschild archives, in the hope of finding documents illustrating that Jewry in the East maintained contacts with both industrialists and Marxist leaders: "We assume," wrote Rosenberg's delegate, "that among the confiscated material in the Rothschild House, some valuable original information on this subject will be found." Hartl's office answered a few weeks later: No material relevant to the exhibition theme could be found in the Rothschild papers.[70] At approximately the same time, SS-Oberführer Albert indicated to his SD colleague, SS-Standartenführer Six, that he was particularly interested in access to the Rothschild archives for "research purposes"; Six assured Albert that the material was accessible, although it had now been moved to several different places; its curators, it should be noted, were not all ordinary archivists: the Frankfurt Rothschild material and the thirty-thousand-volume library that came with it were being kept secure in the SS main region Fulda-Werra.[71]

After the annexation of the Sudetenland, Rosenberg turned to the leader of the Sudeten Germans, Konrad Henlein, with demands for any Marxist, Jewish, and also religious literature that "offers invaluable resources to the library and the scientific research work of the 'Hohe Schule' [institute] that is being established."[72]

It stands to reason that in such a far-flung research drive, some borderline issues presented serious challenges to the Nazi sense of fine distinctions. Thus, on March 9, 1938, Karl Winter, the owner of the Karl Winter University Publishing House in Heidelberg, turned to Rosenberg for advice on a rather delicate matter. In the twenties Winter had published four volumes of a projected five-volume standard edition of Baruch Spinoza's works; the type for the fifth volume had been set in 1932, but the book had not been printed. Winter felt that he could not decide on his own whether to publish the last volume (in his letter he emphasized his long-time party membership and extended involvement in Nazi publishing activities).[73] On March 18 Rosenberg's Main Office for Science (Amt Wissenschaft) authorized publication (probably on the recommendation of party philosopher Alfred Bäumler).[74] Winter, however, was not an old-time party member for nothing: On March 30 he thanked Rosenberg for the authorization and asked whether he could allude to it in the advertisement he was planning to place in the *Bulletin of the German Book Trade*: "I attach importance to it," he added, "in order to protect myself from unjustified attacks." The reaction to Winter's request left immediate traces in the letter's margin: two bold question marks and a "*Nein*" underlined four times.[75] Winter was told the same in no uncertain terms a few days later. To make sure that Winter would not attempt any foul play, the Amt Wissenschaft letter was sent by registered mail.[76]

Sometimes no amount of formal identification helped, and some highly annoying situations arose. Thus, on August 20, 1938, in answer to an inquiry by the political division of the Hesse-Nassau Gauleitung, the woman principal (*Rektorin*) of the Fürstenberger Gymnasium for Girls in Frankfurt had to send a somewhat embarrassed explanation. What had happened could not be denied: A few days before, the two Jewish girls still enrolled at the school had attended the daily flag-raising. Rektorin Öchler tried to explain away the incident by arguing that there had been many changes among the teachers and that the girls had taken advantage of the situation with "a certain Jewish pushiness." Adequate instructions had been given to the teachers and the principal wanted to use the occasion to expel the girls from the school.[77] But the matter did not rest at that. On August 27 the Gauleitung forwarded the file to the Kreisleitung of Greater Frankfurt. Four days later, the Kreisleiter wrote to Mayor Kremmer that what had happened was incomprehensible and inexcusable, despite the principal's explanations: "I ask you to follow up the matter," the Kreisleiter concluded, "and to make sure that the Frankfurt schools are immediately cleansed of Jewish pupils."[78] On September 8 the mayor's office transmitted the case to the city's School Department with an urgent request to clarify the issue, to consider the possibility of cleansing the city schools of their Jewish students, and to prepare a draft answer to the Kreisleiter. The material had to be in by September 18. The School Department reacted to the emergency with calm: Its answer was sent to the mayor on September 26. Basically, it said, the incident had occurred because there had been many changes and replacements among the teachers. Moreover, the presence of Jewish schoolchildren in the city schools was subject to the law of April 25, 1933, against the overcrowding of German schools (that is, Jewish students could be registered up to the limit of 1.5 percent of the overall number, with exemption from the *numerus clausus* for children of front-line veterans and of *Mischling* couples of the first and second degrees).[79]

IV

The anti-Jewish economic campaign started at full throttle in early 1938; laws and decrees followed one another throughout the year, shattering all remaining Jewish economic existence in Germany. As the year began, some 360,000 Jews still lived in the Altreich, most of them in several large cities, mainly in Berlin. Jewish assets, estimated at some ten to twelve billion Reichsmarks in 1933, had been reduced to half that sum by the spring of 1938. This in itself indicates, as Barkai has pointed out, that Aryanization was a gradual process leading to the measures that were to descend on the Jews of Germany throughout 1938.[80]

On April 26 all Jews were ordered to register their property.[81] On June 14 the problem that had defeated the boycott committee on April 1, 1933, was solved. According to the third supplementary decree to the Reich Citizenship Law, "a business was Jewish if the proprietor was a Jew, if a partner was a Jew, or if, on January 1, 1938, a member of the board of directors was a Jew. Also considered Jewish was a business in which Jews owned more than one quarter of the shares or more than one half of the votes, or which was factually under predominantly Jewish influence. A branch of a Jewish business was considered Jewish if the manager of the branch was a Jew."[82]

On July 6, 1938, a law established a detailed list of commercial services henceforth forbidden to Jews, including credit information, real estate brokerage, and so on.[83] On July 25, the fourth supplementary decree to the Reich Citizenship Law put an end to Jewish medical practice in Germany: The licenses of Jewish physicians were withdrawn as of September 30, 1938.[84] As Raul Hilberg indicates, "That was no more than a re-enactment of canon law, but the modern innovation was the provision that leases for apartments rented by Jewish physicians were terminable at the option of either landlord or tenant."[85] The last line of the decree related neither to canon law nor to modern innovations, but was entirely in the spirit of the new Germany: "Those [physicians] who receive an authorization [to give medical services to Jewish patients] are not authorized to use the appellation 'physician,' but only the appellation 'caretakers of the sick.'"[86] Incidentally, the decree was signed and promulgated in Bayreuth: Hitler was attending the festival.

On September 27, 1938, on the eve of the Munich conference, Hitler signed the fifth supplementary decree, forbidding Jews to practice law.[87] The decree was not immediately made public because of the international tension. Finally, on October 13, he allowed the announcement to be made the next day.[88] The decree was to take effect in the Altreich on November 30 and in former Austria (with a partial and temporary exception in Vienna) on December 31.

The final blow that destroyed all Jewish economic life in Germany came on November 12, when, just after the Kristallnacht pogrom, Göring issued a ban on all Jewish business activity in the Reich. Meanwhile, however, National Socialist physicians and lawyers were still not satisfied with having definitively driven the Jews out of their professions. As was usual in the world of Nazi anti-Jewish measures, concrete destruction had to find a symbolic expression as well. On October 3, 1938, the Reich Physicians' Chamber (Reichsärztekammer) had demanded of the Minister of Education that Jewish physicians, now forbidden to practice, should also suffer further deprivation: "I am therefore requesting," Reich physicians leader Wagner concluded his letter to Rust, "that the title 'Doctor' should be taken away from these Jews as soon as possible."[89] The minister of education and the minister of justice consulted on the matter: their common proposal to the Ministry of the Interior was not to cancel the title of doctor in medicine and law only, but rather to consider drafting a law that would strip Jews of all titles, academic degrees, and similar distinctions.[90] On the morrow of the November 9–10 pogrom, the matter was postponed.[91]

The atmosphere permeating German business circles as the forced Aryanization—or more precisely, confiscation of all Jewish property—became law is revealed in a letter from a Munich businessman who had been asked by the authorities to serve as a consultant in the Aryanization transactions. The author of the letter described himself as a National Socialist, a member of the SA, and an admirer of Hitler. He then added: "I was so disgusted by the brutal . . . and extraordinary methods employed against the Jews that, from now on, I refuse to be involved in any way with Aryanizations, although this means losing a handsome fee. . . . As an old, honest and upstanding businessman, I [can] no longer stand by and countenance the way many 'Aryan' businessmen, entrepreneurs and the like . . . are shamelessly attempting to grab up Jewish shops and factories, etc. as cheaply as possible and for a ridiculous price. These people are like vultures swarming down, their eyes bleary, their tongues hanging out with greed, to feed upon the Jewish carcass."[92]

The wave of forced Aryanization swept away the relatively moderate behavior that, as we have seen, major corporations had adhered to until then. The new economic incentives, the pressure from the party, the absence of any conservative ministerial countervailing forces (such as Schacht had represented) put an end to the difference between low-grade grabbing and high-level mannerliness. In some cases Hitler's direct intervention can be traced. Thus, in mid-November 1937, "Herbert Göring and Wilhelm Keppler at Hitler's Chancellery [now] summoned Otto Steinbrinck, [steel magnate] Friedrich Flick's chief operative in Berlin, in order to bribe or bully Flick into leading a drive to 'Aryanize' the extensive mining properties of the Julius and Ignaz Petschek families."[93]

It seems that recently established enterprises were more aggressive than older ones: Flick, Otto Wolf, and Mannesmann, for example, three of the fast-growing new giants of heavy industry, were more energetically involved in the Aryanizations than were Krupp or the Vereinigte Stahlwerke (United Steelworks). The same happened in banking, the most aggressive being the regional banks in search of fast expansion, and some of the private banks (Merck, Fink, Richard Lenz). The Dresdner Bank, in need of capital, took the lead in brokering the takeovers, whereas the

Deutsche Bank showed more restraint, and the 2 percent commission it levied on the sales prices of Jewish businesses accumulated to several millions of Reichsmarks from 1937 to 1940.[94]

Not all these operations were as easy as the Nazis would have wished. Some of the major Aryanization initiatives kept them on tenterhooks for months and even years, without Berlin being able to claim full victory.[95] The most notorious cases involved complex negotiations with the Rothschilds for control of the Witkowitz steel works in Czechoslovakia (the Viennese Rothschild, Baron Louis, was held hostage for the duration of the negotiations), and with the Weinmanns and also with Hitler's targets, the Petscheks, for control of steelworks and coal mines in the Reich. The Nazis were caught in a maze of foreign holdings and property transfers aptly initiated by their prospective victims which, during the Petschek negotiations, led Steinbrinck to write in an internal memorandum: "Eventually we will have to consider the use of violence or direct state intervention."[96]

The Nazis were well aware of the dilemma exacerbated by accelerated Aryanization: The rapid pauperization of the Jewish population and the growing difficulties in the way of emigration were creating a new Jewish social and economic problem of massive proportions. At the outset men like Frick still had very traditional views of what could be done. According to a report of June 14, 1938, entitled "Jews in the Economy," in a discussion held in April of that year, Frick had apparently summed up his views as follows: "Insofar as Jews in Germany are able to live off the proceeds of their commercial and other assets, they require strict state supervision. Insofar as they are in need of financial assistance, the question of the *public* support must be solved. Greater use of the various organizations for social welfare appears to be unavoidable."[97]

In the early fall of 1938, another measure, this time involving locally planned economic extortion, was initiated in Berlin. One of the largest low-rent housing companies, the Gemeinützige Siedlungs- und Wohnungsbaugesellschaft (GSW) Berlin, ordered the registration of all its Jewish tenants and canceled most of their leases. Some of the Jewish tenants left, but others sued the GSW. Not only did the Charlottenburg district court back the housing company, it indicated that similar measures could be more generally applied. The court would probably have reached the same decision without external pressure, but it so happened that pressure was applied upon the Ministry of Justice by Albert Speer, whom, in early 1937, Hitler had appointed general inspector for the construction of Berlin. The eager general inspector was simultaneously negotiating with the capital's mayor for the construction of 2,500 small apartments to which to transfer other Jews from their living quarters.[98] These details seem to have escaped Speer's highly selective memory.[99]

Anti-Jewish violence had erupted again in the Altreich in the spring and early summer of 1938. In June, on Heydrich's orders, some ten thousand "asocials" were arrested and sent to concentration camps: Fifteen hundred Jews with prior sentences were included and shipped off to Buchenwald (which had been set up in 1937).[100] A few weeks before, at the end of April, the propaganda minister (and Gauleiter of Berlin) had asked the Berlin police chief, Count Wolf Heinrich Helldorf, for a proposal for new forms of segregation and harassment of the city's Jews. The result was a lengthy memorandum prepared by the Gestapo and handed to Helldorf on May 17. At the last moment the document was hastily reworked by the SD's Jewish section, which was critical of the fact that the maximal segregation measures proposed by the Gestapo would make the first priority, emigration, even more difficult than it already was. The final version of the proposal was passed on to Goebbels and possibly discussed with Hitler at a meeting on July 24.[101] Some of the measures envisaged were already in preparation, others were to be applied after the November pogrom, and others still after the beginning of the war.

Goebbels simultaneously moved to direct incitement. According to his diary, he addressed three hundred Berlin police officers about the Jewish issue on June 10: "Against all sentimentality. Not law is the motto but harassment. The Jews must get out of Berlin. The police will help."[102] Party organizations were brought into action. Now that Jewish businesses had been defined by the decree of June 14, their marking could finally begin. "Starting late Saturday afternoon," the American ambassador to Germany, Hugh R. Wilson, cabled Secretary of State Hull on June 22, 1938, "Civilian groups, consisting usually of two or three men, were to be observed painting on the windows of Jewish shops the word "JUDE" in large red letters, the star of David and caricatures of Jews. On the Kurfürstendamm and the Tauentzienstrasse, the fashionable shopping districts in the West, the task of the painters was made easy by the fact that Jewish shop-owners had been ordered the day before to display their names in white letters. (This step, which was evidently decreed in anticipation of a forthcoming ruling which will require Jews to display a uniform distinctive sign, disclosed that a surprisingly large number of shops in this district are still Jewish.) The painters in each case were followed by large groups of spectators who seemed to enjoy the proceedings thoroughly. The opinion in informed sections of the public was that the task was being undertaken by representatives of the Labour Front rather than as formerly has been the case by the S.A. or the S.S. It is understood that in the district around the Alexanderplatz boys of the Hitler Youth participated in the painting, making up for their lack of skill by a certain imagination and thoroughness of mutilation. Reports are received that several incidents took place in this region leading to the looting of shops and the beating up of their owners; a dozen or so broken or empty showcases and windows have been seen which lend credence to these reports."[103]

Bella Fromm's diary entry describing the Hitler Youth in action against Jewish retail shops is more graphic. "We were about to enter a tiny jewelry shop when a gang of ten youngsters in Hitler Youth uniforms smashed the shop window and stormed into the shop, brandishing butcher knives and yelling, 'To hell with the Jewish rabble! Room for the Sudeten Germans!'" She continued: "The smallest boy of the mob climbed inside the window and started his work of destruction by flinging everything he could grab right into the streets. Inside, the other boys broke glass shelves and counters, hurling alarm clocks, cheap silverware, and trifles to accomplices outside. A tiny shrimp of a boy crouched in a corner of the window, putting dozens of rings on his fingers and stuffing his pockets with wristwatches and bracelets. His uniform bulging with loot, he turned around, spat squarely into the shopkeeper's face, and dashed off."[104] An SD internal report also briefly described the "Jewish action" (*Judenaktion*) in Berlin, indicating that it had started on June 10. According to the SD, all party organizations participated with the authorization of the city Gauleitung.[105]

The situation soon got out of hand, however, and as the American ambassador was sending his cable, an order was emanating from Berchtesgaden: The Führer wished the Berlin action to stop.[106] And so it did. Wide-scale anti-Jewish violence was not what Hitler needed as the international crisis over the fate of the Sudetenland was reaching its climax.

If Goebbels's diary faithfully reproduced the gist of the views Hitler expressed during their July 24 meeting, then he must have been considering several options: "We discuss the Jewish question. The Führer approves my action in Berlin. What the foreign press writes is unimportant. The main thing is that the Jews be pushed out. Within ten years they must be removed from Germany. But for the time being we still want to keep the Jews here as pawns. . . ."[107] Soon, however, the Sudeten crisis would be over and an unforeseen occurrence would offer the pretext for anti-Jewish violence on a yet unseen level. The Berlin events had merely been a small-scale rehearsal.

V

At the beginning of 1938, Werner Best, Heydrich's deputy as head of the Security Police Main Office, had signed an expulsion decree for approximately five hundred Jews of Soviet nationality living in the Reich.[108] This was a measure requested by the Wilhelmstrasse in retaliation for the expulsion of some German citizens from the Soviet Union. As these Soviet Jews were not granted entry permits into the USSR, the expulsion order was twice prolonged—without any result. On May 28, 1938, Heydrich ordered the incarceration of the male Soviet Jews in concentration camps until they could provide proof of immediately forthcoming emigration. In May expulsion orders were also issued to Romanian Jews living in Germany. All of this was but a prologue to the new expulsion drive that was to start in the fall.

During the months immediately following the Anschluss, however, there was a development that threatened to hamper these Nazi plans for rapid forced emigration: the measures taken by Switzerland. Almost all details of the policy followed by the Swiss Confederation with regard to Jewish refugees, before and during the war, were made available in a 1957 report that had been demanded by the Swiss Federal Assembly and was prepared by Federal Councillor Carl Ludwig.[109] And the 1994 publication of Swiss diplomatic documents of the prewar period has added the finishing touches to the picture.

Two weeks after the Anschluss, in its meeting of March 28, 1938, the Swiss Federal Council (the country's executive branch) decided that all bearers of Austrian passports would be obliged to obtain visas for entry into Switzerland. According to the meeting's minutes: "In view of the measures already taken and being prepared by other countries against the influx of Austrian refugees, we find ourselves in a difficult situation. It is clear that Switzerland can only be a transit country for the refugees from Germany and from Austria. Apart from the situation of our labor market, the present excessive degree of foreign presence imposes the strictest defense measures against a longer stay of such elements. If we do not want to create a basis for an anti-Semitic movement that would be unworthy of our country, we must defend ourselves with all our strength and, if need be, with ruthlessness against the immigration of foreign Jews, mostly those from the East. We have to think of the future and therefore we cannot allow ourselves to let in such foreigners for the sake of immediate advantages; such advantages would undoubtedly soon become the worst disadvantages."[110] This was to remain the basic position of the Swiss authorities during the coming seven years, with one additional point sometimes being added in the various internal memoranda: The Swiss Jews certainly did not want to see their own position threatened by the influx of foreign Jews into the country.

Once all Austrian passports were replaced by German ones, the visa requirement was applied to all bearers of German travel documents. The Swiss knew that their visa requirement would have to be reciprocal, that from then on Swiss citizens traveling to Germany would also have to obtain visas. On both sides the dilemma seemed insoluble. For Germany to avoid having visa requirements imposed on its Aryan nationals traveling to Switzerland would mean inserting some distinctive sign into the passports of Jews, which would automatically make their emigration far more difficult. Various technical solutions were considered throughout the summer of that year. At the end of September 1938, undeterred by the Sudeten crisis, a Swiss delegation headed by the chief of the Police Division at the Ministry of Justice, Heinrich Rothmund, traveled to Berlin for negotiations with Werner Best. According to their own report, the Swiss envoys described to their German colleagues the constant struggle of the federal police against the influx of foreign immigrants, particularly those who did not easily assimilate, primarily the Jews. As a result of the Swiss demands, the Germans finally agreed to stamp the passports of Jews with a "J," which would allow the Swiss police "to check at the border whether the carrier of the passport was Aryan or not Aryan" (these were the terms used in the Swiss report). On October 4 the Bern government confirmed the measures agreed upon by the German and the Swiss police delegates.

The Swiss authorities had not yet solved all their problems: Jews who had received an entrance permit before the stamping of their passports might attempt to make early use of it. On October 4, therefore, all border stations were informed that if "there was uncertainty whether a person traveling with a German passport was Aryan or non-Aryan, an attestation to his being Aryan should be produced. In doubtful cases, the traveler should be sent back to the Swiss consulate of his place of origin for further ascertainment."[111] But would all precautions thereby have been taken? The Swiss thought of one more possible way of cheating. A report from their Federal Center for Printed Matter, dated November 11, 1938, announced that, at Rothmund's request, they had tried to erase the "J" in a German passport acquired for a test. The report on the test was encouraging: "Effacing the 'J' stamped in red did not succeed entirely. One can recognize the remaining traces without difficulty."[112] While this was going on, the Jews of the Sudetenland had come under German control.

Austria had barely been annexed when Hitler turned to Czechoslovakia: Prague must allow the Sudetenland, its mainly German-populated province, to secede and join the German Reich. In May the Wehrmacht had received the order to invade Czechoslovakia on October 1. A general war appeared probable when, formally at least, the French declared their readiness to stand by their Czech ally. After a British mediation effort had come to naught, and after the failure of two meetings between British Prime Minister Neville Chamberlain and Hitler, European armies were mobilized. Then, two days before the scheduled German attack, Mussolini suggested a conference of the main powers involved in the crisis (but without the presence of the Czechs—and of the Soviet Union). On September 29 Britain, France, Germany, and Italy signed an agreement in Munich: By October 10 the Sudetenland was to become part of the German Reich. Peace had been saved; Czecho-Slovakia (the newly introduced hyphen came from a Slovak demand) had been abandoned; its new borders, though, were "guaranteed."

As soon as the Wehrmacht occupied the Sudetenland, Hitler informed Ribbentrop that, in addition to the expulsion of those Sudeten Jews who had not yet managed to flee into truncated Czecho-Slovakia, the expulsion of the 27,000 Czech Jews living in Austria should be considered. But the immediate expulsion measures mainly affected the Jews of the Sudetenland: The Germans sent them over the Czech border; the Czechs refused to take them in. Göring was to describe it with glee a month after the event: "During the night [following the entry of the German troops into the Sudetenland], the Jews were expelled to Czecho-Slovakia. In the morning, the Czechs got hold of them and sent them to Hungary. From Hungary back to Germany, then back to Czecho-Slovakia. Thus, they turned round and round. Finally, they ended up on a riverboat on the Danube. There they camped. As soon as they set foot on the river bank they were pushed back."[113] In fact several thousand of these Jews were finally forced, in freezing weather, into improvised camps of tents situated in the no-man's land between Hungary and Czecho-Slovakia, such as Mischdorf, some twenty kilometers from Bratislava.

In early October 1938 this now commonly used method was planned against some Viennese Jews. An SD memorandum of October 5 indicated that at a meeting of leading party representatives at the local Group "Goldegg," the head announced that, in accordance with instructions from the Gau, a stepped-up operation against the Jews was to take place through October 10, 1938: "Since many Jews have no passports, they will be sent over the Czech border to Prague without a passport. If the Jews have no cash money, they will be given RM40—by the Gau, for their departure. In this operation against the Jews, the impression is to be avoided that it is a Party matter; instead, spontaneous demonstrations by the people are to be caused. There could be use of force where Jews resist."[114]

Throughout the summer and autumn, Austrian Jews attempted to flee illegally to various neighboring countries and farther on, to England. The Gestapo had shipped some groups to Finland, to Lithuania, and to Holland or pushed them over the borders into Switzerland, Luxembourg, and France. Yet, as foreign protests grew, illegal entry or expulsion westward became increasingly difficult.[115] Thus, on September 20, the chief of the Karlsruhe Gestapo informed the regional authorities that Austrian Jews were arriving in great numbers in Baden, often without passports or money. "As the emigration of Austrian Jews has for the time being become practically impossible," the Gestapo chief went on, "due to corresponding defense measures taken by foreign countries, particularly by Switzerland, a prolonged stay of these Jews in Baden . . . can no longer be tolerated." The Gestapo did not suggest that the Jews be forcibly compelled to cross any of the western borders; the order had been given for "the immediate repatriation of the Jews to their former places of residence in Vienna."[116] Within days, however, it was the Jews of Polish nationality living in Germany who became the overriding issue.[117]

The census of June 1933 had indicated that among the 98,747 foreign Jews still residing in Germany, 56,480 were Polish citizens. The Polish Republic showed no inclination to add any newcomers to its Jewish population of 3.1 million, and various administrative measures aimed at hindering the return of Polish Jews living in Germany were utilized between 1933 and 1938. But, as it did in other countries, in Poland too the Anschluss triggered much sharper initiatives. On March 31, 1938, the Polish parliament passed a law establishing a wide array of conditions under which Polish citizenship could be taken away from any citizen living abroad.

The Germans immediately perceived the implications of the new law for their forcible emigration plans. German-Polish negotiations led nowhere, and, in October 1938, a further Polish decree announced the cancellation of the passports of residents abroad who did not obtain a special authorization for entry into Poland before the end of the month. As more than 40 percent of the Polish Jews living in the Reich had been born in Germany, they could hardly hope to liquidate their businesses and homes within less than two weeks. Most of them would therefore lose their Polish nationality on November 1. The Nazis decided to preempt the Polish measure.

Whether or not Hitler was consulted about the expulsion of the Polish Jews is unclear. The general instructions were given by the Wilhelmstrasse, and the Gestapo was asked to take over the actual implementation of the measure. Ribbentrop, Himmler, and Heydrich must have sensed, like everyone else, that given the international circumstances after the Munich agreement—the craving for peace and its consequence, appeasement—no one would lift a finger in defense of the hapless Jews. Poland itself was ultimately dependent on German goodwill; had it not just grabbed the Teschen region of northeastern Czecho-Slovakia in the wake of Germany's annexation of the Sudetenland? The timing of the expulsion could not have been more propitious. Thus, according to Himmler's orders, by October 29 all male Polish Jews residing in Germany were to be forcibly deported over the border to Poland.

The Reichsführer knew that the women and children, deprived of all support, would have to follow. On October 27 and 28, the police and the SS assembled and transported Jews to the vicinity of the Polish town of Zbaszyn, where they sent them over the river marking the border between the two countries. The Polish border guards dutifully sent them back. For days, in pouring rain and without food or shelter, the deportees wandered between the two lines; most of them ended up in a Polish concentration camp near Zbaszyn.[118] The rest were allowed to return to Germany.[119] (In early January Jews who were then in Poland were allowed to return temporarily to sell their homes and businesses.)[120] About 16,000 Polish Jews were thus expelled.[121]

The Grynszpans, a family from Hannover, were among the Jews transported to the border on October 27. Herschel, their seventeen-year-old son, was not with them; at the time he was living clandestinely in Paris, barely subsisting on odd jobs and on some help from relatives. It was to him that his sister Berta wrote on November 3: "We were not permitted to return to our home to get at least a few essential things. So I left with a 'Schupo' [*Schutzpolizei*, the German gendarmerie] accompanying me and I packed a valise with the most necessary clothes. That is all I could save. We don't have a cent. To be continued when next I write. Warm greetings and kisses from us all. Berta."[122]

Young Herschel Grynszpan did not know the details of what was happening to his family near Zbaszyn, but he could well imagine it. On November 7 he wrote a note to his Paris uncle: "With God's help [written in Hebrew] . . . I couldn't do otherwise. My heart bleeds when I think of our tragedy and that of the 12,000 Jews. I have to protest in a way that the whole world hears my protest, and this I intend to do. I beg your forgiveness. Hermann." Grynszpan purchased a pistol, went to the German Embassy, and asked to see an official. He was sent to the office of First Secretary Ernst vom Rath; there he shot and fatally wounded the German diplomat.[123]

CHAPTER 9

The Onslaught

I

"On the morning of November 10, 1938, at eight A.M., the farmer and local SA leader of Elberstadt, Adolf Heinrich Frey, accompanied by several of his cronies, set out for the house of the eighty-one-year-old Jewish widow Susannah Stern. According to Frey, the widow Stern took her time before opening the door, and when she saw him she smiled 'provocatively' and said: "Quite an important visit, this morning." Frey ordered her to dress and come with them. She sat down on her sofa and declared that she would not dress or leave her house; they could do with her whatever they wanted. Frey reported that the same exchange was repeated five or six times, and when she again said that they could do whatever they wanted, Frey took his pistol and shot Stern through the chest. "At the first shot, Stern collapsed on the sofa. She leaned backward and put her hands on her chest. I immediately fired the second shot, this time aiming at the head. Stern fell from the sofa and turned. She was lying close to the sofa, with her head turned to the left, toward the window. At that moment Stern still gave signs of life. From time to time she gave a rattle, then stopped. Stern did not shout or speak. My comrade C. D. turned Stern's head to see where she had been hit. I told him that I didn't see why we should be standing around; the right thing to do was to lock the door and surrender the keys. But to be sure that Stern was dead I shot her in the middle of the brow from a distance of approximately ten centimeters. Thereupon we locked the house and I called Kreisleiter Ullmer from the public telephone office in Elberstadt and reported what had happened." Proceedings against Frey were dismissed on October 10, 1940, as the result of a decision of the Ministry of Justice.[1]

In the course of the prewar anti-Jewish persecutions, the pogrom of November 9 and 10, the so-called Kristallnacht, was in many ways another major turning point. The publication in 1992 of Goebbels's hitherto missing diary accounts of the event added important insights about interactions among Hitler, his closest chieftains, the party organizations, and the wider reaches of society in the initiation and management of the anti-Jewish violence. As for the reactions of German and international opinion to the anti-Jewish violence, these raise a host of questions, not least for their relation to events yet to come.

The idea of a pogrom against the Jews of Germany was in the air. "The SD not only approved the controlled and purposeful use of violence, but explicitly recommended it in a memorandum of January 1937." [2] Early in February 1938 the Zionist leadership in Palestine received information from "a very reliable private source—one which can be traced back to the highest echelons of the SS leadership, that there is an intention to carry out a genuine and dramatic pogrom in Germany on a large scale in the near future."[3] In fact, the anti-Jewish violence of the early summer of 1938 had not entirely died down; A synagogue had been set on fire in Munich on June 9, and another in Nuremberg on August 10.[4] For the American ambassador, the anti-Jewish incidents of the early summer of 1938 indicated, as had been the case in 1935, some forthcoming radical anti-Jewish legislation.[5] Finally, shortly before the pogrom, during an inspection journey to Vienna at the end of October 1938, Hagen discussed the "Jewish situation in Slovakia" with his Vienna colleague SS-Obersturmführer Polte. Hagen instructed Polte to indicate to the representatives of the Slovak government that "this problem had to be solved, and that it seemed advisable to stage an action of the people against the Jews."[6]

By then Hitler's hesitations of June 1938 had disappeared. His totally uncompromising position on Jewish matters found another expression in early November. On November 4, in a letter addressed to Frick, Lammers indicated that due to repeated requests for exemption from diverse anti-Jewish measures (such as additional first names, identification cards, and so on), he himself had raised the fundamental aspect of the issue with Hitler. "The Führer is of the opinion," wrote Lammers, "that exemptions from the special regulations valid for the Jews have to be rejected without any exception. Nor does the Führer himself intend to grant any such marks of personal favor."[7]

On November 8 the *Völkischer Beobachter* published a threatening editorial against the Jews, closing with the warning that the shots fired in Paris would herald a new German attitude regarding the Jewish question.[8] In some places local anti-Jewish violence had started even before the Nazi press brandished its first threats. An SD report of November 9 described events that had taken place in the Kassel and Rotenburg/Fulda districts during the night of November 7–8, presumably as an immediate reaction to the news. In some places Jewish house and shop windows had been smashed. In Bebra a number of Jewish apartments had been "demolished," and in Rotenburg the synagogue's furniture was "significantly damaged" and "objects [were] taken away and destroyed on the street."[9]

One of the most telling aspects of the events of November 7–8 was Hitler's and Goebbels's public and even "private" silence (at least as far as Goebbels's diaries are concerned). In his November 9 diary entry (relating events of November 8), Goebbels did not devote a single word to the shots fired in Paris, although he had spent the late evening in discussion with Hitler.[10] Clearly, both had agreed to act, but had probably decided to wait for the seriously wounded Rath's death. Their unusual silence was the surest indication of plans that aimed at a "spontaneous outburst of popular anger," which was to take place without any sign of Hitler's involvement. And, on that same evening of November 8, in his speech commemorating the 1923 putsch attempt, Hitler refrained from any allusion whatsoever to the Paris event.

Rath died on November 9, at 5:30 in the afternoon. The news of the German diplomat's death was officially brought to Hitler during the traditional "Old Fighters" dinner held at the Altes Rathaus in Munich, at around nine o'clock that evening. An "intense conversation" then took place between Hitler and Goebbels, who was seated next to him. Hitler left the assembly immediately thereafter, without giving the usual address. Goebbels spoke instead. After announcing Rath's death, he added, alluding to the anti-Jewish violence that had already taken place in Magdeburg-Anhalt and Kurhessen, that "the Führer had decided that such demonstrations should not be prepared or organized by the party, but insofar as they erupted spontaneously, they were not to be hampered." As later noted by the chief party judge Walter Buch, the message was clear.[11]

For Goebbels there had been no such occasion to display his leadership talents in action since the boycott of April 1933. The propaganda minister, moreover, badly wanted to prove himself in the eyes of his master. Hitler had been critical of the ineffectiveness, in Germany itself, of the propaganda campaign during the Sudeten crisis.[12] Besides, Goebbels was in partial disgrace as the result of his affair with the Czech actress Lida Baarova, and his intention to divorce his wife, Magda, one of Hitler's very special protégées. Hitler had put an end to the romance and the divorce, but his minister was still in need of some major initiative. Now he held it in his hands.

"I report the matter to the Führer," wrote Goebbels on the tenth, alluding to the conversation at the dinner the evening before. "He [Hitler] decides: demonstrations should be allowed to continue. The police should be withdrawn. For once the Jews should get the feel of popular anger. That is right. I immediately give the necessary instructions to the police and the Party. Then I briefly speak in that vein to the Party leadership. Stormy applause. All are instantly at the phones. Now the people will act."

Goebbels then described the destruction of synagogues in Munich. He gave orders to make sure that the main synagogue in Berlin, on Fasanenstrasse, be destroyed. He continued: "I want to get back to the hotel and I see a blood-red [glare] in the sky. The synagogue burns. . . . We extinguish only insofar as is necessary for the neighboring buildings. Otherwise, should burn down. . . . From all over the Reich information is now flowing in: 50, then 70 synagogues are burning. The Führer has ordered that 20–30,000 Jews should immediately be arrested. . . . In Berlin, 5, then 15 synagogues burn down. Now popular anger rages. . . . It should be given free rein."

Goebbels went on: "As I am driven to the hotel, windowpanes shatter [they are being smashed]. Bravo! Bravo! The synagogues burn like big old cabins. German property is not endangered. At the moment nothing special remains to be done."[13] The main Munich synagogue, on Herzog-Max Strasse, was not among those Goebbels saw burning. Its demolition had been started a few months before, on Hitler's explicit orders.[14]

At approximately the same time as the propaganda minister was gleefully contemplating a good day's work, Hitler gave his instructions to Himmler and informed him that Goebbels was in overall charge of the operation. On that same night Himmler summed up his immediate reaction in writing: "I suppose that it is Goebbels's megalomania—something I have long been aware of—and his stupidity which are responsible for starting this operation now, in a particularly difficult diplomatic situation."[15]

The Reichsführer was certainly not opposed to the staging of a pogrom; what must have stung Himmler was the fact that Goebbels had been the first to exploit the shots fired at Rath to organize the action and obtain Hitler's blessing. But he may indeed also have thought that the timing was not opportune.

The propaganda chief concluded his November 10 diary entry by alluding to some of the events of that morning: "Throughout the morning, a shower of new reports. I consider with the Führer what measures should be taken now. Should one let the beatings continue or should they be stopped? That is now the question."[16]

Still in Munich on the eleventh, Goebbels kept writing about the previous day: "Yesterday: Berlin. There, all proceeded fantastically. One fire after another. It is good that way. I prepare an order to put an end to the actions. It is just enough by now. . . . Danger that the mob may appear on the scene. In the whole country the synagogues have burned down. I report to the Führer at the Osteria [a Munich restaurant; Hitler later left for the Obersalzberg]. He agrees with everything. His views are totally radical and aggressive. The action itself took place without the least hitch. 100 dead. But no German property damaged."

What follows shows that some of the most notorious orders given by Göring during the conference that was about to take place on November 12 were decisions made by Hitler on the tenth: "With small changes, Hitler agrees to my decree concerning the end of the actions," wrote Goebbels, and he added: "The Führer wants to take very sharp measures against the Jews. They must themselves put their businesses in order again. The insurance companies will not pay them a thing. Then the Führer wants a gradual expropriation of Jewish businesses. . . . I give appropriate secret orders. We now await the foreign reactions. For the time being, they are silent. But the uproar will come. . . ."[17]

"Information arrives from Berlin about enormous anti-Semitic riots. Now the people move ahead. But now one has to stop. I give the requisite instructions to the police and the Party. Then everything will be calm."[18]

The pogrom was much less coordinated than Goebbels claimed. According to one reconstruction of the sequence of events, after Goebbels's initial order "the Gauleiters started around 10:30 P.M. They were followed by the SA at 11:00, by the police shortly before midnight, by the SS at 1:20 in the morning, and again by Goebbels at 1:40."[19] Heydrich's orders to the Gestapo and the SD were precise. No measures endangering German life or property were to be taken, in particular when synagogues were burned down; Jewish businesses or apartments could be destroyed but not looted (looters would be arrested); foreigners (even when identified as Jews) were not to be molested; synagogue archives were to be seized and transferred to the SD. Finally, "inasmuch as in the course of the events of this night the employment of officials used for this purpose would be possible, in all districts as many Jews, especially rich ones, are to be arrested as can be accommodated in the existing jails. For the time being only healthy men not too old should be arrested. Upon their arrest, the appropriate concentration camps should be contacted immediately, in order to confine them in these camps as fast as possible. Special care should be taken that the Jews arrested in accordance with these instructions are not mistreated."[20]

The November 10 telephone report from SA Brigade 151 in Saarbrücken was concise and to the point: "During the past night, the synagogue in Saarbrücken was set on fire; the synagogues in Dillingen, Merzig, Saarlautern, Saarwillingen, and Broddorf were also destroyed. The Jews were taken into custody. The fire brigade is engaged in extinguishing the fires. In the area of Brigade 174, all synagogues were destroyed."[21]

On November 17 Hitler attended Rath's funeral in Düsseldorf.

Only a few hundred Jews lived in the Gau Tyrol-Vorarlberg. Like all other Jews of the Austrian province, they had to leave the country by mid-December or move to Vienna. In October, Eichmann had arrived in Tyrol's main city, Innsbruck, and issued a personal warning to the community's three leading Jews: Karl Bauer and Alfred and Richard Graubart. Gauleiter Franz Hofer and the local SD office meant to fulfill Himmler's orders and have the Gau *judenrein* within weeks. The night of November 9 to 10 offered an unexpected opportunity. Hofer rushed back from the Alte Kämpfer dinner in Munich and set the tone: "In response to the cowardly Jewish assassination of our embassy counsellor vom Rath in Paris, in the Tyrol too the seething soul of the people should, this night, rise against the Jews."[22]

The SS had been put on alert by Heydrich's message. After the midnight swearing-in ceremony of the new SS recruits which on that same night had taken place in Innsbruck as in all other major cities of the Reich, the men reassembled in civilian clothes around 2:30 in the morning, under the command of SS-Oberführer Hanns von Feil. Within minutes a special SS murder commando, divided into three groups, was on its way to No. 4–5 Gänsbacherstrasse, where some of the more prominent Jewish families of Innsbruck still lived. According to SS-Obersturmführer Alois Schintlholzer, he "received instructions at the Hochhaus in Innsbruck from regional leader Feil to kill the Jews on Gänsbacherstrasse silently."

At No. 4 Gänsbacherstrasse the engineer Richard Graubart was stabbed to death in the presence of his wife and daughter. On the second floor of the same building, Karl Bauer was dragged into the hall, stabbed, and beaten with pistol butts; he died on the way to the hospital. On Anichstrasse, the turn of Richard Berger, the president of the Jewish community in Innsbruck, came approximately at the same time. Berger was taken out in pajamas and winter coat and pushed into an SS car that was supposedly taking him to Gestapo headquarters. But the car started off in a different direction. According to SS-Untersturmführer Walter Hopfgartner: "We drove west through Anichstrasse, over the university bridge, in the direction of Kranebitten. During the trip, Berger asked where we were going, since this was not the way to the Gestapo. Berger, who, understandably, was somewhat nervous, was calmed down by the men in the back of the car. . . . Suddenly Lausegger announced, in a voice sufficiently loud so that all could hear him, that 'no firearms are to be used.' This upset Berger again and he asked what we wanted from him, but he was quieted down again. . . . After Lausegger's statement, I realized immediately that Berger was to be killed."

At a bend of the Inn River, Berger was dragged out of the car, battered with pistols and stones, and thrown into the river. Against instructions he was shot at, but the subsequent Gestapo investigation established that by then he was already dead.

All the SS men involved in the Innsbruck murders were old-timers fanatically devoted to Hitler, extreme anti-Semites and exemplary members of the order. Gerhard Lausegger, the man in charge of the squad that killed Berger, had been a member of a student corporation and had "headed the federation of all dueling companies at the University of Innsbruck." On March 11 he had been among the men who, just before the arrival of the Wehrmacht, seized the provincial administration hall of the city.

Heydrich's report of November 11 indicated that thirty-six Jews had been killed and the same number seriously injured throughout the Reich.

"One Jew is still missing, and among the dead there is one Jew of Polish nationality and two others among those injured."[23] The real situation was worse. Apart from the 267 synagogues destroyed and the 7,500 businesses vandalized, some ninety-one Jews had been killed all over Germany and hundreds more had committed suicide or died as a result of mistreatment in the camps.[24] "The action against the Jews was terminated quickly and without any particular tensions," the mayor of Ingolstadt wrote in his monthly report on December 1. "As a result of this measure a local Jewish couple drowned themselves in the Danube."[25]

For the Würzburg Gestapo nothing was self-evident. In an order issued on December 6 to the heads of the twenty-two administrative districts of Gau Main-Franken (Franconia) as well as to the mayors of Aschaffenburg, Schweinfurt, Bad Kissingen, and Kitzingen, the secret police demanded immediate details about Jews who had committed suicide "in connection with the action against the Jews"; question no. 3 required information about "the presumed motive."[26]

In a secret letter addressed on November 19 to the Hamburg Prosecutor General about the events of November 9–11, the Ministry of Justice stated that the destruction of synagogues and Jewish cemeteries, as well as of Jewish shops and dwellings, if not committed for purposes of looting, were *not* to be prosecuted. The murder of Jews and the infliction of serious bodily damage were to be prosecuted "only if committed for selfish reasons."[27]

The decisions of the courts and the various administrative decrees regarding the (lack of) culpability of the murderers were given their adequate "conceptual framework" in the report prepared by the Supreme Court of the NSDAP of February 13, 1939. The report stated that on November 10, at 2 A.M., Goebbels had been informed of the first killing, that of a Polish Jew. He was told that something had to be done to stop what could become a dangerous development. According to the report, Goebbels's answer was in terms of "not getting upset because of a dead Jew." The report then adds the following comment: "At this point in time, most of the killings could still have been hindered by an additional order. As it was not given, this very fact as well as the remarks expressed [by Goebbels] lead to the conclusion that the final result was intended or at least taken into account as possible and desirable. This being the case, the individual perpetrator has put into action not merely what he assumed to be the intention of the leadership, but what he rightly recognized as such, even though it was not clearly stated. For this he cannot be punished."[28]

Was the Nazi action perceived by its perpetrators as a step that could hasten the emigration of the Jews from the Reich or possibly as an initiative aimed at furthering some other, more encompassing policy? After the pogrom Göring, on Hitler's orders, would make the most of the Paris shooting. But despite prior SD plans about the use of violence, nothing systematic seems to have been considered before the unleashing of the action of November 9. At that moment total, abysmal hatred appears as the be-all and end-all of the onslaught. The only immediate aim was to hurt the Jews as badly as the circumstances allowed, by all possible means: to hurt them and to humiliate them. The pogrom and the initiatives that immediately followed have quite rightly been called "a degradation ritual."[29] An explosion of sadism threw a particularly lurid light on the entire action and its sequels; it burst forth at all levels, that of the highest leadership and that of the lowliest party members. The tone of Goebbels's diary entries was unmistakable; the same tone would suffuse the November 12 conference.

An uncontrollable lust for destruction and humiliation of the victims drove the squads roaming the cities. "Organized parties moved through Cologne from one Jewish apartment to another," the Swiss consul reported. "The families were either ordered to leave the apartment or they had to stand in a corner of a room while the contents were hurled from the windows. Gramophones, sewing machines, and typewriters tumbled down into the streets. One of my colleagues even saw a piano being thrown out of a second-floor window. Even today [November 13] one can still see bedding hanging from trees and bushes.[30] Even worse was reported from Leipzig: "Having demolished dwellings and hurled most of the movable effects to the streets," the American consul in Leipzig reported, "the insatiably sadistic perpetrators threw away many of the trembling inmates into a small stream that flows through the Zoological Park, commanding the horrified spectators to spit at them, defile them with mud and jeer at their plight. . . . The slightest manifestation of sympathy evoked a positive fury on the part of the perpetrators, and the crowd was powerless to do anything but turn horror-stricken eyes from the scene of abuse, or leave the vicinity. These tactics were carried out the entire morning of November 10 without police intervention and they were applied to men, women and children."

The same scenes were repeated in the smallest towns: the sadistic brutality of the perpetrators, the shamefaced reactions of some of the onlookers, the grins of others, the silence of the immense majority, the helplessness of the victims. In Wittlich, a small town in the Moselle Valley in the western part of Germany, as in most places, the synagogue was destroyed first: "The intricate lead crystal window above the door crashed into the street and pieces of furniture came flying through doors and windows. A shouting SA man climbed to the roof, waving the rolls of the Torah: 'Wipe your asses with it, Jews,' he screamed while he hurled them like bands of confetti on Karnival." Jewish businesses were vandalized, Jewish men beaten up and taken away: "Herr Marks, who owned the butcher shop down the street, was one of the half dozen Jewish men already on the truck. . . . The SA men were laughing at Frau Marks who stood in front of her smashed plate-glass window [with] both hands raised in bewildered despair. 'Why are you people doing this to us?' She wailed at the circle of silent faces in the windows, her lifelong neighbors. 'What have we ever done to you?'"[31]

Soon the Jewish masses of occupied Poland would offer the choicest targets to the unquenchable rage that, stage by stage, propelled the Greater German Reich against the hapless European Jews.[32]

Once again Hitler had followed the by-now familiar pattern he had displayed throughout the 1930s. Secretly he gave the orders or confirmed them; openly his name was in no way to be linked with the brutality. Having refrained from any open remark about the events on November 7–8, Hitler also avoided any reference to them in his midnight address to SS recruits in front of the Feldherrnhalle on November 9. At the time of his address, synagogues were already burning, shops being demolished, and Jews wounded and killed throughout the Reich. A day later, in his secret speech to representatives of the German press, Hitler maintained the same rule of silence regarding events that could not but be on the mind of every member of the audience;[33] he did not even speak at Rath's funeral. The fiction of a spontaneous outburst of popular anger imposed silence. Any expression of Hitler's wish or even any positive comment would have been a "Führer order." Of Hitler's involvement the outside world—including trustworthy party members—was, at least in principle, to know nothing.

However, knowledge of Hitler's direct responsibility quickly trickled out from the innermost circle. According to the diaries of Ulrich von

Hassell, the former German ambassador to Rome and an early opponent of the regime, many conservatives were outraged by the events, and the minister of finance of Prussia, Johannes Popitz, protested to Göring and demanded the punishment of those responsible for the action. "My dear Popitz, do you want to punish the Führer?" was Göring's answer.[34]

At the low end of the party hierarchy some justifications were rapidly concocted. On November 23 a Hüttenbach Blockleiter (block leader), who was also the chronicler of party history in his town, was ordered by his district party leader to collect incriminating evidence against the local Jews. Only two days later he had completed his research and could report that the task had been accomplished: "Herewith," the Blockleiter wrote, "I am sending some material about the Jews in Hüttenbach. I don't know whether I hit upon the right things. I couldn't do it more quickly, if what was wanted was an overview of these racial foreigners and about how they behaved in Hüttenbach." At that point the Blockleiter, with engaging openness, voiced some doubts about his own qualifications as a full-fledged historian: "I may have more material here, but I cannot become a historian along with my professional work and in any case the necessary documentation is missing."[35]

Incidentally, the same local historian had not yet exhausted his efforts, or his worries, regarding the events of November 9 and 10. On February 7, 1939, he announced to his district party leader that he had completed the chronicle for the year 1938. The November events he memorialized as follows: "During the night of November 9, 1938, Party member v. Rath died in Paris as a result of the cowardly aggression perpetrated by the Jew Grünspan. During the same night all the Jews' synagogues went up in flames all over Germany; Party member Ernst v. Rath was avenged. Early, at 5 in the morning, District Party Leader Party Comrade Waltz and Mayor Party Comrade Herzog, District Propaganda Leader Party Comrade Büttner and Sturmführer Brand, arrived and set the Jewish temple on fire. Party members from the local section gave energetic help. But this sentence was criticized by a few Party members: it should not be written that Party members Walz, Herzog, and Büttner/Brand set the synagogue on fire, but the people did it. That's right. But as the writer of a chronicle I should and I must report the truth. It would easily be possible to take this page out and to prepare another entry. I ask you, my District Party Leader, how should I prepare this entry and how should it be worded? Heil Hitler."[36]

II

On the morning of November 12, Goebbels summed up the events of the previous days in the *Völkischer Beobachter*: "The Jew Grynszpan," so the last paragraph ran, "was the representative of Jewry. The German vom Rath was the representative of the German people. Thus in Paris Jewry has fired on the German people. The German government will answer legally but harshly."[37] The German government's legal answers were hurled at the Jews throughout the remaining weeks of 1938; they were accompanied by three major policy interpretations: the first on November 12, at the top-echelon conference convened by Göring; the second on December 6, in Göring's address to the Gauleiters; the third on December 28, in a set of new rules also announced by Göring. All of Göring's initiatives and interpretations were issued on Hitler's explicit instructions.

It has often been assumed that Göring had lost much of his influence since the summer of 1938 as a result of his relatively moderate attitude in the Sudeten crisis.[38] The new star among Hitler's underlings was Ribbentrop, the presumptuous new foreign minister, who was convinced that the British would continue to back down from crisis to crisis. It may well be, however, that within the new scheme of things following Munich, Göring's role as coordinator of Jewish matters had become essential for Hitler's plans. Göring was to orchestrate all measures that would make Jewish life in Germany untenable and accelerate Jewish emigration. The constant threat of further anti-Jewish violence and the need to find places of refuge for the fleeing Jews would mobilize "world Jewry" and convince it to tone down its anti-German incitement; this in turn would persuade the Western governments that compromise solutions to Hitler's new demands were a necessity. In other words, Hitler may well have thought that anti-Jewish pressure in Germany would ensure the success of Nazi foreign aggressions, due to what he believed to be the influence of world Jewry on the policies of the Western democracies. And Hitler's forthcoming declarations about settling Jews in some Western colonies, and his public threats to exterminate them in case of war, also indicate that his belief in the influence of world Jewry in Paris, London, and Washington was an essential part of his worldview.

The conference of high-ranking officials that Göring convened on November 12 at the Air Transport Ministry has become notorious. "Gentlemen," Göring began, "today's meeting is of decisive importance. I received a letter that Bormann, the Führer's Deputy's chief of staff, wrote

to me on instruction from the Führer, according to which the Jewish question should now be dealt with in a centralized way and settled in one form
or another. In a telephone call which I received from the Führer yesterday,
I was once again instructed to centralize the decisive steps to be taken
now."[39]

The concrete discussions that took place on November 12 at Göring's
headquarters dealt not only with various additional ways of harassing the
Jews and further economic steps to be taken against the Jews but also, and
at length, with the immediate problem of insurance compensation for the
damages inflicted on Jewish property during the pogrom. A representative
of the German insurance companies, Eduard Hilgard, was called in. The
windowpanes alone destroyed in Jewish shops were insured for about six
million dollars, and because the glass was Belgian, at least half of this
amount would have to be paid in foreign currency. That prompted an aside
by Göring to Heydrich: "I wish you had killed two hundred Jews and not
destroyed such property." Heydrich: "Thirty-five were killed."[40] Göring
issued the orders secretly given by Hitler two days before: The Jews would
bear all the costs of repairing their businesses; the Reich would confiscate
all payments made by German insurance companies. "The Jews of German
citizenship will have to pay as a whole a contribution of 1,000,000,000 RM
to the German Reich."[41]

On the same day Göring ordered the cessation of all Jewish business
activity as of January 1, 1939. The Jews had "to sell their enterprises, as well
as any land, stocks, jewels, and art works. They could use the services of
'trustees' to complete these transactions within the time limit. Registration
and deposit of all shares was compulsory."[42] Göring's main policy statement, again delivered after consultations with Hitler, was yet to come, in a
meeting with the Gauleiters on December 6. But more than for its major
executive decisions, the November 12 conference remains significant for its
sadistic inventiveness and for the spirit and tone of the exchanges.

Still carried away by the flurry of his activities the days before, the propaganda minister had a whole list of proposals: The Jews should be compelled to demolish the damaged synagogues at their own expense; they
should be forbidden public entertainments ("I am of the opinion that it is
impossible to have Jews seated next to Germans at variety shows, cinemas,
or theaters; one could eventually envisage later that here in Berlin one or
two movie houses be put at the disposal of the Jews in which they could
present Jewish films").[43] At that point a notorious debate arose between

Goebbels and Göring on how to segregate Jews on trains. Both agreed on the necessity of separate compartments for Jews but, Goebbels declared, there should be a law forbidding them to claim a seat even in a Jewish compartment before all Germans had secured one. The mere existence of a separate compartment would have the undesirable effect of allowing some Jews to sit at their ease in an overcrowded train. Göring had no patience for such formalities: "Should a case such as you mention arise and the train be overcrowded, believe me, we won't need a law. We will kick him [the Jew] out and he will have to sit all alone in the toilet all the way!" Goebbels insisted on a law, to no avail.[44]

This minor setback did not paralyze Goebbels's brainstorming: the Jews, he demanded, should absolutely be forbidden to stay in German resorts. The propaganda minister also wondered whether German forests should not be made out of bounds for them ("Nowadays, packs of Jews run around in Grunewald; it is a constant provocation, we constantly have incidents. What the Jews do is so annoying and provoking that there are brawls all the time"). This gave Göring an idea of his own: Some sections of the forests should be open to Jews, and animals that resembled Jews—"the elk has a crooked nose like theirs"—should be gathered in those sections. Goebbels continued; he demanded that parks should also be forbidden to Jews, as Jewish women, for instance, might sit down with German mothers and engage in hostile propaganda ("There are Jews who do not look so very Jewish"). There should also be separate benches for Jews, with special signs: FOR JEWS ONLY! Finally, Jewish children should be excluded from German schools ("I consider it as out of the question that my son be seated next to a Jew in a German school and [that the Jew] be given a German history lesson.").[45]

At the end of the debate on the economic issues, Göring made it clear that the decisions taken would have to be "underpinned by a number of police, propaganda, and cultural measures, so that everything should happen right away and that, this week, slam-bang, the Jews should have their ears slapped, one slap after the other."[46]

It was Heydrich who reminded those present that the main problem was to get the Jews out of Germany. The idea of setting up a central emigration agency in Berlin on the Viennese model was broached (Eichmann had been specially summoned from Vienna for the occasion). But in Heydrich's opinion at the current rate it would take some eight to ten years to achieve a solution of the problem, which, it will be remembered, was also

Hitler's assessment in the meeting with Goebbels on July 24. How, then, should the Jews be isolated in the meantime from the German population without their losing all possibility of a livelihood? Heydrich was in favor of a special badge to be worn by all those defined as Jews by the Nuremberg Laws ("A uniform!" Göring exclaimed. Heydrich repeated: "A badge.") Göring was skeptical: He himself was in favor of establishing ghettos on a large scale in the major cities. For Heydrich ghettos would become "hiding places for criminal activities" uncontrollable by the police, whereas a badge would allow surveillance by "the vigilant eye of the population." The debate on the introduction of a badge or the creation of ghettos went on, concentrating on the ways the Jews would pursue their daily life ("You can't let them starve!" Göring argued).[47] The difference of opinion remained unresolved, and, three weeks later, Hitler was to reject both badges and ghettos.

Like Goebbels earlier, Heydrich had more suggestions on his list: no driver's licenses, no car ownership ("the Jews could endanger German life"), no access to areas of national significance in the various cities, no access to cultural institutions—along the lines of Goebbels's suggestion—none to resorts and not even to hospitals ("a Jew cannot lie in a hospital together with an Aryan Volksgenosse"). When the discussion moved to what the Jews could do to counter the financial measures about to be taken against them, Göring was sure that they would do nothing whatsoever. Goebbels concurred: "At the moment, the Jew is small and ugly and he will remain at home."[48]

Shortly before the last exchange, Göring commented, as if an afterthought: "I would not like to be a Jew in Germany." The General-feldmarschall then mentioned that on November 9 Hitler had told him of his intention to turn to the democracies that were raising the Jewish issue and to challenge them to take the Jews; the Madagascar possibility would also be brought up, as well as that of "some other territory in North America, in Canada or anywhere else the rich Jews could buy for their brethren." Göring added: "If in some foreseeable future an external conflict were to happen, it is obvious that we in Germany would also think first and foremost of carrying out a big settling of accounts with the Jews."[49]

On the same day that Goebbels forbade Jews access to cultural institutions, he also banned the Jewish press in Germany. Shortly afterward, Erich Liepmann, director of the *Jüdische Rundschau*, which by then had been closed down, was summoned to the propaganda minister's office: "'Is the Jew here?' Goebbels yelled by way of greeting," Liepmann recalled. "He was sitting at his desk; I had to stand some eight meters away. He yelled: 'An informational paper must be published within two days. Each issue will be submitted to me. Woe to you if even one article is published without my having seen it. That's it!'"[50] Thus the *Jüdisches Nachrichtenblatt* was born: It was designed to inform the Jews of all the official measures taken to seal their fate.

But sometimes, it seems, even Goebbels's eye wasn't sharp enough. In early December, some six weeks after Kristallnacht, the *Nachrichtenblatt* reviewed the American film *Chicago*: "A city goes up in flames and the fire-fighters stand by without taking any action. All the hoses are poised, the ladders have been prepared . . . but no hand moves to use them. The men wait for the command, but no command is heard. Only when the city has burned down and is lying in cinders and ashes, an order arrives; but the firefighters are already driving away. A malicious invention? An ugly tale? No. The truth. And it was revealed in Hollywood."[51]

The law of November 12 compelling the Jews to sell all their enterprises and valuables, such as jewels and works of art, inaugurated the wholesale confiscation of art objects belonging to them. The robbery that had already taken place in Austria now became common practice in the Reich. In Munich, for example, the procedure was coordinated by Gauleiter Wagner himself who, in the presence of the directors of state collections, gave the orders for "the safekeeping of works of art belonging to Jews." This "safe-keeping" was implemented by the Gestapo: An inventory was duly taken in the presence of the owners (or their "delegates") and a receipt issued to them. One of these documents reads: "25 November 1938. Protocol, recorded in the residence of the Jew Albert Eichengrün, Pilotystrasse 11/1, presently in protective custody. The housekeeper, Maria Hertlein, b. 21/10/1885, in Wipolteried, B.V., Kempten, was present. Dr. Kreisel, Director, Residenzmuseum, and criminal investigators Huber and Planer officiated."[52]

On November 15 all Jewish children still remaining in German schools were expelled.[53] In a letter the same day addressed to all state and party agencies, Secretary of State Zschintsch explained the minister of education's decision. "After the heinous murder in Paris one cannot demand of any German teacher to continue to teach Jewish children. It is also self-evident that it is unbearable for German schoolchildren to sit in the same classroom with Jewish children. Racial separation in schooling has already been accomplished in general over the last few years, but a remnant of Jewish children has stayed in German schools, for whom school attendance together with German boys and girls cannot be permitted anymore. . . . I therefore order, effective immediately: Attendance at German schools is no longer permitted to Jews. They are allowed to attend only Jewish schools. Insofar as this has not yet happened, all Jewish schoolchildren who at this time are still attending a German school must be dismissed."[54]

On November 19 Jews were excluded from the general welfare system. On November 28 the minister of the interior informed all the federal state presidents that some areas could be forbidden to Jews and that their right of access to public places could also be limited to a few hours a day.[55] It did not take long for the Berlin police chief to move ahead. On December 6 the city's Jews were banned from all theaters, cinemas, cabarets, concert and conference halls, museums, fairs, exhibition halls, and sports facilities (including ice-skating rinks), as well as from public and private bathing facilities. Moreover Jews were banned from the city districts where most government offices and major monuments and cultural institutions were located: "the Wilhelmstrasse from the Leipzigerstrasse to Unter den Linden, including the Wilhelmsplatz, the Vossstrasse from the Hermann Göring-Strasse to the Wilhelmstrasse, the Reich Commemorative Monument including the northern pedestrian way on Unter den Linden from the University to the Arsenal." The announcement indicated that in the near future the banning of Jews would probably be applied to "a great number of Berlin streets."[56]

On December 3, on Himmler's orders, the Jews were deprived of their driver's licenses. The access of Jewish scholars who possessed a special authorization to university libraries was cancelled on December 8. On December 20 Jews were no longer allowed to train as pharmacists, and a day later they were excluded from midwifery.[57] On the twenty-eighth, besides further measures of segregation (access was prohibited that day to dining and sleeping cars on trains, and also to public swimming pools and hotels that usually catered to party members), the first indications of a potential physical concentration of the Jews (to be discussed further on) appeared.[58] On November 29 the minister of the interior forbade Jews to keep carrier pigeons.[59]

In the meantime, following Heydrich's order of November 9, the

Gestapo started to impound all Jewish communal archives. The police and SA did the preliminary work, even in the smallest towns. In Memmingen the criminal police arrested the local Jewish religion teacher who also dealt with all of the community's official correspondence. He was forced to lead the inspectors to the archives, which were kept in "three old closets" in the synagogue and in "a wooden case" in the attic of his house. The closets and the case were locked and sealed, the key of the synagogue deposited at the police station.[60] In large cities the procedure was basically the same. According to a report by the state archives director in Frankfurt, he was ordered by the mayor on November 10 to take over all Jewish communal archives. When he arrived at the Fahrgasse synagogue, he found "broken windowpanes, unhinged gates, paintings cut to pieces, smashed exhibit cases, files and books scattered all over the floors, and so on." On the twelfth, a small fraction of the files were removed to the state archives for examination by the Gestapo. On the fifteenth, two Gestapo officials started cataloging, with the historical material destined to be added to the ever-growing collection of plundered Judaica being assembled in Frankfurt for the new Research Institute on the Jewish Question. In passing, the state archives director mentioned that among the files there was a complete list of the (some 23,000) Jews living in Frankfurt;[61] for the Gestapo such a list must have been of particular interest.

Göring's main policy statement was delivered on December 6, following the instructions given him by Hitler on December 4. This time he was addressing the inner core of the party, the Gauleiters, and although the speech, in the usual Göring style, was relaxed in tone, there could have been no doubt in the minds of the audience that he was conveying clear orders backed by Hitler's authority. These were to be followed strictly to the letter. As Göring put it (regarding Hitler's decision that the Jews would not be marked by any special sign), "Here, gentlemen, the Führer has forbidden it, he has expressed his wish, he has given the order, and I think that this should entirely suffice for even the lowliest employee not to get the idea that the Führer actually wishes it but maybe he wishes even more that I do the opposite. In terms of the Führer's authority, it is clear that there is nothing to change and nothing to interpret."[62]

What is striking in Göring's address is his constant reference to the fact that these were Hitler's orders, that all the steps mentioned had been discussed with Hitler and had his complete backing. The most likely reason for this repeated emphasis was that some of the measures announced would not be popular with the assembly, since they would put an end to the profits party members of all ranks, including some Gauleiters, had derived from their seizure of Jewish assets. It seems that this was why Göring repeatedly linked the Jewish issue to the general economic needs of the Reich. Party members were to be fully aware that any transgression of the new orders was harmful to the Reich's economy and an outright violation of the Führer's orders. In concrete terms, after stressing the fact that the party and the Gaue had taken Jewish assets, Göring made it clear that, on Hitler's orders, such unlawfully acquired property would have to be transferred to the state: "The Party should not engage in any business. . . . A Gau leadership cannot set up an Aryanization office. The Gau leadership has no authority to do this, it is not allowed to do it. . . . The Führer has issued the following guidelines: obviously, Aryanization has to take place locally, because the state itself cannot do it . . . but the benefits from all the Aryanization measures belong exclusively and solely to the Reich, i.e., to its authorized representative, the Reich Finance Minister, and to no one else in the whole Reich; it is only thus that the Führer's rearmament program can be accomplished."[63] Previously Göring had made it clear that deals already made by party members in order to enrich themselves were to be cancelled. It was not the fate of the Jews that mattered, Göring added, but the reputation of the party inside and outside Germany.[64] The other internal party issue dealt with at some length was that of punishment for deeds committed on November 9 and 10: Whatever was undertaken on purely ideological grounds, out of a justified "hatred for the Jews," should go unpunished; purely criminal acts of various kinds were to be prosecuted as they would be prosecuted under any other circumstances, but all publicity liable to cause scandal was to be strictly avoided.[65]

As for the main policy matters regarding the Jews, the recurring two issues—two facets of the same problem—reappeared once again: measures intended to further Jewish emigration, and those dealing with the Jews remaining in the Reich. In essence the life of the Jews of Germany was to be made so unpleasant that they would make every effort to leave by any means; however, those Jews still remaining in the Reich had to feel that they had something to lose, so that none of them would take it into their heads to make an attempt on the life of a Nazi leader—possibly the highest one of all.[66]

Forced emigration was to have top priority: "At the head of all our considerations and measures," Göring declared, "there is the idea of transferring the Jews as rapidly and as effectively as possible to foreign countries, of accelerating the emigration with all possible pressure and of pushing aside anything that impedes this emigration." Apparently Göring was even willing to refrain from stamping Jewish passports with a recognizable sign (the letter "J") if a Jew had the means to emigrate but would be hindered from doing so by such identification.[67] Göring informed the Gauleiters that the money needed to finance the emigration would be raised by an international loan (precisely the kind of loan that, as we shall see, Schacht was soon to be discussing with the American delegate of the Intergovernmental Committee for Refugees); Hitler, Göring stated, was very much in favor of this idea. The guarantee for the loan, presumably to be raised by "world Jewry" and by the Western democracies, was to consist of the entire assets still belonging to the Jews in Germany—one reason why Jewish houses were not to be forcibly Aryanized at that stage,[68] even though many party members were particularly tempted by that prospect.

From world Jewry Göring demanded the bulk not only of the loan but also the cessation of any economic boycott of Germany, so that the Reich could obtain the foreign currency needed to repay the principal and the interest on the international loan. In the midst of these practical explanations, Göring mentioned that he wanted the Jews to promise that the "international department store corporations, which in any case are all in Jewish hands, should commit themselves to take millions worth of German goods."[69] The myth of Jewish world power loomed again.

Regarding the Jews still remaining in Germany, Göring announced Hitler's rejection of any special identifying signs, and of excessively drastic travel and shopping restrictions. Hitler's reasons were unexpected: Given the state of mind of the populace in many Gaue, if Jews wore identifying signs they would be beaten up or refused any food. The other limitations would make their daily life so difficult that they would become a burden on the state.[70] In other words, the Gauleiters were indirectly warned not to launch any new actions of their own against the Jews in their Gaue. Jewish-owned houses, as has been seen, were the last Jewish assets to be Aryanized.

While discussing the measures that would induce the Jews to leave Germany, Göring assured his listeners he would make sure that the rich Jews would not be allowed to depart first, leaving the mass of poor Jews behind. This remark probably explains what followed three days later.

*　　　*　　　*

In the Würzburg Gestapo files there is an order issued December 9, on Göring's instructions, to the twenty-two Franconia district offices; that order must have been issued on a national scale. In it, the State Police demanded the "immediate forwarding of a list of 'influential Jews' living in each of the districts." The criteria of influence were spelled out as follows: wealth and relations with foreign countries (of "economic, family, personal, or other kinds"). The regional officers were to give the reason for every influential Jew's inclusion on the list. The matter was so urgent that the lists had to be sent in by express mail on the next day, the tenth, so as to reach Würzburg Gestapo headquarters by 9 A.M. on Saturday, December 11. Each regional office director was made "personally responsible for strict adherence to the deadline."[71]

There is no explanation in the files regarding Göring's intentions, nor any record of further action; it may have been a short-lived start at taking rich and influential Jews hostage to guarantee the departure of the poor ones.

A few days before the Würzburg Gestapo transmitted Göring's orders, Frick informed the federal state presidents and interior ministers that "by expressly highest order"—a formula used only for orders from Hitler—no further anti-Jewish measures were to be taken without explicit instructions from the Reich government.[72] The echo of Göring's announcement to the Gauleiters is clearly perceptible. On December 13 it was the Propaganda Ministry's turn to inform its agencies that "the Führer had ordered that all political broadcasts deal exclusively with the Jewish problem; political broadcasts on other topics were to be avoided so as not to diminish the effect of the anti-Jewish programs."[73] In short, German opinion had yet to be convinced that the November pogrom had been amply justified.

After the series of internal meetings of party and state officials that aimed at clarification of the goals and limits of post-pogrom anti-Jewish policies, one additional conference took place on December 16. Convened by Frick, that meeting was held in the presence of Funk, Lammers, Helldorf, Heydrich, Gauleiters and various other party and state representatives. In the main Frick and Funk took up Göring's explanations, exhortations, and orders. Yet it also became apparent that throughout the Reich, party organizations such as the German Labor Front had put pressure on shopkeepers not to sell to Jews. And, mainly in the Ostmark, *Mischlinge* were being treated as Jews, both in terms of their employment and of their business activities. Such initiatives were unacceptable in Hitler's eyes. Soon no Jewish businesses would be left, and the Jews would have to be allowed to buy in German stores. As for the *Mischlinge*, the policy, according to Frick, was to absorb them gradually into the nation (strangely enough Frick did not distinguish the half- from the quarter-Jews), and the current discrimination against them contravened the distinctions established by the Nuremberg Laws. On the whole, however, the main policy goal was emphasized over and over again: Everything had to coincide to expedite the emigration of the Jews.[74]

Yet another set of measures descended on the Jews toward the end of December. On the twenty-eighth Göring, again referring to orders explicitly given by Hitler, both at the beginning of the document and in its conclusion, established the rules for dealing with dwellings belonging to Jews (they should not be Aryanized at this stage, but Jewish occupants should gradually move to houses owned and inhabited only by Jews) and defined the distinction between two categories of "mixed marriages." Marriages in which the husband was Aryan were to be treated more or less as regular German families, whether or not they had children. The fate of mixed marriages in which the husband was Jewish depended on whether there were children. The childless couples could eventually be transferred to houses occupied only by Jewish tenants, and in all other respects as well, they were to be treated as full Jewish couples. Couples with children—whereby the children were *Mischlinge* of the first degree—were temporarily shielded from persecution.[75] "The government appears, if one is prepared to accept the government's point of view, to treat correctly 'Aryan' husbands of Jewish wives," Jochen Klepper noted in his diary. "Many of them hold important positions in the army and the economy. They are not forced to divorce and are able to transfer their possessions. But much worse off are 'Aryan' wives of Jewish men. They are expelled to Jewry; on their heads fall all the misfortunes that we others are spared according to the present regulations and conditions."[76] Klepper mentions that Aryan husbands of Jewish wives "are not forced to divorce." Aryan wives of Jewish husbands were not forced to divorce either, but the law of July 6, 1938 (already mentioned in chapter 4), had made divorce on racial grounds possible, and Göring's decree of December 28 clearly encouraged Aryan women to leave their Jewish husbands: "If the German wife of a Jew divorces him," said the decree, "she again joins the community of German blood, and all the disadvantages [previously imposed on her] are eliminated."[77]

Why did Hitler oppose the yellow badge and outright ghettoization in December 1938? Why did he create a category of "privileged mixed marriages" and also decide to compensate *Mischlinge* who had suffered damages during the pogrom? In the first case, wariness about German and international opinion was probably the main factor. As for the mixed marriages and compensation decision for *Mischlinge*, it seems evident that Hitler wanted to circumscribe as tightly as possible the potential zones of discontent that the persecution of mixed marriages and of *Mischlinge* in general could create within the population.

Göring's decree of December 28 aiming at the concentration of Jews in "Jews' houses" became more easily applicable when, on April 30, 1939, further regulations allowed for the rescinding of leasing contracts with Jews.[78] The Aryan lessor could not annul a contract of Jewish tenants prior to obtaining a certificate from the local authorities that alternative quarters in a Jewish-owned house had been secured. But, as noted by the American chargé d'affaires Alexander Kirk, these new regulations allowed municipal and communal authorities to "compel Jewish house-holders, or Jewish tenants in a Jewish owned house, to register with them vacant rooms, or space which they would not seem to require for their own needs. The latter may then be forced, even against their will, to lease these quarters to other Jews who are liable to eviction from 'Aryan' houses. The local authorities may draw up the terms of these involuntary contracts and collect a fee for this service."[79]

On January 17, 1939, the eighth supplementary decree to the Reich Citizenship Law forbade Jews to exercise any paramedical and health-related activities, particularly pharmacy, dentistry, and veterinary medicine.[80] On February 15 members of the Wehrmacht, the Labor Service, party functionaries, and members of the SD were forbidden to marry "*Mischlinge* of the second degree,"[81] and on March 7, in answer to a query from the Justice Minister, Hess decided that Germans who were considered as such under the Nuremberg Laws but who had some Jewish blood were not to be hired as state employees.[82]

During the crucial weeks from November 1938 to January 1939, the measures decided upon by Hitler, Göring, and their associates entirely destroyed any remaining possibility for Jewish life in Germany or for the life of Jews in Germany. The demolition of the synagogues' burned remains symbolized an end; the herding of the Jews into "Jewish houses" intimated a yet-unperceived beginning. Moreover, the ever-present ideological obsession that was to receive a paroxysmic expression in Hitler's Reichstag speech of January 30, 1939, continued unabated: A stream of bloodthirsty statements flowed from the pages of *Das Schwarze Korps*, and an address by Himmler to the SS top-echelon leadership of the elite unit SS-Standarte "Deutschland," on November 8, 1938, carried dire warnings.

Himmler made no mention of the Paris shooting the day before, and most of his speech dealt with the organization and tasks of the SS. But the Jewish question was ominously there. Himmler warned his audience that, within ten years, the Reich would face unprecedented confrontations "of a critical nature." The Reichsführer referred not only to national confrontations but, in particular, to the clash of worldviews in which the Jews stood behind all other enemy forces and represented the "primal matter of all that was negative." The Jews—and the forces they directed against the Reich— knew that "if Germany and Italy were not annihilated, they themselves would be annihilated." "In Germany," Himmler prophesied, "the Jew will not be able to maintain himself; it is only a matter of years." How this would be achieved was obvious: "We will force them out with an unparalleled ruthlessness." There followed a description of how anti-Semitism was intensifying in most European countries, as a result of the arrival of Jewish refugees and the efforts of Nazi propaganda.

Then Himmler launched into his own vision of the final phase. Trapped, the Jews would fight the source of all their troubles, Nazi Germany, with all the means at their disposal. For the Jews the danger would be averted only if Germany were burned down and annihilated. There should be no illusions, said Himmler, and repeated his warning that in case of a Jewish victory, there would be total starvation and massacre; not even a reservation of Germans would remain: "Everybody will be included, the enthusiastic supporters of the Third Reich and the others; speaking German and having had a German mother would suffice."[83] The implicit corollary was clear.

In October 1935, in the immediate wake of the promulgation of the Nuremberg Laws, Goebbels had issued a decree according to which the names of fallen Jewish soldiers would not be inscribed on any memorial erected in Germany from then on.[84] It so happened, however, that when on June 14, 1936, a memorial was unveiled in the small town of Loge, in Eastern Friesland, the name of the Jewish soldier Benjamin was inscribed among those who had fallen in 1915. Loge's Gruppenleiter took the initiative of having Benjamin's name deleted and replaced (to fill the conspicuous void) by that of a local soldier who had died of his wounds soon after the war's end. Local protests, including those of Dutch citizens living in this border town, such as the retired ambassador Count van Wedel, led to the removal of the new name. Was, then, the Jewish soldier Benjamin to be reinscribed? The Gauleiter of Weser-Ems decided that such a move would be "intolerable."[85]

The overall problem remained unsolved until the pogrom of November 1938. On November 10 Paul Schmitthenner, rector of Heidelberg University, wrote to the Baden minister of education in Karlsruhe: "In view of the struggle of world Jewry against the Third Reich, it is intolerable that the names of members of the Jewish race remain on plaques of the war dead. The students," Schmitthenner continued, "were demanding the removal of the plaque, but this was not done out of respect for the German dead." The rector therefore asked the ministry to find an immediate solution to the problem in cooperation with the Reich student leader: "I consider removal of the Jewish names necessary," Schmitthenner concluded. "It should take place in an orderly and dignified way in the spirit of the regulations I am asking for."[86]

The minister of education of Baden forwarded Schmitthenner's letter to the Reich minister of education with the following comment: "In my opinion, as the question is of fundamental significance, it should be submitted to the Führer for decision."[87] Rust did so, and on February 14 he was able to announce Hitler's decision: Names of Jews on existing memorials would not be removed. Newly erected memorials would not include names of Jews.[88]

Schmitthenner's resolve to eliminate the names of fallen Jewish soldiers from the halls of Heidelberg was echoed by the no less determined action of Friedrich Metz, the Freiburg University rector, who thereby preempted a decision that would be taken in Berlin on December 8. "I have been informed," Metz wrote to the university library director on November 17, "that the library of the university and the academic reading room are still being visited by Jews. I have already instructed former members of the faculty Professor Jonas Cohn and Professor Michael, who are in question in this matter, to abstain from using any services of the Albert-Ludwig University in order to avoid unpleasantness. I authorize you herewith to act in the same spirit if the university library or the academic reading room are visited by other Jews."[89]

III

"Regarding your request concerning a residence permit for your wife, I have to inform you of the following." Thus started a letter that Party Comrade Seiler, the Kreisleiter of Neustadt on the Aisch, near Nuremberg, addressed on November 21, 1938, to the German farmer and grocery store owner Fritz Kestler of Ühlfeld. Kestler's wife, the mother of his four children and the family member in charge of the grocery store, had been expelled from Ühlfeld during the November pogrom and was temporarily staying with relatives in Nuremberg.

"Your wife, born Else Rindeberg," the letter continued, "is a full-blooded Jewess. That is why she has repeatedly shown to all members of her race, through personal contact and all possible help, that she feels that she totally belongs to them. That is why, for instance, she took the reponsibility for the reimbursements of debts owed to Ühlfeld Jews. Moreover, she has given shelter to Jews who felt threatened. Further, she allowed Volksgenossen who have not learned a thing and who wished to buy at the Jew Schwab's to walk through her store and enter Schwab's premises from the back. Your wife has proved thereby that she considers herself a Jew and that she thinks she can make fools of the political leadership and the authorities.

"I am not astonished that you were not enough of a man to put an end to this, since someone who admits that he has been happily married to a Jewess for twenty-five years shows that he is badly contaminated by this evil Jewish spirit. If, at the time, you were oblivious enough of your race to marry a Jewess against the warnings of your parents, you cannot expect today to have the right to ask that an exception be made for your Jewish wife." After warning Kestler that his wife should not try to return, Kreisleiter Seiler ended his letter with the appropriate flourish: "Your question regarding what should now happen to your wife is of as little interest to me as twenty-five years ago it was of interest to you what would become of the German people if everybody entered a marriage that defiled the race."[90]

Seiler's anti-Jewish fury was not shared by the majority of Germans. On November 10 a clear difference emerged from the outset between activists and onlookers on the streets of the large cities: "I myself," the counselor of the British Embassy reported to his foreign minister a few days later, "and members of the staff were witnesses of the later stages of the excesses in Berlin, which lasted until well into the night of the 10th. Gangs of youths in plain clothes and armed with poles, hammers and other appropriate weapons were visiting the Jewish shops and completing the work of destruction, done in the early morning. In some cases the premises had been entirely looted, in others the stock in trade was only mishandled and scattered. And at one or two places a crowd was gaping in silent curiosity at the efforts of the owners to tidy up the *débris*. I especially noted the demeanor of the groups which followed each band of marauders. I heard no expression of shame or disgust, but, in spite of the complete passiveness of many of the onlookers, I did notice the inane grin which often inadvertently betrays the guilty conscience."[91]

Whereas the British diplomat recognized the signs of a troubled conscience on the onlookers' faces, the French chargé d'affaires perceived "silent condemnation" in the attitude of the people on the streets.[92]

The SD reports show widespread popular criticism of the violence and the damage caused during the pogrom. Some of the criticism, expressed even by people usually favorable to the regime, was motivated by practical considerations: the wanton destruction of property and the losses thus incurred not only by all Germans but also by the state. When news of the billion-mark fine imposed on the Jews was announced, and when official propaganda stressed the immense wealth still possessed by the Jews, the general mood improved.[93] Sometimes, however, the reactions of the population were not negative at all. Thus, according to a SOPADE report of December 1938, "the broad mass of people has not condoned the destruction, but we should nevertheless not overlook the fact that there are people among the working class who do not defend the Jews. There are certain circles where you are not very popular if you speak disparagingly about the recent incidents. The anger was not, therefore, as unanimous as all that. Berlin: the population's attitude was not fully unanimous. When the Jewish Synagogue was burning . . . a large number of women could be heard saying, 'That's the right way to do it—it's a pity there aren't any more Jews inside, that would be the best way to smoke out the whole lousy lot of them.' No one dared to take a stand against these sentiments. . . . If there has been any speaking out in the Reich against the Jewish pogroms, the excesses of arson and looting, it has been in Hamburg and the neighboring Elbe district. People from Hamburg are not generally anti-Semitic, and the Hamburg Jews have been assimilated far more than the Jews in other parts of the Reich. They have intermarried with Christians up to the highest levels of officialdom and the wholesale and shipping trades."[94]

How did people closer to Hitler who were neither committed party members nor "old-fashioned" conservatives react? In his memoirs, Albert Speer indicates a measure of unease, if only because of the material destruction and the "disorder": "On November 10, driving to the office, I passed by the still smoldering ruins of the Berlin synagogues. . . . Today this memory is one of the most doleful of my life, chiefly because what really disturbed me at the time was the aspect of disorder that I saw on Fasanenstrasse: charred beams, collapsed façades, burned-out walls, . . . The smashed panes of shop windows offended my sense of middle-class order."[95] But even this lack of any human empathy compounded with later pseudo-candor demands some qualification. According to Speer's recent biographer, Gitta Sereny, there was nothing about Kristallnacht in the early draft of Speer's book, and it was only after the proddings of his publisher, Wolf Jobst Siedler, and of Hitler's biographer Joachim Fest that Speer came up with his feelings of annoyance at the material damage.[96] Thus, even a questionable but clever sincerity may have been entirely faked: Speer may simply not have felt anything at all, as was probably the case when he planned the eviction of Jewish tenants from their Berlin apartments. As for Speer's secretary, Annemarie Kempf, she knew nothing and saw nothing: "I just never knew about it," she declared, "I remember that someone was shot in an embassy abroad, and Goebbels gave speeches, and there was a lot of anger. But that's all."[97] Again, however, even among these young technocrats the reactions were not all the same. Consider one of "Speer's men," Hans Simon. "When [Kristallnacht] happened," another witness later told Sereny, "Simon said: for people like that, I don't work. And he resigned from the GBI [Generalbauinspektorat, or Construction Inspectorate General]."[98]

No criticism of the pogrom was publicly expressed by the churches. Only a month after the events, in a message to the congregations, did the Confessing Church make an oblique reference to the most recent persecutions, albeit in a peculiar way. After declaring that Jesus Christ was the "propitiation of our sins" and "also the propitiation for the sins of the Jewish people," the message continued with the following words: 'We are bound together as brethren with all the believers in Christ of the Jewish race. We will not separate ourselves from them, and we ask them not to separate themselves from us. We exhort all members of our congregations to concern themselves with the material and spiritual distress of our Christian brothers and sisters of the Jewish race, and to intercede for them in their prayers to God.' The Jews as such were excluded from the message of compassion and, as has been noted, "the only reference to the Jewish people as a whole was a mention of their sin."[99]

Some individual pastors did protest; we know of them mainly from brief mentions in surveillance reports. Thus the monthly report for November 1938 for Upper and Mid-Franconia notes laconically: "Pastor Seggel of Mistelgau, administrative district Bayreuth, expressed himself critically on the Day of Prayer and Repentance regarding the actions against the Jews. The State Police of Nuremberg-Fürth was informed."[100]

The overall attitude of the Catholic Church was no different. Apart from Provost Bernhard Lichtenberg of Berlin's St. Hedwig Cathedral, who declared on November 10 that "the temple which was burnt down outside is also the House of God," and who later was to pay with his life for his public prayers for the Jews deported to the East,[101] no powerful voice was raised. Quite to the contrary, Cardinal Faulhaber found it necessary to proclaim in his New Year's Eve sermon, less than two months after the pogrom: "That is one advantage of our time; in the highest office of the Reich we have the example of a simple and modest alcohol- and nicotine-free way of life."[102]

No open criticism (or even indirect protest) came from the universities. Some strong condemnations of the pogrom were committed to private correspondence and, probably, to the privacy of diaries. On November 24, 1938, the historian Gerhard Ritter wrote to his mother: "What we have experienced over the last two weeks all over the country is the most shameful and the most dreadful thing that has happened for a long time."[103] Ritter's indignation, however, and the initiative that followed, paradoxically shed some light on the anti-Semitism that underlay the attitudes of the churches and the universities.

Following the pogrom, and certainly in part as a result of it, an opposition group was formed at Freiburg University. The Freiburg Circle (*Freiburger Kreis*) was composed mainly of university members close to the Confessing Church (and also of some Catholics); Gerhard Ritter, Walter Eucken, Franz Böhm, Adolf Lampe, and Constantin von Dietze were its leading figures.[104] The group's discussions resulted in the drafting of the "Great Memorandum," which offered a social, political, and moral basis for a post–National Socialist Germany. The fifth and last appendix to the Memorandum, completed by Dietze in late 1942, listed "Proposals for a Solution of the Jewish Question in Germany."[105] Present-day German historians still find it hard to explain these proposals, and refer to the "schizoid atmosphere" that engendered them.[106] The Freiburg group—which had come into being after the pogrom and by the time of this last appendix was also fully aware of the extermination of the Jews (which is mentioned explicitly in Dietze's "Proposals")—suggested nonetheless that after the war the Jews be internationally subjected to a special status. Moreover, although the "Proposals" rejected the Nazis' racial theories, they recommended caution regarding close contacts and intermarriage between German Christians and other races—the allusion to the Jews is clear. [107] It seems that even in one of the most articulate groups of anti-Nazi academics, there was explicit and deep-seated anti-Jewish prejudice. One of the best-informed historians on the subject of the Freiburg Circle, Klaus Schwabe, rejects the conclusion that Dietze was motivated by anti-Semitism.[108] Yet, in his program, Dietze accepted and recommended some of the traditional German conservative anti-Semitic positions, despite what he knew of the Jews' fate. The logical corollary is obvious: If a university resistance group, consisting mostly of members of the Confessing Church or the Catholic Church, could come up with such proposals even though they had knowledge of the extermination, the evidence of prevalent anti-Semitism among Germany's elites must be taken into account as a major explanation of their attitudes during the Third Reich.

In an indirect way, however, the pogrom created further tension between the German Catholic Church and the state. On November 10 the National Socialist Association of Teachers decided not only to expel all remaining Jewish pupils from German schools but also to stop providing (Christian) religious education—as had been the rule until then—under the pretext that "a glorification of the Jewish murderers' nation could no longer be tolerated in German schools." Cardinal Bertram sent a vigorous protest to Rust in which he stated that "whoever has the least familiarity with the Catholic faith and certainly every believing teacher knows that this assertion [that the Christian religion glorified the Jews] is false and that the contrary is true."[109]

I V

"The foreign press is very bad," Goebbels noted on November 12. "Mainly the American. I receive the Berlin foreign correspondents and explain the whole issue to them. . . . This makes a big impression."[110] Press comments were scathing indeed. "There happen in the course of time," said the

Danish *Nationaltidende* on November 12, "many things on which one must take a stand out of regard for one's own human dignity, even if this should involve a personal or national risk. Silence in the face of crimes committed may be regarded as a form of participation therein—equally punishable whether committed by individuals or by nations. . . . One must at least have the courage to protest, even if you feel that you do not have power to prevent a violation of justice, or even to mitigate the consequences thereof. . . . Now that it has been announced that after being plundered, tortured and terrorized, this heap of human beings [the Jews of Germany] will be expelled and thrown over the gate of the nearest neighbor, the question no longer remains an internal one and Germany's voice will not be the only one that will be heard in the council of nations."[111]

The American press was particularly vehement. "In the weeks following *Kristallnacht*, close to 1,000 different editorials were published on the topic. . . . Practically no American newspaper, irrespective of size, circulation, location, or political inclination failed to condemn Germany. Now even those that, prior to *Kristallnacht*, had been reluctant to admit that violent persecution was a permanent fixture in Nazism criticized Germany."[112] President Roosevelt recalled Ambassador Hugh Wilson for consultation.

But despite such emotional outpourings, basic attitudes and policies did not change. In the spring of 1939, Great Britain, increasingly worried by the pro-Axis shift in the Arab world—a trend with possibly dire consequences for Britain in case of war—reneged on its commitments and for all practical purposes closed the doors of Palestine to Jewish immigration. No alternative havens were even envisaged by the British colonial authorities. As A. W. G. Randall of the Foreign Office stated on June 1: "The proposed temporary solution of Cyprus has, I understand, been firmly rejected by the Governor; it is unthinkable that a miscellaneous crowd of Jews could be admitted to any other part of the Empire."[113]

After slightly liberalizing its immigration policy in 1937, the United States did not even fill the quotas for Germany and Austria in 1938.[114] In July 1938 the Wagner-Rogers Child Refugee Bill, which would have allowed twenty thousand Jewish refugee children to enter the country, was not passed by the Senate,[115] and, at the same time, despite all entreaties, the 936 hapless Jewish emigrants from Germany who had sailed on the soon-to-become-notorious *St. Louis*, after being denied entry to Cuba, their destination, were not admitted into the United States.[116] Their voyage back to Europe became a vivid illustration of the overall situation of Jewish refugees from Germany. After Belgium, France, and England finally agreed to give asylum to the passengers, the London *Daily Express* echoed the prevalent opinion in no uncertain terms: "This example must not set a precedent. There is no room for any more refugees in this country. . . . They become a burden and a grievance."[117]

By then even some relatively well-known Jews had not the least certainty of reaching the the United States. In February 1939 Thomas Mann intervened in favor of Kafka's friend and biographer Max Brod with H. M. Lyndenberg, the director of the New York Public Library: "Dr. Max Brod, the German-Czechoslovakian novelist and dramatist . . . is anxious to leave Czechoslovakia and come to the United States. He fears he will not survive the period of fifteen months to two years which he would have to wait to enter this country as an ordinary immigrant. . . . He writes that he is willing to give his collection of books and manuscripts of Franz Kafka to any institution of repute which would accept it and in return offer him a position to act as assistant or curator of the collection, and so make possible his entry into this country. . . . Perhaps you will agree with me that the possibility of acquiring the manuscripts and books of so well known a writer as Franz Kafka is an opportunity deserving of consideration quite apart from the human tragedy of the individual for whom the collection represents the one real chance of escape from an intolerable situation."[118] Ultimately Brod managed to escape to Palestine.

France was neither more nor less inhospitable than other countries, but it did not volunteer even a symbolic gesture of protest against the anti-Jewish pogrom. It was the only major democratic country that did not react.[119] Most newspapers expressed their outrage, but neither Prime Minister Édouard Daladier nor Foreign Minister Georges Bonnet did so. On the contrary, Bonnet continued with the planning for Ribbentrop's visit to Paris, which was to lead to a Franco-German agreement.

In a way the official French attitude demonstrated that Hitler did not have to worry too much about international reactions when he unleashed the pogrom. But the outcry that immediately followed the events of November and the criticism now directed at the French attitude confirmed that the Munich atmosphere was quickly dissipating. No less a supporter of appeasement than the London *Times* was taken aback by Bonnet's eagerness to go ahead with the agreement, the pogrom notwithstanding. The American secretary of state rejected Bonnet's request that the American government express its approval of the agreement, even if only in the form of a press statement. In view of the strained United States–German relations following Kristallnacht, the secretary deemed such approval entirely inappropriate. Even the Italian government expressed surprise that "the recrudescence of anti-Semitic persecutions in Germany did not lead to the ruin of the project of Franco-German declaration."[120]

The German foreign minister arrived in Paris on December 6. According to the German version of the second discussion between Ribbentrop and Bonnet, which took place on December 7, the French foreign minister told Ribbentrop "how great an interest was being taken in France in a solution of the Jewish problem," and he added that "France did not want to receive any more Jews from Germany." Bonnet then supposedly asked whether Germany could not take measures to prevent further German Jewish refugees from coming to France, since France itself would have to ship ten thousand Jews somewhere else. (They were actually thinking of Madagascar for this.) Ribbentrop then told Bonnet, "'We all wish to get rid of our Jews,' but the difficulty lay in the fact that no country wished to receive them and, further, in the shortage of foreign currency." [121]

Bonnet's oft-quoted remarks to Ribbentrop were not an isolated occurence. In fact, less than two weeks before the Franco-German meeting, on November 24, the prime ministers and foreign ministers of Great Britain and France met in Paris in order to coordinate their countries' policies. The problem of the Jewish refugees from Germany was raised. Daladier complained that there were some forty thousand of them in France and that no more could be taken in. The possibility of sending the refugees to the colonies was discussed. It was agreed that the French would ask Ribbentrop if the German measures making it almost impossible for the refugees to take along some of their belongings could be alleviated.[122] Whether this issue was mentioned at all during Ribbentrop's visit to Paris is unclear.

Yet another sequel to the events of November took place—at least for a time—in the French capital: preparations for the trial of Herschel Grynszpan. The forthcoming event attracted worldwide attention. Hitler dispatched Professor Friedrich Grimm, one of Goebbels's deputies, to Paris in order to follow the work of the prosecution, while an international committee headed by the American journalist Dorothy Thompson collected money to pay for Grynszpan's defense. Grynszpan's lawyer, Vincent

Moro-Giafferi, was one of the most respected criminal lawyers in France and an ardent antifascist.[123]

The beginning of the war interrupted the preparations of both prosecution and defense. When the Germans occupied France, the Vichy government duly delivered to them the young Jew they were searching for. Grynszpan was incarcerated in Germany, and Goebbels started to plan a huge show trial in which Herschel Grynszpan would have stood for "international Judaism." Nothing came of it, as in 1942 the accused suddenly announced that he had had a homosexual relationship with Rath. Such a line of defense, if presented in public, would have been disastrous in the eyes of the Nazis. Grynszpan did not survive the war; the circumstances of his death remain unknown.[124]

During these early months of 1939, the expulsion of the Jews from the Reich continued to follow the pattern inaugurated in 1938; the Jews were sent over the borders, but usually to no avail. In February 1939 a SOPADE report described a scene witnessed in the west of the country, near the border with France. The Jews were taken from their homes and herded together in the city square. In the evening they were transported to the border, only to be brought back the next day, as the French would not let them through. Later they were shipped off to Dachau.

The report described the jeering and the insults coming from youths and "hysterical women." But "most of the older people who accidentally came upon this scene could not hide their indignation over this spectacle. Words were exchanged with people who wanted to defend the measures against the Jews. People said: 'They [the Jews] are no worse than other businessmen; and those who took over their businesses are more expensive and have poorer quality goods.' The excitement was so great that nothing could be undertaken [by the authorities] against these dissidents. A large segment of those previously transported are here again, and have been received kindly by the public. People ask them sympathetically if they have no possibilities of emigrating. Some answer that they are trying, and others point to the great difficulties. Now it has reached the point where children confront Jews and demand money. Some give it to them and create the impression that they themselves have become childish."[125]

On December 23, 1938, very strict orders had been issued by Gestapo headquarters to all stations on the western borders of the Reich to prevent illegal crossings of Jews into neighboring countries, due to increasing complaints. However, as the SOPADE report indicates, and as a further Gestapo order of March 15, 1939, confirms, such illegal crossings, mostly initiated, it seems, by local authorities, must have continued well into the spring of that year.[126] On exceptionally rare occasions, officials on the non-German side of the borders took the risk of aiding the illegal entry of Jews into their countries, whether the refugees were pushed over the frontier by the Germans or were trying to cross on their own. Such was the case of Paul Grüninger, the commander of the border police in the Swiss canton Saint Gall. By predating visas and falsifying other documents, he helped some 3,600 Jewish refugees to enter Switzerland in late 1938 and early 1939. Grüninger's activities were discovered. In April 1939 he was dismissed and, later, sentenced to a heavy fine and to the loss of his pension rights.[127] As the result of a lengthy public campaign, Grüninger was rehabilitated—fifty-four years after his sentencing, twenty-three years after his death.[128]

One escape route was still open, but only for a very short time. An interministerial conference held in Tokyo on December 6, 1938, decided on a lenient policy toward Jewish refugees, making Japanese-occupied Shanghai accessible to them and even permitting prolonged transit stays in Japan itself. The Japanese seem to have been moved by their distrust of Germany and possibly by humane considerations, but undoubtedly too, as accounts of the conference show, by their belief in Jewish power—a belief reinforced by Nazi propaganda and by study of the *Protocols of the Elders of Zion*—and its possible impact on Japanese interests in Great Britain and the United States. Be that as it may, Shanghai, where no visa was required, became an asylum for desperate German and Austrian Jews. By the end of 1938, fifteen hundred refugees had arrived; seven months later the number had reached fourteen thousand, and if the Japanese had not begun curtailing access to the city because of local conditions, the total would have mushroomed. On the eve of the war, the Jews who had reached the safe shores of the China Sea numbered between seventeen and eighteen thousand.[129] This influx triggered a fear of economic competition among some of the earlier Jewish settlers who had not yet established themselves, as well as among the large community of White Russian exiles. Some aspects of the European pattern reappeared with uncanny similarity. But there were very few reactions among the great majority of the Shanghai population, the Chinese themselves, because their standard of living was too low for any sort of competition.[130]

Thus some tens of thousands of Jews managed to leave Germany for neighboring European countries, North, Central, and South America, and remote Shanghai. Tiny groups were driven over Germany's borders. And finally, despite British policy, Jewish emigrants managed to reach Palestine by way of illegal transports organized secretly both by the majority Zionist leadership and by its right-wing rivals, the Revisionists. These illegal operations were backed by Heydrich and all branches of the SD and the Gestapo, with the full knowledge of the Wilhelmstrasse. On the occasion of the first working session of the newly established Reich Central Office for Jewish Emigration, on February 11, 1939, Heydrich was quite explicit: "He [Heydrich] stated that any illegal emigration should be opposed on principle, to be sure. In the case of Palestine, however, matters were such that illegal transports were already going there at the present time from many other European countries, which were themselves only transit countries, and in these circumstances this opportunity could also be utilized in Germany, though without any official participation. Senior Counselor Walter Hinrichs and Minister Ernst Eisenlohr from the Foreign Ministry had no objection to this and expressed the viewpoint that every possibility for getting a Jew out of Germany ought to be taken advantage of."[131]

The illegal road first led through Yugoslavia, then down the Danube to the Romanian harbor of Constantsa. The main problem was not for the emigrants to leave the Greater Reich, but for the Zionist organizations to find the money to bribe officials and buy ships, and then to avoid the British patrols along the Palestine coast. Some seventeen thousand illegal immigrants reached Palestine from early 1939 to the outbreak of the war.[132] On September 2, 1939, off the beach at Tel-Aviv, a Royal Navy ship fired at the *Tiger Hill*, which was carrying fourteen hundred Jewish refugees, two of whom were killed. As Bernard Wasserstein ironically noted, "these were probably the first hostile shots fired by British forces after the [previous day's German] attack on Poland."[133]

On March 15, 1939, the Wehrmacht had occupied Prague. Czecho-Slovakia ceased to exist. Slovakia became a German satellite; Bohemia-Moravia was turned into a protectorate of the Reich. The crisis had started in the early days of the month. Enticed and supported by the Germans, the Slovaks seceded from the already truncated Czecho-Slovakia. The elderly Czech President, Emil Hacha, was summoned to Berlin, threatened with the bombing of Prague, and bullied into acceptance of all the German demands. But before he even signed the document of his country's submission, the first German units had crossed the border. Some 118,000 more Jews were now under German domination. Stahlecker was transferred from Vienna to Prague to become inspector of the security police and the SD in the new protectorate, and Eichmann soon followed; imitating the Viennese model, he set up a Central Office for Jewish Emigration in Prague.[134]

"At home for breakfast, I found that I myself had a refugee, a Jewish acquaintance who had worked many years for American interests," the American diplomat George F. Kennan, who had been posted to the Prague legation a few months earlier, wrote in a March 15 memorandum. "I told him that I could not give him asylum, but that as long as he was not demanded by the authorities he was welcome to stay here and to make himself at home. For twenty-four hours he haunted the house, a pitiful figure of horror and despair, moving uneasily around the drawing room, smoking one cigarette after another, too unstrung to eat or think of anything but his plight. His brother and sister-in-law had committed suicide together after Munich, and he had a strong inclination to follow suit. Annelise pleaded with him at intervals throughout the coming hours not to choose this way out, not because she or I had any great optimism with respect to his chances for future happiness but partly on general Anglo-Saxon principles and partly to preserve our home from this sort of unpleasantness."[135]

CHAPTER 10

A Broken Remnant

I

"Guests of the Jewish race," read the "welcoming" card at the Hotel Reichshof in Hamburg sometime in early 1939, "are requested not to lounge in the lobby. Breakfast will be served in the rooms and the other meals in the blue room next to the breakfast hall on the mezzanine. The Management." These words were addressed to lucky emigrants still managing to flee the Reich through its major northern harbor. On the back of the card was an advertisement for the travel agency located in the hotel lobby, where "you may obtain boat tickets." The advertisement carried the slogan: "Travel is pleasant on the ships of the Hamburg-Amerika Line."[1]

Through a process of interpretation and innovation, party, state, and society gradually filled in the remaining blanks of the ever harsher code regulating all relations with Jews. What party agencies and the state bureaucracy left open was dealt with by the courts, and what the courts did not rule on remained for *Volksgenossen* (such as the Reichshof managers) to figure out.

Sometimes court decisions may have appeared improbable or even paradoxical, but only at first glance. More closely considered, they expressed the essence of the system. Thus, on June 30, 1939, a Frankfurt district court ordered a language- school director to refund advance payments received from a Jew for English lessons not provided in full; the court then followed by ruling that a German woman had to pay (in monthly installments, with interest) for goods she had bought and not paid for when her husband, a party member, insisted on immediate cessation of the transaction on discovery of the seller's Jewish identity. In both cases the German defendants also had to bear the court costs.[2]

There was a slight twist, however, to this unexpected show of justice. The rulings most probably resulted from instructions regarding the legal status of Jews issued by the Ministry of Justice on June 23, 1939, to all presidents of regional higher courts; the guidelines had been agreed upon by the ministers concerned at the beginning of the year and had already been communicated orally at the end of January. Thus the courts were well aware of their "duty."

The opening paragraph of the memorandum conveyed the gist of the ministry's position: "The exclusion of the Jews from the German economy must be completed according to plan and in stages on the basis of the existing regulations. . . . Businesses and other properties in the possession of Jews, which would allow for economic influence, will become German property in accordance with the prescribed ways." There is no possible mistake about the goals here defined. At this point, though, the bureaucracy sets the "limits," requiring that, beyond the aforesaid measures, the Jews (whether plaintiffs or defendants) be treated by the courts according to all accepted legal norms in any financial litigation: "Intervention in the economic situation of the Jews by the use of measures devoid of any explicit legal basis should be avoided. Therefore, the Jews should be able to turn to the courts with claims stemming from their [economic] activity and to have the rulings enforced when cases are decided in their favor." The ministry did not conceal the reason for this sudden legal concern: "It is undesirable, on public welfare grounds alone, to let the Jews become totally impoverished." In a prior paragraph this rather crass reasoning had been preempted by a declaration of high principles: "The enforcement [of rulings] . . . is not only a matter for the parties involved but also serves . . . as an expression of the authority of the state." Even judges who were party members could not avoid the application of the law to Jews, because in their function as judges they were also part of an administrative organ.[3]

This text represents a classic example of Nazi thinking. There is an absolute cleavage between the apparent significance of the text and the reality to which it alludes. The apparent significance here was that the Jews were entitled to their share of justice so that they would not become a burden on the state and because the enforcement of justice was the ultimate expression of state authority. But this declaration came after the Jews had been dispossessed of all their rights and of all possibilities of material subsistence by the very state authorities that were ordering that justice be enforced.

Up to that point, there had been a measure of consonance between the significance of decrees, as brutal as they were, and the facts they dealt with, as calamitous as *they* were. The laws of exclusion were explicit and led to the dismissal of the Jews from public office and official life; the segregation edicts led to complete separation between Germans and Jews; the expropriation decrees dealt with the destruction of the concrete economic situation of the Jews in Germany. But the edict of June 1939 was calling for a measure of justice in a situation in which day in, day out, the Nazi authority that was demanding such justice was imposing ever harsher injustices, a situation in which court decisions on individual claims had become irrelevant in practice, given the public burden (the impoverishment of the Jews) the same authority had itself already created.

Although the instructions given to the courts in January (and June) 1939 were unknown to the litigants, they introduced a new dimension within the administration itself: the double language that increasingly characterized all measures taken against the Jews—the internal camouflage that was to contribute to the success of the "Final Solution." And, whereas concrete measures were increasingly disguised by a new form of language and concepts, open statements, particularly the utterances of the leadership and of the Nazi press, attained unequaled degrees of violence. Hitler threatened extermination; the Ministry of Justice enjoined abiding by the rules.

I I

As in every year since 1933, the Reichstag was convened in festive session on January 30, 1939, to mark the anniversary of Hitler's accession to power. Hitler's speech started at 8:15 in the evening and lasted for more than two and a half hours. The first part of the speech dealt with the history of the Nazi movement and the development of the Reich. Hitler then castigated some of the main British critics of appeasement, whom he accused of calling for a war against Germany. Since the Munich agreement Hitler had already twice lashed out in public against his English enemies, Winston Churchill, Anthony Eden, Alfred Duff Cooper, and on one occasion at least, in his speech of October 9, he had explicitly mentioned the Jewish wire pullers he perceived behind the anti-German incitement.[4] The same rhetoric unfurled on January 30. Behind the British opponents of Munich, the Führer pointed to "the Jewish and non-Jewish instigators" of that campaign. He promised that when National Socialist propaganda went on the offensive, it would be as successful as it had been within Germany, where "we knocked down the Jewish world enemy . . . with the compelling strength of our propaganda."[5]

After referring to the American intervention against Germany during World War I, which, according to him, had been determined by purely capitalistic motives, Hitler—probably infuriated by the American reactions to the November pogrom and to other Nazi measures against the Jews—thundered that nobody would be able to influence Germany in its solution of the Jewish problem. He sarcastically pointed to the pity expressed for the Jews by the democracies, but also to the refusal of these same democracies to help and to their unwillingness to take in the Jews to whom they were so sympathetic. Hitler then abruptly turned to the principle of absolute national sovereignty: "France to the French, England to the English, America to the Americans, and Germany to the Germans." This allowed for a renewed anti-Jewish tirade: The Jews had attempted to control all dominant positions within Germany, particularly in culture. In foreign countries there was criticism of the harsh treatment of such highly cultured people. Why then weren't the others grateful for the gift Germany was giving to the world? Why didn't they take in these "magnificent people"?

From sarcasm Hitler moved to threat: "I believe that this [Jewish] problem will be solved—and the sooner the better. Europe cannot find peace before the Jewish question is out of the way. . . . The world has enough space for settlement, but one must once and for all put an end to the idea that the Jewish people have been chosen by the good Lord to exploit a certain percentage of the body and the productive work of other nations. Jewry will have to adapt itself to productive work like any other nation or it will sooner or later succumb to a crisis of unimaginable dimensions." Up to that point, Hitler was merely rehashing an array of anti-Jewish themes that had become a known part of his repertory. Then, however, his tone changed, and threats as yet unheard in the public pronouncements of a head of state resonated in the Reichstag: "One thing I would like to express on this day, which is perhaps memorable not only for us Germans: In my life I have often been a prophet, and I have mostly been laughed at. At the time of my struggle for power, it was mostly the Jewish people who laughed at the prophecy that one day I would attain in

Germany the leadership of the state and therewith of the entire nation, and that among other problems I would also solve the Jewish one. I think that the uproarious laughter of that time has in the meantime remained stuck in German Jewry's throat." Then came the explicit menace: "Today I want to be a prophet again: If international finance Jewry inside and outside Europe again succeeds in precipitating the nations into a world war, the result will not be the Bolshevization of the earth and with it the victory of Jewry, but the annihilation of the Jewish race in Europe."[6]

Over the preceding weeks and months Hitler had mentioned any number of possibilities regarding the ultimate fate of the German (and more often than not, of the European) Jews. On September 20, 1938, he had told the Polish ambassador to Berlin, Jósef Lipski, that he was considering sending the Jews to some colony in cooperation with Poland and Romania. The same idea, specifying Madagascar, had come up in the Bonnet-Ribbentrop talks and, earlier, in Göring's addresses of November 12 and December 6. (The Generalfeldmarschall had explicitly referred to Hitler's ideas on this issue.) To South African Defense Minister Oswald Pirow, Hitler declared on November 24, 1938, that "some day, the Jews will disappear from Europe." On January 5, 1939, Hitler stated to Polish Foreign Minister Beck that had the Western democracies had a better understanding of his colonial aims, he would have allocated an African territory for the settlement of the Jews; in any case, he made it clear once more that he was in favor of sending the Jews to some distant country. Finally, on January 21, a few days before his speech, Hitler told Czech Foreign Minister František Chvalkovsky that the Jews of Germany would be "annihilated," which in the context of his declaration seemed to mean their disappearance as a community; he added again that the Jews should be shipped off to some distant place. A more ominous tone appeared in this conversation when Hitler mentioned to Chvalkovsky that if the Anglo-Saxon countries did not cooperate in shipping out and taking care of the Jews, they would have their deaths on their consciences.[7] If Hitler was mainly thinking in terms of deporting the Jews from Europe to some distant colony, which at this stage was clearly a completely vague plan, then the threats of extermination uttered in the January 30 speech at first appear unrelated. But the background needs to be considered once more.

On the face of it, Hitler's speech seems to have had a twofold context. First, as already mentioned, British opposition to the appeasement policy, and the strong American reactions to Kristallnacht, would have sufficed to explain his multiple references to Jewish-capitalist warmongering. Second, it is highly probable that in view of his project of dismembering what remained of Czecho-Slovakia, and of the demands he was now making on Poland, Hitler was aware of the possibility that the new international crisis could lead to war (he had mentioned this possibility in a speech given a few weeks before, in Saarbrücken).[8] Thus Hitler's threats of extermination, accompanied by the argument that his past record proved that his prophecies were not to be made light of, may have been aimed in general terms at weakening anti-Nazi reactions at a time when he was preparing for his most risky military-diplomatic gamble. More precisely the leader of Germany may have expected that these murderous threats would impress the Jews active in European and American public life sufficiently to reduce what he considered to be their warmongering propaganda.

The relevance of Hitler's speech to the immediate international context appears to be confirmed by a Wilhelmstrasse memorandum sent on January 25, 1939, to all German diplomatic missions, regarding "the Jewish question as a factor of foreign policy during the year 1938." The memorandum linked the realization of "the great German idea," which had occurred in 1938 (the annexation of Austria and the Sudetenland), with steps for the implementation of a solution of the Jewish question. The Jews were the main obstacles to the German revival; the rise of German strength was therefore necessarily linked to the elimination of the Jewish danger from the German national community. The memorandum, which reaffirmed Jewish emigration as the goal of German policy, identified the United States as the headquarters of Jewish international action and President Roosevelt, notoriously surrounded by Jews, as the force attempting to organize international pressure on Germany both in general political terms and also in order to ensure that Jewish emigrants from Germany could benefit from the full recovery of Jewish assets.[9] Thus it seems that for the Wilhelmstrasse and for Hitler, the Western democracies and the United States in particular were temporarily taking the place of Bolshevik Russia as the seat of international Jewish power and therefore of militant hostility to the rise of German power.

It was precisely because Hitler believed in Jewish influence in the capitalist world that, in its immediate context, his speech may be considered as yet another exercise in blackmail. The Jews of Germany (and of Europe) were to be held hostage in case their warmongering brethren and assorted governments were to instigate a general war. This idea, which had been aired by *Das Schwarze Korps* on October 27, 1938, in an article entitled "An Eye for an Eye, a Tooth for a Tooth," was circulating in Germany during these very months. On November 3 *Das Schwarze Korps* returned to the same theme: "If the Jews declare war on us—as they have already done [in the past]—we will treat the Jews who live among us as the citizens of a belligerent state. . . . The Jews of Germany are part of world Jewry, and they partake in the responsibility for everything that world Jewry initiates against Germany, as they are a guarantee against the harm that world Jewry causes to us and still wants to inflict upon us."[10] The idea of holding the Jews hostage did not necessarily contradict the urgent desire to expel them from Germany. As has been seen, Hitler himself evoked this idea in his conversation with Goebbels on July 24, 1938. In his December 6 address to the Gauleiters, Göring returned to it as part of his emigration plan. Moreover, during the negotiations between Schacht and Rublee, which will be discussed below, the plan submitted by the Reichsbank president foresaw the departure of 150,000 Jews with their dependents over the following three years, whereas some 200,000 Jews, mainly the elderly, would stay behind in order to ensure international Jewry's positive behavior toward the Reich.

It would be a mistake, however, to consider Hitler's January 30 speech merely in its short-term, tactical context. The wider vistas may have been part calculated pressure, part uncontrolled fury, but they may well have reflected a process consistent with his other projects regarding the Jews, such as their transfer to some remote African territory. This was, in fact, tantamount to a search for radical solutions, a scanning of extreme possibilities. Perceived in such a framework, the prophecy about extermination becomes one possibility among others, neither more nor less real than others. And—like the hostage idea—the possibility of annihilation was in the air.

Himmler's speech of November 8, 1938, and its implicit corollaries have already been mentioned. A few weeks later, in an article published on November 24, *Das Schwarze Korps* was far more explicit. After announcing the need for the total segregation of the Jews of Germany in special areas and special houses, the SS periodical went one step further: The Jews could not continue in the long run to live in Germany: "This stage of development [of the situation of the Jews] will impose on us the vital necessity to exterminate this Jewish sub-humanity, as we exterminate all criminals in our ordered country: by the fire and the sword! The outcome will be the final catastrophe for Jewry in Germany, its total annihilation."[11]

It is not known if it was this article in *Das Schwarze Korps* that incited the American consul general in Berlin, Raymond Geist, to write in early December that the Nazi objective was the "annihilation" of the Jews,[12] or whether foreign observers sensed, at the inner core of the regime, the utter hatred that a few weeks later found its expression in Hitler's speech. Significantly, a few days before the Reichstag declaration, Heydrich, in an address to high-ranking SS officers, defined the Jews as "subhuman" and pointed to the historical mistake of expelling them from one country to another, a method that did not solve the problem. The alternative, although not expressed, was not entirely mysterious, and after the speech, Himmler entered a rather cryptic remark in his notes: "inner martial spirit."[13]

How far the reality of the Jews as a "threatening world power" had been internalized at all levels of the Nazi apparatus is possibly best illustrated by a text entitled "International Jewry," prepared by Hagen for Albert Six, the head of II 1. In its final version it was forwarded to Six on January 19, 1939, for a lecture at Oldenburg (probably at a meeting of the higher SS leadership) on the Jewish question.[14]

The opening paragraph of Hagen's memorandum was unequivocal: The Jewish question was "the problem, at the moment, of world politics." After showing that the Western democracies (including the United States) had no intention of solving the "Jewish problem" because the Jews themselves had no intention of leaving the countries of which they had taken hold, and were planning to use Palestine only as some sort of "Jewish Vatican," the text described the links between Jewish organizations in various countries and the channels through which they were exercising a determining influence on the politics and the economies of their host countries. Hagen's production bristled with the names of personalities and groups whose visible and invisible ties were uncovered in a mighty crescendo: "All the organizational and personal ties of Jewry, established from country to country, come together in the summit organizations of the Jewish International." These summit organizations were the World Jewish Congress, the World Zionist Organization—and B'nai B'rith. The mastermind at the center of it all was Chaim Weizmann, whose collected essays and speeches, published in Tel Aviv in 1937, were repeatedly quoted. Hagen's memorandum was no mere exercise in cynicism. "The Jewish 'experts' of the SD believed in their constructs . . . [for them,] anti-Semitism, which they pretended was matter of fact, scientific, and rational, was the basis of their action."[15]

Himmler, Heydrich, and *Das Schwarze Korps* illustrate the constant dichotomy of Nazi thinking regarding the Jews during the last months of peace: On the one hand, emigration by all means was the concrete aim and the concrete policy, but there was also the realization that, given its world-threatening nature, the Jewish problem could not be solved by mere practicalities, that something infinitely more radical was necessary. This was the gist of Hitler's "prophecy," even if tactically his threats were aimed at intimidating the British and American "warmongers." One way or another, through every available channel, the regime was convincing itself and was conveying the message that the Jews, as helpless as they may have looked on the streets of Germany, were a demonic power striving for Germany's perdition. On January 11 and 13 it was Walter Frank's turn to have his say, in a two-part radio broadcast entitled "German Science in Its Struggle Against World Jewry." After emphasizing that scientific research on the Jewish question could not be pursued in isolation but had to be integrated into the totality of national and world history, Frank plunged into deeper waters: "Jewry is one of the great negative principles of world history and thus can only be understood as a parasite within the opposing positive principle. As little as Judas Iscariot with his thirty silver coins and the rope with which he ultimately hanged himself can be understood without the Lord whose community he betrayed with a sneer, but whose face haunted him to his last hour—that night side of history called Jewry cannot be understood without being positioned within the totality of the historical process, in which God and Satan, Creation and Destruction confront each other in an eternal struggle."[16]

Thus, alongside and beyond obvious tactical objectives, some other thoughts were emerging on the eve of the war. No program of extermination had been worked out, no clear intentions could be identified. A bottomless hatred and an inextinguishable thirst for a range of ever-harsher measures against the Jews were always very close to the surface in the minds of Hitler and of his acolytes. As both he and they knew that a general war was not excluded, a series of radical threats against the Jews were increasingly integrated into the vision of a redemptive final battle for the salvation of Aryan humanity.

Throughout the weeks during which Hitler was hinting, in his conversations with foreign dignitaries, at the dire fate in store for the Jews and publicly threatening them with extermination, he was kept informed of the negotiations taking place between German representatives and the Intergovernmental Committee for Refugees set up at Evian to formulate an overall plan for the emigration of the Jews from Germany. The negotiations were in line with the general instructions given by Göring on November 12 and December 6, 1938. Although Hitler was fully cognizant of the progress of the discussions, it was Göring who was in charge of the actual steps.[17]

At an early stage, in November 1938, Ribbentrop had tried to play a part in these negotiations, which he had at first entirely opposed, issuing orders to Hans Fischböck, the former Austrian Nazi minister of the economy, to initiate contacts with the Intergovernmental Committee. The Ribbentrop-Fischböck intermezzo did not last long, and in December, Schacht, by now president of the Reichsbank, took over the negotiations with Rublee, first in London and then in Berlin. On January 16, 1939, in a conversation with the Hungarian foreign minister, Count Csáky, Hitler mentioned the possibility of solving the Jewish emigration issue by way of a financial plan.

Schacht was dismissed by Hitler from his position as president of the Reichsbank on January 20, 1939—for reasons entirely unrelated to the negotiations with Rublee (mainly in response to a memorandum warning Hitler of the financial difficulties resulting from the pace of military expenditures); Rublee, a political appointee, had resigned in mid-February 1939, in order to return to private law practice. The contacts continued nonetheless: Helmut Wohltat, one of the highest officials of the Four-Year Plan administration, took over on the German side, and the British diplomat Sir Herbert Emerson henceforth represented the Intergovernmental Committee. An agreement in principle between Wohltat and Rublee had been achieved on February 2. As has been seen, it envisaged that some 200,000 Jews over the age of forty-five would be allowed to stay in the Greater German Reich, whereas some 125,000 Jews belonging to the younger male population would emigrate, with their dependents. (The numbers varied slightly from one proposal to another.) The emigration process was to be spread over a period of three to five years, with its financing to be ensured by an international loan mainly taken out by Jews all over the world and secured by the assets still belonging to the Jews of Germany (approximately six billion RM, less the billion-mark fine imposed after the pogrom). As in the Haavarah Agreement, the Germans made sure that various arrangements included in the plan would enhance the export of

German goods and thus ensure a steady flow of foreign currency into the Reich. The agreement was nothing less than Germany's use of hostages in order to extort financial advantages in return for their release.

The concrete significance of the agreement depended on the successful floating of the loan and, in particular, on the designation of the countries or areas to which the Jews leaving Germany were to emigrate. Each of the Western powers involved had its preferred territorial solution, usually involving some other country's colony or semicolony: Angola, Abyssinia, Haiti, the Guianas (now Guyana, French Guiana, and Surinam), Madagascar, and so on. In each case some obstacle arose or, more precisely, was raised as a pretext; even on paper no refuge zone was agreed upon before the outbreak of the war put an end to all such pseudo-planning.

Thus by means of pressure, threats, and grand schemes Hitler may have imagined that "the Jews of the world" would become pawns in his plans for aggression, because the Jews of Germany were now hostages in his hands.

On November 7, 1938, while the German Foreign Ministry was still refusing to have any contact with the Intergovernmental Committee and its representative, George Rublee, State Secretary Ernst von Weizsäcker received the British chargé d'affaires, Sir George Ogilvie-Forbes, to discuss the issue. "As Ogilvie-Forbes indicated that he personally knew Rublee well from Mexico," Weizsäcker wrote in a memorandum to Undersecretary Ernst Woermann, the chief of the political division, "I asked him to what percentage was Rublee an Aryan. Ogilvie-Forbes believes that Rublee does not have any Jewish blood."[18] Three days later Woermann himself inquired about Rublee's racial origins, this time of an American diplomat; the answer was the same: Rublee was unquestionably an Aryan. When, on November 15, the American ambassador, Hugh Wilson, came to take leave of Ribbentrop, the foreign minister felt the need to ask once more: Wilson had to state emphatically that Rublee was of French Huguenot origin and that not a drop of Jewish blood flowed in his veins.[19]

III

According to the German census of May 1939 and to various computations made since the war, 213,000 full Jews were living in the Altreich at the time of the census.[20] By the end of 1939, the number had been reduced to 190,000.[21] Strangely enough, a June 15, 1939, SD report indicated that at the end of December 1938, 320,000 full Jews were still living in the Altreich.[22] There is no explanation for the inflated numbers produced by the SD (the numbers do not tally with what is known even if accelerated emigration during 1939 is taken into account). Whatever the reasons for these discrepancies, the demographic data provided by the Jewish Section of the SD are nonetheless significant. Only 16 percent of the Jewish population (on December 31, 1938) were under age twenty; 25.93 percent were between twenty and forty-five, and 57.97 percent over forty-five.[23] These indications correspond to other known estimates: The Jewish population in Germany was rapidly becoming a community of elderly people. And it was also becoming hopelessly impoverished. Whereas in 1933, for example, there had been more than 6,000 "Jewish" small businesses in Berlin, by April 1, 1938, their number had been reduced to 3,105. By the end of that year, 2,570 had been liquidated and 535 been "sold" to Aryans.[24] More than two centuries of Jewish economic activity in the Prussian and German capital had come to an end.

The daily situation of these Jews was described in a memorandum sent in February 1939 by Georg Landauer, director of the Central Bureau for the Settlement of German Jews in Palestine, to his Jerusalem colleague Arthur Ruppin: "Only the employees of Jewish organizations," wrote Landauer, "and some people who rent rooms or cater meals are still earning something. . . . In West Berlin [a Jew] can get a coffee only in the waiting room of the Zoo [Railroad] Station and a meal in a Chinese or some other foreign restaurant. As the Jews' leases are constantly being rescinded in buildings inhabited by a 'mixed population,' they increasingly move in with each other and brood over their fate. Many of them have not yet recovered from the 10th of November and are still fleeing from place to place in Germany or hiding in their apartments. Travel agencies, mainly in Paris, get in touch with consulates that can be bribed—this is mainly true of Central and South American republics—and purchase visas to foreign countries for high prices and enormous commissions. It has often happened that, having suddenly granted several hundred visas, consuls pocketed the money and were then dismissed by their governments. After that, the chances of Jews to enter the countries concerned disappear for a long time. Early in the morning, Jews appear at travel agencies and stand in long lines waiting to ask what visas one can obtain that day."[25]

Landauer's description found an uncanny echo in an SD report two months later: "The defense measures taken by the Party and the state, which follow each other in quick succession, no longer allow the Jews to catch their breath; a real hysteria has set in among both Jewish women and men. Their mood of helplessness is possibly best expressed by the words of a Ludwigsburg Jewess, who declared 'that if she didn't have children, she would long ago have committed suicide.'"[26]

For some time the Nazis had been aware that, in order to expedite the emigration of the Jews, they had to hold them in an even tighter organizational grip than before, and that they themselves also needed to set up a centralized emigration agency on the Viennese model, so as to coordinate all the emigration measures in the Reich.

The establishment of the new body that henceforth was to represent the Jews in Germany was initiated in the summer of 1938. By the beginning of 1939, its shape and function were clear. According to a February memorandum from the Düsseldorf Gestapo, "the Jewish organizations must be associated with all measures taken to prepare for the emigration of the Jews. To further that aim, it is necessary to bring together in one single organization for the whole Reich the means dispersed among the various organizations. The Reichsvertretung has therefore been given the task of building a so-called Reichvereinigung [Reich Association of the Jews in Germany] and of ensuring that all existing Jewish organizations disappear and put all their installations at the disposal of the Reichsvereinigung."[27]

The Reichsvereinigung was finally established on July 4, 1939, by the tenth supplementary decree to the Reich Citizenship Law. Its main function was clearly defined in Article 2: "The purpose of the Reichsvereinigung is to further the emigration of the Jews."[28] But despite the Nazis' clear priorities, the bulk of the decree dealt with the other functions, such as education, health, and especially welfare: "The Reichsvereinigung is also the independent Jewish welfare system." And the minister of the interior was entitled to add further responsibilities to the new organization.[29] Thus the structure of the decree clearly conveyed the impression that the Nazis themselves did not believe in the success of the emigration drive. For all practical purposes, the Reichsvereinigung was becoming the first of the Jewish Councils, the Nazi-controlled Jewish organizations that, in most parts of occupied Europe, were to carry out the orders of their German masters regarding life and death in their respective communities.

A few months earlier, on January 24, Göring informed the minister of the interior that a Reich Central Office for Jewish Emigration (Reichszentrale für Jüdische Auswanderung) was being set up within the framework of the ministry, but as Heydrich's sole responsibility: "The Reich Central Office will have the task of devising uniform policies as follows: (1) measures for the *preparation* of increased emigration of Jews; (2) the *channeling* of emigration, including, for instance, preference for emigration of the poorer Jews . . . ; (3) the speeding up of emigration in *individual* cases."[30] Heydrich appointed the head of the Gestapo, SS-Standartenführer Heinrich Müller, chief of the new Reich Central Office.

On October 30, 1938, the local party leader in Altzenau (Franconia) wrote to the district party office in Aschaffenburg that two houses belonging to different members of a Jewish family named Hamburger were being acquired by party members, each for half its market value of 16,000 RM. The local party section requested the right to acquire one of these two houses. The authorization was granted in June 1939 and the price established by the party district office at 6,000 RM, slightly more than a third of the real value. In December 1938 the same Altzenau party chief informed his district leader that Jews who—as of January 1, 1939—would no longer be allowed to engage in business, were selling off their goods at rock-bottom prices. The local population was asking whether it could buy the Jewish merchandise, the ban on commerce with Jews notwithstanding.[31]

The Jews of Germany who had not managed to flee were increasingly dependent on public welfare. As noted in the previous chapter, from November 19, 1938, on, Jews were excluded from the general welfare system: They had to apply to special offices, and they were subjected to different and far more stringent assessment criteria than the general population. The German welfare authorities attempted to shift the burden onto the Jewish welfare services, but there too the available means were overstrained by the increasing need. The solution to the problem soon became evident, and on December 20, 1938, the Reich Labor Exchange and Unemployment Insurance issued a decree ordering all unemployed Jews who were fit for work to register for compulsory labor. "It was obvious that only carefully chosen hard and difficult work was to be assigned to the Jews. Building sites, road and motorway work, rubbish disposal, public toilets and sewage plants, quarries and gravel pits, coal merchants and rag and bone works were regarded as suitable."[32] But from a Nazi point of view, the decree created a series of new problems.

For instance, some of the tasks alloted to the Jews had a special national significance or were linked to the name of the Führer, an unacceptable outrage for some party members. "The assignment of Jews to work on the *Reichsautobahnen* [Reich freeways], the inspector-general of German roads wrote to the Reich Labor Minister on June 22, 1939, "cannot in my opinion be in accord with the prestige given to the *Reichsautobahnen* as Roads of the Führer." The general inspector suggested that Jews be used only in work indirectly related to the construction or repair of the motorways, such as in stone quarries and the like.[33]

The December 1938 decree had imposed strict segregation on Jewish workers: They had to be kept "separated from the community."[34] But in many cases, mostly on farms, contact was unavoidable. The reactions from party activists were foreseeable. On April 13, 1939, a party district leader in Baden wrote to a local labor exchange: "Peasants who are still employing Jews are those who know the Jews very well, who did business with them and possibly still owe them money. An honest German peasant with but a minimum of National Socialist consciousness would never take a Jew into his house. If, on top of that, Jews were allowed to stay overnight, our race laws would be worthless."[35]

Even more serious concern was expressed in a letter from a party district leader in Mannheim to the director of the Labor Exchange in that city. The subject was the employment of "the Jew Doiny" by a local bakery. The district leader was unable to understand how a Jew could be employed in food-related business. Should the Volksgenossen patronize a bakery in which bread is baked by a Jew?[36] At times such dangerous contacts could be eliminated in a summary fashion. On August 29, 1939, the district governor of Hildesheim could inform all of the area's heads of administrative regions and mayors of rather momentous news: "In the district of Hildesheim, all business activity of Jewish barbers and Jewish undertakers is terminated."[37]

In the meantime, throughout the prewar months of 1939, the concentration of Jews in Jewish-owned dwellings continued; it was made easier, as has been noted, by the April 30, 1939, decree allowing rescinding of leases with Jews. In Berlin the entire operation was spurred by Speer's agency, and the municipal authorities, supported by the party, started pressuring Aryan landlords to put an end to their contracts with Jewish tenants. Pressure was

indeed necessary, according to an official report, "since for political reasons, the Jews were the quietest and the most unassuming tenants" and did not "cause any trouble" to their landlords.[38] Once the transfers had taken place, it became clear that the areas cleared of Jews coincided exactly with the those designated by Speer's offices to be "Jew free."[39]

At some stage the Propaganda Ministry discovered that 1,800 window openings belonging to Jewish inhabitants would face the planned huge avenue called the East-West Axis. As that could be dangerous, Hitler was to be asked what appropriate measures should be taken.[40]

Even the most brutal systems sometimes make exceptions among their designated victims. In Nazi Germany such exceptions never applied to "full" Jews but only to some *Mischlinge* who were deemed unusually useful (Milch, Warburg, Chaoul) or especially well connected (Albrecht Haushofer). But in the rarest of cases, exceptions could also apply to *Mischlinge* of the first degree who were so insignificant and so persistent that both state and party bureaucracies were finally worn down. This was to be the unlikely conclusion of the story of Karl Berthold, the Chemnitz civil servant whose struggle to keep his job has been followed in these pages since its beginning, in 1933.

In her January 23, 1936, letter to the Reich minister of labor, Ada Berthold, Karl Berthold's wife, had expressed only desperation: her husband's three-year struggle had left both of them devastated in health and spirit. For Ada Berthold, there was now only one hope: A meeting with Hitler.[41] The appointment was not granted, and, at that very same time, Berthold was ordered by the Ministry of the Interior's Reich Office for Kinship Research to undergo a racial examination at the Institute for Racial Science and Ethnology at the University of Leipzig.[42] Meanwhile the Office for Kinship Research had found the presumed Jewish father in Amsterdam; but the man denied being Karl Berthold's father. The racial examination, however, was not in the subject's favor: "A number of indices point to a Jewish begetter."[43] In November 1938 the verdict came down: Berthold must be dismissed from his job. It was then that he played his last card: a personal petition to Hitler. In it Berthold very clearly summed up his situation: "Since April 1924, I have been a permanent employee of the Social Benefits Office in Chemnitz, where, for almost five years now [actually more than five], a procedure for my dismissal has been pending because of my inability to prove my Aryan descent. Since then, there has

been a search for my father (he is totally unknown to me, as I was an illegitimate child). No paternity was ever recognized in court. It is only
because of the circumstance that my deceased mother mentioned a Jewish
name that this matter has become fateful to me, without any objective
proof. In consequence of the fact that, as already mentioned, the begetter
could never be identified, I was ordered to undergo an examination at the
Racial Science Institute in Leipzig, with which I complied. Then, it was
supposedly ascertained that I showed Jewish characteristics. On the basis
of this attestation of origins of May 23, 1938, the Minister of Labor has
ordered my dismissal from the Social Benefits Office in Chemnitz."

After describing the tragic consequences of this situation for himself
and his family, Berthold continued: "I feel myself to be a true German,
with a true German heart, who has never seen or heard anything of Jews
and who has no desire ever to know them." He listed the events of German
nineteenth-century history in which his maternal ancestors had taken part
and all the national duties he and his mother had fulfilled in the war. He
had been a party member since March 1933 and had "foolishly" resigned
his membership in 1936 because of the ongoing investigation. Of his three
sons, the youngest was a member of the Jungvolk, the next oldest a Hitler
Youth, and the oldest in his third year of military service.

"Such are my circumstances," Berthold added. "They certainly are to
be considered as normal, and it can be derived from them that I have nothing to do in any way with the Jewish rabble."[44]

Berthold's petition was forwarded by Hitler's Chancellery to that of
Deputy Führer Hess. In February 1939 it appeared that the answer would
be negative. However, on August 16, 1939, a letter from the Deputy Führer
to the labor minister "concerning the continuing employment of a Jewish
Mischling in public service" announced the verdict: Karl Berthold was to be
allowed to keep his position as an employee of the Chemnitz Social
Benefits Office.[45]

Karl Berthold's story throughout the first six years of the regime shows
in microcosm how a modern bureaucracy could be the efficient purveyor of
exclusion and persecution and, at the same time, could be slowed down by
an individual's use of the system's loopholes, the ambiguity of decrees, and
the immense variety of individual situations. Since, the party and the state
during the thirties, decided to deal with every issue related to the Jews in
the most minute detail, and, in particular, to resolve each case of legal or
administrative exception, the entire policy might have ground to a halt as a

result of the very complexity of the task. That this did not happen is possibly the most telling proof of the relentless obstinacy of the anti-Jewish effort, a kind of determination that mere bureaucratic routine alone could not have mobilized.

It is difficult to obtain a clear picture of the attitudes of ordinary Germans toward the increasingly more miserable Jews living in their midst in the spring of 1939. As we saw from the SOPADE report about the populace's responses to the group of Jews being sent back and forth over the western border during those weeks, hatred and sympathy were mixed, possibly according to differences in age. One obtains the same mixed impression from memoirs, such as those of Valentin Senger, a Jew who survived the Nazi period in Frankfurt,[46] or from Klemperer's diaries. There is no doubt that, at least in smaller towns and villages, some people were still patronizing Jewish stores, although in principle no Jewish business (unless it was an exporter or belonged to foreign Jews) was allowed to function after January 1, 1939. How else to explain the confidential report addressed on February 6 by the Bernburg district party leadership to its counterpart in Rosenheim, regarding "lists of clients of Jewish stores in the Bernburg district"? The report not only gives the list of "verified customers of Jews" but also indicates the store owners' names and the dates of the purchases and amounts paid.[47]

On May 5, 1939, the Fischbach police station informed the Labor Office in Augsburg of its attempt to send three men of a local Levi family (Manfred Israel, Sigbert Israel, and Leo Israel) to compulsory work at the Hartmann brick factory in Gebelbach. Whereas Manfred Levi was in Altona (a suburb of Hamburg) attending a Zionist professional training school to prepare for his emigration to Palestine, Sigbert and Leo's German employers had come to the police station to request permission to retain the services of their Jewish carpenter and gardener.[48]

Gestapo surveillance of the churches reveals the same mixed attitudes. Thus, in January 1939, at a meeting of the Evangelical Church in Ansbach, one Knorr-Köslin, a physician, declared that in present-day Germany the sentence "all salvation comes from the Jews" should be deleted from the Bible; the report indicates that Knorr-Köslin's outburst caused a protest from the audience; the protest might have been only on purely religious grounds.[49] When, on the other hand, Pastor Schillfarth of Streitberg declared that "after baptism, Jews become Christians," one of his young students retorted ("in a strong and well-deserved way," says the report), "But Pastor, even if you pour six pails of water on a Jew's head, he still remains a Jew."[50]

In small towns some municipal officials avoided the mandatory forms of addressing Jews. When, in early 1939, the town officials of Goslar negotiated with the head of the local Jewish community for the acquisition of the synagogue building, their letters were addressed "Herrn Kaufmann W. Heilbrunn" (Mr. W. Heilbrunn, merchant), without using the obligatory "Israel."[51]

And yet . . . In a December 1938 diary entry, Victor Klemperer told of a policeman who in the past had been friendly to him, even encouraging. When he encountered him that month, in the municipal office of the small town where the Klemperers owned a country house, the same policeman passed by him "looking fixedly ahead, as distant as could be. In his behavior," Klemperer commented, "the man probably represents 79 million Germans."[52]

Looking back over the first six years of the regime, this much can be said with a measure of certainty: German society as a whole did not oppose the regime's anti-Jewish initiatives. Hitler's identification with the anti-Jewish drive, along with the populace's awareness that on this issue the Nazis were determined to push ahead, may have reinforced the inertia or perhaps the passive complicity of the vast majority about a matter that most, in any event, considered peripheral to their main interests. It has been seen that economic and religious interests triggered some measure of dissent, mainly among the peasantry and among Catholics and members of the Confessing Church. Such dissent did not, however, except in some individual instances, lead to open questioning of the policies. Yet, during the thirties, the German population, the great majority of which espoused traditional anti-Semitism in one form or another, did not demand anti-Jewish measures, nor did it clamor for their most extreme implementation. Among most "ordinary Germans" there was acquiescence regarding the segregation and dismissal from civil and public service of the Jews; there were individual initiatives to benefit from their expropriation; and there was some glee in witnessing their degradation. But outside party ranks, there was no massive popular agitation to expel them from Germany or to unleash violence against them. The majority of Germans accepted the steps taken by the regime and, like Klemperer's policeman, looked the other way.[53]

From within the party ranks hatred flowed in an ever more brutal and open way. Sometimes, as with anonymous informers, it is not known whether it originated in the party or among unaffiliated citizens. In any case, denunciations reached such proportions on the eve of the war that Frick, on orders from Göring, had to intervene, addressing a January 10, 1939, letter to the whole array of civilian and police authorities.

Marked CONFIDENTIAL, Frick's letter tersely indicated his subject: "The Jewish Question and Denunciations." It related that—on the occasion of a conference with Göring regarding the necessity of eliminating the Jews from the German economy and of using their assets for the goals of the Four-Year Plan—the Generalfeldmarschall had mentioned "that it had been recently noticed that German Volksgenossen were being denounced because they had once bought in Jewish shops, inhabited Jewish houses, or had some other business relations with Jews." Göring considered that a very unpleasant development, which, in his opinion, could hurt the realization of the Four-Year Plan: "The General-Feldmarschall wishes therefore that everything be done to put an end to this nuisance."[54]

Frick's order probably did not reach party member Sagel of Frankfurt. On January 14, 1939, a grocer named Karl Schué complained to his local group leader that female party member Sagel had berated him for having sold butter to a Jew (the last one, wrote Schué, "who still buys his butter in my store") and told him that she had informed the local [party] leader accordingly. Schué used the occasion to unfold the tale of his economic woes as a the owner of a small store and then returned to Sagel: "Maybe you could inform female Party Comrade Sagel that I do not wear any uniform, as she told me that I should take off my uniform. It is really sad," he concluded, "that even today, in Greater Germany, such incidents could occur, instead of help being provided to a struggling businessman to allow him to get on his feet and spare his family serious worry."[55]

It could be that denunciations were forbidden only when they concerned events in the distant past. Recent occurrences were another matter. On Sunday, June 25, 1939, Fridolin Billian, a local party cell leader and teacher in Theilheim, in the Schweinfurt district of Main-Franken, reported to the local police station that a sixteen-year-old Jew, Erich Israel Oberdorfer, a horse dealer's son, had perpetrated indecent acts on Gunda Rottenberger, a workingman's ten-year-old daughter. The story had been told to him by Gunda's mother, supposedly because Gunda had admitted that Erich Oberdorfer had lured her to the stable and told her that she would get five *Pfennig* if she took off her underpants. Oberdorfer denied the accusation; Gunda herself said that he had made the offer, but that nothing happened when she refused, except that they had eaten cherries in the stable and, in order to explain their prolonged absence from home, decided to tell Gunda's mother they had been counting the hens.[56]

After the Theilheim police proved unable to obtain confirmation of a sexual misdeed from Gunda Rottenberger herself, the Gestapo took over and produced one Maria Ums, who readily admitted that some years (she could not remember how many) earlier, Erich, who was her own age, had touched her genitals and even inserted his member into her "sexual parts." Then a certain Josef Schäfner came forward. He remembered that Siegfried Oberdorfer, Erich's father, had told him that during the war he had hit a lieutenant with his pistol butt (because the lieutenant had called him a dirty Jew) and killed him. Siegfried Oberdorfer denied it all; according to him, it was a tale invented by Schäfner, who spread it in the local inns when he was drunk.[57]

The hearings in young Erich Oberdorfer's case were over by 1940: He was sentenced to one year in prison. In 1941, on his release from the Schweinfurt jail, he was sent to Buchenwald as a race defiler.[58] His dossier was closed and his short life, too, possibly reached its end.

In April 1939 the Ministry of Religious Affairs reached an agreement with the Evangelical Church Leaders' Conference on further relations between the Protestant churches and the state. The agreement was strongly influenced by German-Christian ideology, but nonetheless not opposed, at least not formally, by a majority of German pastors; the Godesberg Declaration of the same month gave its full weight to this new statement.

"What is the relation between Judaism and Christianity?" it asked. "Is Christianity derived from Judaism and has therefore become its continuation and completion, or does Christianity stand in opposition to Judaism? We answer: Christianity is in irreconcilable opposition to Judaism."[59]

A few weeks later, the signatories of the Godesberg Declaration met at the Wartburg near Eisenach, a site sacred to the memory of Luther and hallowed by its connection with the German student fraternities, to inaugurate the Institute for Research on Jewish Influence on the Life of the German Church. According to a historian of the German churches, "a surprisingly large number of academics put themselves at the disposal of the institute, which issued numerous thick volumes of proceedings and prepared a revised version of the New Testament (published in an edition of 200,000 copies in early 1941). It omitted terms such as "Jehovah," "Israel," "Zion," and "Jerusalem" which were considered to be Jewish.[60]

Cleansing Christianity of its Jewish elements was a Sisyphean task indeed. Just at the time of the Godesberg Declaration, when the Eisenach Institute was being set up, an urgent query was addressed to the SD by the party's Education Office: could it be that Philipp Melanchthon, possibly the most important German figure of the Reformation after Martin Luther himself, was of non-Aryan origin? The Education Office had discovered this piece of unwelcome news in a book by one Hans Wolfgang Mager, in which, on page 16, the author stated: "Luther's closest collaborator and confidant, Philipp Melanchthon, was a Jew!" The SD answered that it could not deal with this kind of investigation; the Reich Office for Ancestry Research would possibly be the right address.[61]

Whether or not Melanchthon's case underwent further scrutiny, it seems that the great reformer was not excluded from the fold. It was easier to eliminate lesser servants of the church, such as pastors and the faithful of Jewish origin. On February 10, 1939, the Evangelical Church of Thuringia forbade its own baptized Jews access to its temples. Twelve days later the Saxon Evangelical Church followed suit; the ban then spread to the churches of Anhalt, Mecklenburg, and Lübeck. In the early summer, all pastors of non-Aryan ancestry were dismissed. The letter sent on July 11, 1939, to Pastor Max Weber of Neckarsteinach in Hesse-Nassau by the president of the Land Church Office used a standard formula: "The mandate you received on January 10, 1936—No. 941—to administer the parish of Neckarsteinach, under condition of a possible cancellation at any time, is hereby revoked; you are dismissed from your position as of the end of July this year. The director of the German Evangelical Church Office has ordered on May 13, 1939—K.K. 420/39—that the provisions of the German Civil Service Law of January 26, 1937 [excluding all *Mischlinge* from the civil service], be administratively applied to all clergymen and employees of the Church. According to the provisions of the German Civil Service Law, only a person of German or related blood can be a civil servant (see para. 25). As you are a *Mischling* of the second degree [one Jewish grandparent], not of German or related blood and thereby according to the meaning of the German Civil Service Law cannot be a clergyman or remain one, your dismissal has had to be declared."[62]

The Eisenach Institute dealt with Jews and traces of Judaism in

Christianity; the project to establish a research institute on Jewish affairs in Frankfurt, on the other hand, was concerned with the comprehensive task of submitting all matters Jewish to scientific Nazi scrutiny. The existence of a large research library on Jewish affairs at the University of Frankfurt, along with the rift between Walter Frank and Wilhelm Grau—which led to Grau's dismissal from his position as director of the Jewish section of the Institute for the History of the New Germany—enabled the mayor of Frankfurt, Fritz Krebs, to suggest, in the fall of 1938, that the new institute be set up, with Grau as its director.[63] The minister of education and Hess approved the project and preparations began: The festive opening was to take place two years later, in 1941.

Goebbels was also active in this effort to identify non-Aryans in the various cultural areas—and in the purges that followed. Since 1936, the Propaganda Ministry had been compiling and publishing lists of Jewish, mixed, and Jewish-related figures active in cultural endeavors[64] and prohibiting their membership in non-Jewish organizations and the exhibition, publication, and performance of their works. But Goebbels evidently felt that he had not yet achieved total control. Thus, throughout 1938 and early 1939, the propaganda minister harassed the heads of the various Reich chambers to obtain updated and complete lists of Jews who had been excluded from pursuing their professions.[65] One list after another was sent to the Ministry of Propaganda with the admission that it was still incomplete. (A sample of one such, sent by the Reich Music Chamber on February 25, 1939: "Ziegler, Nora, piano teacher; Ziffer, Margarete, private music teacher; Zimbler, Ferdinand, conductor; Zimmermann, Artur, pianist; Zimmermann, Heinrich, clarinetist; Zinkower, Alfons, pianist; Zippert, Helene, music teacher; . . . Zwillenberg, Wilhelm, choir conductor.")[66]

The Rosenberg files contain similar lists. One document contains part 6 of a list of Jewish authors—those with names beginning with the letters *S* through *V*—including three Sacher-Masochs and six Salingers, followed by Salingré and Salkind, and ending with Malea Vyne, who, according to the compiler, is the same person as Malwine Mauthner.[67]

I V

In the fall of 1938, when Tannenhof, an institution for mentally ill patients (belonging to the Evangelical Kaiserswerth Association) was formulating its new statutes, the board decided that they "must take into account the changed attitude of the German Volk to the race question by excluding the admission of patients of Jewish origin. . . . The institution's administration is instructed that from now on it should not admit patients of Jewish origin and . . . with the aim of freeing itself as soon as possible of such patients . . . it should give notice to private patients of Jewish origin at the earliest possible date and, in the case of regular patients [of Jewish origin], should ask the district administration to transfer them to another institution."[68]

Other Evangelical institutions had already started practicing such selection several months earlier. Thus, on March 7, 1938, Dr. Oscar Epha, director of the Evangelical Inner Mission in Schleswig-Holstein, wrote to Pastor Lensch in Alsterdorf: "I have informed the Hamburg public welfare authorities that we can no longer take in any Jewish patients, and we have asked [them] to transfer to Hamburg the four Jewish patients we still have."[69] The Inner Mission's initiative thus preceded the Interior Ministry decree of June 22, 1938, according to which "the accommodation of Jews in medical institutions is to be executed in such a way that the danger of race defilement is avoided. Jews must be accommodated in special rooms." [70] How this decree was to be carried out was not always clear: "We ask you to inform us," the hospital administration in Offenburg wrote to its sister institution in Singen on December 29, 1938, "whether you accept Jews and, in case you do, whether you put them together with Aryan patients or whether special rooms are kept ready for them." The Singen colleagues answered promptly: "As there is no Jewish hospital in this region, and as to this day we have not received any instructions in this matter, we cannot refuse to accept Jewish emergency patients. But, as there are only a few of them, we accommodate the Jewish patients separately."[71] In the Hamburg area, on the other hand, the instructions from the Health Office were unambiguous: "Because of the danger of race defilement, special attention should be devoted to the accommodation of Jews in institutions for the sick. They must be separated spatially from patients of German or related blood. Insofar as Jews who are not bedridden have to remain in institutions for the sick, their accommodation and arrangements regarding their movements inside or on the grounds must make certain to exclude any danger of race defilement. . . . I therefore demand that this danger be prevented under all circumstances."[72]

Dead Jews were no less troublesome than sick ones. On March 17, 1939, the Saxon office of the German Association of Municipalities wrote to the head office in Berlin that, since the Jews had their own cemetery nearby, the mayor of Plauen intended to forbid the burial or cremation of racial Jews in the municipal cemetery.[73] The letter writer wanted to be assured of the legality of this decision, which was obviously directed against converted Jews or those who had simply left their religious community. In his answer two months later, Bernhard Lösener wrote that "the burial of Jews can be forbidden in a municipal cemetery when there is a Jewish cemetery in the same district. The definition of a Jew has been established by the Nuremberg Laws and is also applicable to converted Jews. . . . The owner of the Jewish cemetery is not allowed to forbid the burial of a converted Jew." Lösener also informed the Association of Municipalities that a cemetery law was in preparation. Whether access to a municipal cemetery could be refused to Jews who had already acquired graves there or who wished to take care of the tombs of deceased relatives was, according to Lösener, still under consideration.[74]

V

The Polish crisis had unfolded throughout the spring and summer of 1939. This time, however, the German demands were met by an adamant Polish stand and, after the occupation of Bohemia and Moravia, by new British resolve. On March 17, in Birmingham, Chamberlain publicly vowed that his government would not allow any further German conquests. On March 31 Great Britain guaranteed the borders of Poland, as well as those of a series of other European countries. On April 11 Hitler gave orders to the Wehrmacht to be ready for "Operation White," the code name for the attack on Poland.

On May 22, Germany and Italy signed a defense treaty, the Steel Pact. Simultaneously, while Great Britain and France were conducting hesitant and noncommittal negotiations with the Soviet Union, Hitler made an astounding political move and opened negotiations of his own with Stalin. The Soviet dictator had subtly indicated his readiness for a deal with Nazi Germany in a speech in early March and by a symbolic act: on May 2, he dismissed Foreign Minister Maxim Litvinov and replaced him with Vyacheslav Molotov. Litvinov had been the apostle of collective security— that is, of a common front against Nazism. Moreover, he was a Jew.

The German-Soviet Nonaggression Pact was signed on August 23; an attached secret protocol divided a great part of Eastern Europe into areas to be eventually occupied and controlled by the two countries in case of war. Hitler was now convinced that, as a result of this coup, Great Britain and France would be deterred from any military intervention. On September 1, the German attack on Poland started. After some hesitation the two democracies decided to stand by their ally, and on September 3, France and Great Britain were at war with Germany. World War II had begun.

In the meantime other events were occurring in Hitler's Reich. Soon after the handicapped Knauer baby had been put to death in Leipzig, Hitler instructed his personal physician, Karl Brandt (who had performed the euthanasia), and the head of his personal chancellery, Philipp Bouhler, to see to the identification of infants born with a variety of physical and mental defects. These preparations were undertaken, in the strictest secrecy, during the spring of 1939. On August 18, doctors and midwives were ordered to report any infants born with the defects that had been listed by a committee of three medical experts from the Reich Committee for Hereditary Health Questions. These infants were to die.[75]

Another initiative was taken at the same time; it was, as we have seen, one about which religious authorities at first kept prudent silence. Sometime prior to July 1939, in the presence of Bormann and Lammers, Hitler instructed State Secretary Leonardo Conti to begin preparations for adult euthanasia. Brandt and Bouhler quickly succeeded in getting Conti out of the way and, with Hitler's assent, took over the entire killing program. Both the mass murder of handicapped children and of mentally ill adults had been decided upon by Hitler, and both operations were directed under cover of the Führer's Chancellery.[76]

None of this could yet have had any impact on the popular fervor surrounding Hitler or on the public's ardent adherence to many of the regime's goals. Hitler's accession to power would be remembered by a majority of Germans as the beginning of a period of "good times." The chronology of persecution, segregation, emigration, and expulsion, the sequence of humiliations and violence, of loss and bereavement that molded the memories of the Jews of Germany from 1933 to 1939 was not what impressed itself on the consciousness and memory of German society as a whole. "People experienced the breakneck speed of the economic and foreign resurgence of Germany as a sort of frenzy—as the common expression has it," writes German historian Norbert Frei. "With astonishing rapidity, many identified themselves with the social will to construct a Volksgemeinschaft that kept any thoughtful or critical stance at arm's length. . . . They were beguiled by the esthetics of the Nuremberg rallies and enraptured by the victories of German athletes at the Berlin Olympic Games. Hitler's achievements in foreign affairs triggered storms of enthusiasm. . . . In the brief moments left between the demands of a profession and those of the ever-growing jungle of Nazi organizations, they enjoyed modest well-being and private happiness."[77]

It was in this atmosphere of national elation and personal satisfaction that, on April 20, 1939, some four months before the war, eighty million Germans celebrated Hitler's fiftieth birthday. During the following weeks hundreds of theaters showed avid audiences the pageantry and splendor of the event. Newsreel No. 451 was a huge success. Terse comments introduced the various sequences: "Preparations for the Führer's fiftieth birthday/The entire nation expresses its gratitude and offers its wishes of happiness to the founder of the Greater German Reich/Gifts from all the Gaue of the Reich are continuously brought to the Reich Chancellery/Guests from all over the world arrive in Berlin/On the eve of the birthday, the Inspector General for the Construction of the Capital of the Reich, Albert Speer, presents to the Führer the completed East-West Axis/The great star of the newly erected victory column shines/Slovak Premier Dr. Josef Tiso, President of the Protectorate of Bohemia and Moravia Emil Hacha, and the Reich Protector Freiherr von Neurath . . . /The troops prepare for the parade/The Third Reich's greatest military spectacle begins/For four and a half hours, formations from all branches of the armed forces march by their Supreme Commander . . . !"[78]

Resuming its activities—briefly curtailed after the November 1938 pogrom—early in the year on orders from above, the Kulturbund in its Berlin theater that April staged *People at Sea*, a drama by English writer J. B. Priestley. An American correspondent, Louis P. Lochner of the Associated Press, covered the April 13 opening: "Because . . . the British playwright has renounced all claims to royalties from German Jews, the Jewish Kulturbund was able tonight to present a beautiful premiere rendition in German of *Men at Sea* [sic]. The translation was by Leo Hirsch, the stage setting by Fritz Wister. Almost 500 attentive, art-loving Jews witnessed the performance and applauded generously. Outranking all others in the depth of her emotional acting was Jennie Bernstein as Diana Lissmore. Alfred Berliner, with his face made up to look much like Albert Einstein's, also scored signally with his interpretation of the role of

Professor Pawlet. The audience wistfully nodded when Fritz Grüne as Carlo Velburg complained again and again that he had no passport. Thirty-nine performances of the Priestley play are planned for the coming weeks."[79]

The play tells of the terrors and hopes of twelve people on a ship in the Caribbean disabled by fire, adrift, and in danger of sinking. The characters depicted on the stage are saved at the end. Most of the Jews seated in the Charlottenstrasse theater that night were doomed.

Notes

Introduction

1. Clearly sharing no common ground with us is the small group of historians of the same generation whose apologetic interpretations of Nazism and the Holocaust were sharply confronted during the "historians' controversy" of the mid-1980s. For that specific debate see Charles S. Maier, *The Unmasterable Past: History, Holocaust, and German National Identity* (Cambridge, Mass., 1988), and Richard J. Evans, *In Hitler's Shadow: West German Historians and the Attempt to Escape from the Nazi Past* (New York, 1989); for a particularly perceptive discussion of the issues, see Steven E. Aschheim, *Culture and Catastrophe: German and Jewish Confrontations with National Socialism and Other Crises* (New York, 1996). For this and other debates on the historical representation of the Holocaust, see the essays included in Peter Baldwin, ed., *Reworking the Past: Hitler, the Holocaust and the Historians* (Boston, 1990), and in Saul Friedländer, ed., *Probing the Limits of Representation: Nazism and the "Final Solution"* (Cambridge, Mass., 1992).

2. One of the earliest examples of the first approach is Raul Hilberg, *The Destruction of the European Jews* (Chicago, 1961); the best illustration of the second is Lucy S. Dawidowicz, *The War Against the Jews, 1933-1945* (New York, 1975).

3. In representing the life of the victims and some attitudes of surrounding society I have drawn most of my illustrations from everyday life. In this respect, and with regard to some other issues brought forth in this book, I have accepted some of Martin Broszat's insights that I criticized in my debate with him in the late 1980s. Yet, I have attempted to avoid some of the pitfalls of the historicization of National Socialism precisely by emphasizing the everyday life of the victims rather than that of the *Volksgemeinschaft*. For the debate, see Martin Broszat, "A Plea for the Historicization of National Socialism" in Baldwin, *Reworking the Past*; Saul Friedländer, "Some Thoughts on the Historicization of National Socialism," ibid.; Martin Broszat and Saul Friedländer, "A Controversy about the Historicization of National Socialism," ibid.

4. For the importance of this wider context, see Omer Bartov, *Murder in Our Midst: The Holocaust, Industrial Killing, and Representation*, New York, 1996. For the impact of modernity as such on the genesis of the "Final Solution," see, among many other studies, Detlev J. K. Peukert, "The Genesis of the 'Final Solution' from the Spirit of Science," in Thomas Childers and Jane Caplan, eds., *Reevaluating the Third Reich* (New York, 1993); Zygmunt Bauman, *Modernity and the Holocaust* (New York, 1989); Götz Aly and Susanne Heim, *Vordenker der Vernichtung: Auschwitz und die deutschen Pläne für eine neue europäische Ordnung* (Hamburg, 1991). For an excellent presentation of related issues in the history of Nazism see Michael Burleigh, ed., *Confronting the Nazi Past: New Debates on Modern German History* (London, 1996).

5. For internal competition as the basis of Nazi radicalization, see mainly the works of Hans Mommsen, particularly "The Realization of the Unthinkable," in *From Weimar to Auschwitz* (Princeton, N.J., 1991). For the cost-benefit calculations of technocrats as incentives for the "Final Solution," see Aly and Heim, *Vordenker der Vernichtung*.

6. Redemptive anti-Semitism is different, as I shall indicate, from the "eliminationist anti-Semitism" referred to by Daniel Jonah Goldhagen in *Hitler's Willing Executioners: Ordinary Germans and the Holocaust* (New York, 1996). Moreover, it represented an ideological trend shared at the outset by a small minority only, and, in the Third Reich, by a segment of the party and its leaders, not by the majority of the population.

7. Because of my emphasis upon the interaction between Hitler, his ideological motivations, and the constraints of the system within which he acted, I hesitate to identify my approach as "intentionalist." Moreover, whereas during the thirties Hitler decided on all major anti-Jewish steps and intervened in the details of their implementation, later his guidelines left much greater leeway to his subordinates in the implementation of the concrete aspects of the extermination. As for Hitler's impact on the Germans, it has been the subject of countless studies and the basic theme of major biographies. For a complex approach both to Hitler's charismatic effect and to his interaction with the populace, see in particular J. P. Stern, *Hitler, The Fuehrer and the People* (Glasgow, 1975), and Ian Kershaw, *Hitler* (London 1991).

8. This point is made both in Michael Wildt, *Die Judenpolitik des SD 1935 bis 1938* (Munich 1995) and in Ulrich Herbert, *Best. Biographische Studien über Radikalismus, Weltanschauung und Vernunft 1903–1989* (Bonn, 1996). For a discussion of this theme, see chapter 6.

9. The reference here is to the opposed theses of Christopher R. Browning, *Ordinary Men: Reserve Police Battalion 101 and the Final Solution in Poland* (New York, 1992), and of Goldhagen, *Hitler's Willing Executioners*. The issue will be discussed at length in volume 2. The impact of Nazi ideology on various Wehrmacht units and its relation to the extreme barbarization of warfare on the Eastern front must also be considered in that context. For this issue see mainly Omer Bartov, *Hitler's Army: Soldiers, Nazis and War in the Third Reich* (New York, 1991).

10. See Martin Broszat, "A Plea," in Baldwin, *Reworking the Past*.

11. The issue is thoroughly discussed in Dominick LaCapra, *Representing the Holocaust: History, Theory, Trauma* (Ithaca, N.Y., 1994).

Chapter 1 Into the Third Reich

1. Walter Benjamin, *The Correspondence of Walter Benjamin*, ed. Gershom Scholem and Theodor Adorno (Chicago, 1994), p. 406.

2. Lion Feuchtwanger and Arnold Zweig, *Briefwechsel 1933–1958*, vol. 1 (Frankfurt am Main, 1986), p. 22.

3. Erik Levi, *Music in the Third Reich* (New York, 1994), p. 42; Sam H. Shirakawa, *The Devil's Music Master: The Controversial Life and Career of Wilhelm Furtwängler* (New York, 1992), pp. 150–51.

4. Alan E. Steinweis, "Hans Hinkel and German Jewry, 1933–1941," *Leo Baeck Institute Yearbook* [hereafter *LBIY*] 38 (1993): 212.

5. Shirakawa, *The Devil's Music Master*, p. 151.

6. Joseph Goebbels, *Die Tagebücher von Joseph Goebbels*, ed. Elke Fröhlich, part 1, *1924–1941*, vol. 2, *1.1.1931–31.12.1936* (Munich, 1987), p. 430.

7. Fred K. Prieberg, *Musik im NS-Staat* (Frankfurt am Main, 1982), pp. 41–42. For a more thorough discussion of the dismissal of Jewish musicians, see Levi, *Music in the Third Reich*, pp. 41ff.

8. Ibid., p. 41.

9. Lawrence D. Stokes, *Kleinstadt und Nationalsozialismus: Ausgewählte Dokumente zur Geschichte von Eutin 1918–1945*, (Neumünster, 1984), p. 730. (Initials are used instead of full names as indicated in the source.)

10. Klaus Mann, *Mephisto* (New York, 1977), p. 157. (Klaus Mann was one of Thomas Mann's sons. The original German edition was published in Amsterdam in 1936; Mann describes Höfgens's happiness at not being Jewish as it found expression in 1933, soon after the *Machtergreifung*.)

11. On the details of this issue see Peter Stephan Jungk, *Franz Werfel: A Life in Prague, Vienna, and Hollywood* (New York, 1990), p. 140.

12. Quoted and excerpted in Golo Mann, *Reminiscences and Reflections: A Youth in Germany* (New York, 1990), p. 144.

13. Jungk, *Franz Werfel*, pp. 141–44.

14. Ibid., p. 145.

15. Joseph Wulf, ed., *Die bildenden Künste im Dritten Reich: Eine Dokumentation* (Gütersloh, 1963), pp. 36, 81ff.

16. Ibid., p. 36.

17. Thomas Mann, *The Letters of Thomas Mann 1889–1955* (London, 1985), p. 170.

18. Ibid., p. 191.

19. Ronald Hayman, *Thomas Mann: A Biography* (New York, 1995), pp. 407–8.

20. Thomas Mann, *Tagebücher 1933–1934,* ed. Peter de Mendelssohn (Frankfurt am Main, 1977), p. 46.

21. Ibid., p. 473.

22. On Thomas Mann's anti-Jewish stance see Alfred Hoelzel, "Thomas Mann's Attitudes toward Jews and Judaism: An Investigation of Biography and Oeuvre," *Studies in Contemporary Jewry* 6 (1990): 229–54.

23. Thomas Mann, *Tagebücher 1933–1934*, p. 473.

24. After the death of the publisher Samuel Fischer, his son-in-law, Gottfried Bermann, took steps to transfer at least part of the firm out of the Reich. S. Fischer Verlag would remain in Germany, in Aryan hands. The new Bermann Fischer publishing house—and with it some of the most prestigious names of contemporary German literature (Mann, Döblin, Hofmannsthal, Wassermann, Schnitzler)—was ready to start activities in Zurich. This, however, was a serious misjudgment of Swiss hospitality on Bermann's part. The main Swiss publishers opposed the move, and the literary editor of the *Neue Zürcher Zeitung*, Eduard Korrodi, did not mince words: The only German literature that had emigrated, he wrote in January 1936, was Jewish ("the hack-writers of the novel industry"). Bermann Fischer moved to Vienna. This time Thomas Mann reacted. His open letter to the newspaper was his first major public stand since January 1933: Mann drew Korrodi's attention to the obvious: Both Jews and non-Jews were to be found among the exiled German writers. As for those who remained in Germany, "being völkisch is not being German. But the German or the German rulers' hatred of the Jews is in the higher sense not directed against Europe and all loftier Germanism; it is directed, as becomes increasingly apparent, against the Christian and classical foundations of Western morality. It is the attempt . . . to shake off the ties of civilization. That attempt threatens to bring about a terrible alienation, fraught with evil potentialities, between the land of Goethe and the rest of the world . . ." Mann, *The Letters*, p. 209. Within a few months all members of the Mann family who had not yet been deprived of their German citizenship lost it, and on December 19, 1936, the dean of the Faculty of Philosophy of Bonn University announced to Thomas Mann that his name had been "struck off the roll of honorary doctors." Nigel Hamilton, *The Brothers Mann: The Lives of Heinrich and Thomas Mann, 1871–1950* (London, 1978), p. 298.

25. Frederic Spotts, *Bayreuth: A History of the Wagner Festival* (New Haven, Conn., 1994), p. 168.

26. Quoted in Moshe Zimmermann, "Die aussichtslose Republik— Zukunftsperspektiven der deutschen Juden vor 1933," in *Menora: Jahrbuch für deutschjüdische Geschichte 1990* (Munich, 1990), p. 164. This did not mean, however, that Jewish votes shifted to extremist parties. After the disappearance of the German Democratic Party (DDP), Jewish votes in the crucial elections of 1932 probably led to

the election of two Social Democratic deputies and one deputy from the Catholic Center. Ernest Hamburger and Peter Pulzer, "Jews as Voters in the Weimar Republic," *LBIY* 30 (1985): 66.

27. Kurt Jakob Ball-Kaduri, *Das Leben der Juden in Deutschland im Jahre 1933: Ein Zeitbericht* (Frankfurt am Main, 1963), p. 34.

28. Quoted in Wolfgang Benz, ed., *Das Exil der kleinen Leute: Alltagserfahrung deutscher Juden in der Emigration* (Munich, 1991), p. 16.

29. Ibid., p. 17.

30. According to the June 16, 1933, census, 499,682 persons of the "Mosaic faith" lived in Germany (the Saar territory not included) on that date, which amounts to 0.77 percent of the total German population. See Ino Arndt and Heinz Boberach, "Deutsches Reich" in Wolfgang Benz, ed., *Dimension des Völkermords: Die Zahl der jüdischen Opfer des Nationalsozialismus* (Munich, 1991), p. 23. It is plausible that approximately 25,000 Jews had fled Germany between January and June 1933.

31. For the petition and the other details, see *Akten der Reichskanzlei: Die Regierung Hitler, 1933–1938*, part 1, *1933–1934*, ed. Karl-Heinz Minuth, vol. 1 (Boppard am Rhein, 1983), pp. 296–98, 298n.

32. Zimmermann, "Die aussichtslose Republik," p. 160.

33. Rüdiger Safranski, *Ein Meister aus Deutschland: Heidegger und seine Zeit* (Munich, 1994), p. 271.

34. Wolfgang Benz, ed., *Die Juden in Deutschland 1933–1945: Leben unter nationalsozialistischer Herrschaft* (Munich, 1988), p. 18.

35. Nahum N. Glatzer and Paul Mendes-Flohr, eds., *The Letters of Martin Buber* (New York, 1991), p. 395.

36. Jews were also shipped off to the new concentration camps: Four were killed in Dachau on April 12. Both in Dachau and in Oranienburg "Jewish units" were set up from the outset. See Klaus Drobisch, "Die Judenreferate des Geheimen Staatspolizeiamtes und des Sicherheitsdienstes der SS 1933 bis 1939," *Jahrbuch für Antisemitismusforschung* 2 (1933): 231.

37. To this day the most thorough study of the Nazi takeover during the years 1933 and 1934 remains Karl Dietrich Bracher, Wolfgang Sauer, and Gerhard Schulz, *Die nationalsozialistische Machtergreifung* (Cologne, 1962).

38. Drobisch, "Die Judenreferate," p. 231.

39. Martin Broszat, Elke Fröhlich, and Falk Wiesemann, eds., *Bayern in der NS-Zeit: Soziale Lage und politisches Verhalten der Bevölkerung im Spiegel vertraulicher Berichte* (Munich, 1977), p. 432.

40. District President, Hildesheim, to Local Police Authorities of the District, 31.3.1933, Aktenstücke zur Judenverfolgung, Ortspolizeibehörde Göttingen, microfilm MA–172, Institut für Zeitgeschichte, Munich (hereafter IfZ).

41. Local Police Authority, Göttingen, to District President, Hildesheim, 1.4.33, ibid.

42. Deborah E. Lipstadt, *Beyond Belief: The American Press and the Coming of the Holocaust 1933–1945* (New York, 1986), pp. 44–45. On Walter Lippmann's positions see mainly Ronald Steel, *Walter Lippmann and the American Century* (Boston, 1980), particularly pp. 330–33.

43. *Akten der Reichskanzlei: Die Regierung Hitler*, part 1, vol. 1, p. 251.

44. Zimmermann, "Die aussichtslose Republik," pp. 155, 157–58.

45. Avraham Barkai, *From Boycott to Annihilation: The Economic Struggle of German Jews 1933–1943* (Hanover, N.H., 1989), p. 15.

46. Heinz Höhne, *Die Zeit der Illusionen: Hitler und die Anfänge des Dritten Reiches 1933–1936* (Düsseldorf, 1991), p. 76.

47. For a description of various components of this radical tendency, see Dietrich Orlow, *The History of the Nazi Party 1933–1945*, vol. 2 (Pittsburgh, 1973), pp. 40ff.

48. Richard Bessel, *Political Violence and the Rise of Nazism: The Storm Troopers in Eastern Germany, 1925–1934* (New Haven, Conn., 1984), p. 107.

49. David Bankier, "Hitler and the Policy-Making Process on the Jewish Question," *Holocaust and Genocide Studies* 3, no. 1 (1988): 4.

50. *Akten der Reichskanzlei: Die Regiergung Hitler,* part 1, vol. 1, p. 277.

51. Memoranda of telephone conversations between the State Department and the U.S. Embassy in Berlin, March 31, 1933, *Foreign Relations of the United States, 1933,* vol. 2 (Washington, D.C., 1948), pp. 342 ff.

52. Henry Friedlander and Sybil Milton, eds., *Archives of the Holocaust,* vol. 17, *American Jewish Committee New York,* ed. Frederick D. Bogin (New York, 1993), p. 4. In May 1933 a trilingual German, English, and French collection of various Jewish declarations was printed (probably in Berlin) by an ostensibly Jewish publisher, "Jakov Trachtenberg," under the title *Atrocity Propaganda Is Based on Lies, the Jews of Germany Themselves Say. (Die Greuel-Propaganda ist eine Lügenpropaganda, sagen die deutschen Juden selbst.)* The book was probably meant for worldwide distribution. I am indebted to Hans Rogger for drawing my attention to this publication.

53. Yoav Gelber, "The Reactions of the Zionist Movement and the Yishuv to the Nazis' Rise to Power," *Yad Vashem Studies* 18 (1987): 46. From Gelber's text it is not clear whether the telegram was sent before or after April 1.

54. About the quandary of the American Jewish leadership see Gulie Ne'eman Arad, *The American Jewish Leadership and the Nazi Menace* (Bloomington, Ind., forthcoming [1997]).

55. Gelber, "The Reactions of the Zionist Movement," pp. 47–48. On the American Jewish boycott see mainly Moshe R. Gottlieb, *American Anti-Nazi Resistance, 1933–1941: An Historical Analysis* (New York, 1982).

56. Goebbels, *Die Tagebücher,* vol. 2, pp. 398–99.

57. Ibid., p. 400.

58. Ibid.

59. Barkai, *From Boycott to Annihilation,* p. 2.

60. For a detailed account of the concrete problems encountered by the Nazis, see Karl A. Schleunes, *The Twisted Road to Auschwitz: Nazi Policy toward German Jews 1933–1939* (Urbana, Ill., 1970), pp. 84–90.

61. Ibid., p. 94.

62. Peter Hanke, *Zur Geschichte der Juden in München zwischen 1933 und 1945* (Munich, 1967), p. 85.

63. Ibid., pp. 85–86.

64. For Martha Appel's memoirs see Monika Richarz, ed., *Jüdisches Leben in Deutschland: Selbstzeugnisse zur Sozialgeschichte 1918–1945* (Stuttgart, 1982), pp. 231–32.

65. Broszat, Fröhlich, and Wiesemann, *Bayern in der NS-Zeit,* vol. 1, p. 435.

66. Helmut Genschel, *Die Verdrängung der Juden aus der Wirtschaft im Dritten Reich* (Göttingen, 1966), p. 58.

67. On April 5 the German ambassador to France reported to Berlin: "How unfavorable the effects of the action against the Jews had been in France was best shown by the sympathy expressed by high-ranking Catholic and Protestant clergy at the French-

Jewish demonstrations against the anti-Jewish movement in Germany. . . . There was no doubt . . . that the operation had been fully exploited by French circles antagonistic to Germany for material or political reasons and that they had fully attained their purpose of painting again in the darkest of colors, even to the rural population, the danger from a Germany inclining to deeds of violence." Koester to Foreign Minister, 5 April 1933. *Documents on German Foreign Policy, Series C (1933–1937)*, vol. 1 (Washington, D.C., 1957), p. 251.

68. Ernst Noam and Wolf-Arno Kropat, *Juden vor Gericht, 1933–1945: Dokumente aus hessischen Justizakten* (Wiesbaden, 1975), pp. 84–86.

69. Files of the NSDAP Main Office, microfiche 581 00181, IfZ. (Parteikanzlei der NSDAP)

70. David Bankier, "The German Communist Party and Nazi Anti-Semitism, 1933–1938," *LBIY* 32 (1987): 327.

71. Barkai, *From Boycott to Annihilation*, p. 17.

72. Ibid., p. 72. As a result, part of the shares of Tietz were acquired by major German banks. In 1934 the Tietz brothers sold the remainder; the firm was Aryanized and renamed Hertie AG.

73. National Socialist enterprise cell of Ullstein Verlag to Reich Chancellor, 21.6.1933, Max Kreuzberger Research Papers, AR 7183, Box 10, Folder 1, Leo Baeck Institute [hereafter LBI], New York.

74. Ron Chernow, *The Warburgs* (New York, 1993), p. 377.

75. Harold James, "Die Deutsche Bank und die Diktatur 1933–1945," in Lothar Gall et al., eds., *Die Deutsche Bank 1870–1995* (Munich, 1995), p. 336.

76. Ibid.

77. The overall argument and a wealth of supporting archival material is presented in Peter Hayes, "Big Business and 'Aryanisation' in Germany, 1933–1939," in *Jahrbuch für Antisemitismusforschung* 3 (1994), pp. 254ff.

78. Peter Hayes, *Industry and Ideology: IG Farben in the Nazi Era* (New York, 1987), p. 93.

79. Chernow, *The Warburgs*, pp. 379–80.

80. Jeremy Noakes and G. Pridham, eds., *Nazism: A History in Documents and Eyewitness Accounts 1919–1945*, vol. 1 (New York, 1983), pp. 14–16.

81. Herbert Michaelis and Ernst Schraepler, eds., *Ursachen und Folgen: Vom deutschen Zusammenbruch 1918 und 1945 bis zur staatlichen Neuordnung Deutschlands in der Gegenwart: Eine Urkunden- und Dokumentsammlung zur Zeitgeschichte*, vol. 9 (Berlin, n.d.), p. 383.

82. Joseph Walk, ed., *Das Sonderrecht für die Juden im NS-Staat* (Heidelberg, 1981), p. 4.

83. Hermann Graml, *Anti-Semitism in the Third Reich* (Cambridge, Mass., 1992), p. 97.

84. Walk, *Das Sonderrecht*, p. 3.

85. Ibid., p. 36.

86. For a detailed description of these laws, see in particular Schleunes, *The Twisted Road*, pp. 102–4.

87. To this day the best overall analysis of the Civil Service Law is still to be found in Hans Mommsen, *Beamtentum im Dritten Reich* (Stuttgart, 1966), pp. 39ff.

88. Walk, *Sonderrecht*, pp. 12–13. The party program of 1920 excluded Jews from party membership. After 1933 most organizations directly affiliated with the party, such as the German Labor Front, for instance, excluded membership for anyone with

Jewish ancestry after 1800. See Jeremy Noakes, "Wohin gehören die Judenmischlinge? Die Entstehung der ersten Durchführungsverordnungen zu den Nürnberger Gesetzen," in Ursula Büttner, Werner Johe, and Angelika Voss, eds., *Das Unrechtsregime: Internationale Forschung über den Nationalsozialismus*, vol. 2, *Verfolgung, Exil, belasteter Neubeginn* (Hamburg, 1986), p. 71.

89. Hilberg, *The Destruction of the European Jews*, p. 54. For Hilberg there was a straight line between the first definition and the ultimate extermination.

90. For details regarding the origins of the anti-Jewish paragraph of the Civil Service Law, see Günter Neliba, *Wilhelm Frick: Der Legalist des Unrechtsstaates: Eine Politische Biographie* (Paderborn, 1992), pp. 168ff.

91. Ibid., p. 171; see also Mommsen, *Beamtentum*, pp. 48, 53.

92. Hans-Joachim Dahms,"Einleitung" in Heinrich Becker, Hans-Joachim Dahms, Cornelia Wegeler, eds., *Die Universität Göttingen unter dem Nationalsozialismus: verdrängte Kapitel ihrer 250 jährigen Geschichte* (Munich, 1987), pp. 17–18.

93. Achim Gercke, "Die Lösung der Judenfrage," *Nationalsozialistische Monatshefte* 38 (May 1933): 196. Gercke did not simply write that the laws were "educational" but that they were "educational insofar as they indicated a direction."

94. Walk, *Das Sonderrecht*, p. 12.

95. Comité des Délégations Juives, ed., *Das Schwarzbuch: Tatsachen und Dokumente: Die Lage der Juden in Deutschland 1933* (Paris, 1934; reprint,Berlin, 1983), p. 105.

96. Uwe Dietrich Adam, *Judenpolitik im Dritten Reich* (Düsseldorf, 1972), pp. 50ff., 65ff. For Schlegelberger's report to Hitler on April 4, see *Akten der Reichskanzlei*, part 1, vol. 1, p. 293n. For Hitler's statement, see the protocol of the cabinet meeting of April 7, 1933, ibid., p. 324.

97. Dirk Blasius, "Zwischen Rechtsvertrauen und Rechtszerstörung: Deutsche Juden 1933–1935," in Dirk Blasius and Dan Diner, eds., *Zerbrochene Geschichte: Leben und Selbstverständnis der Juden in Deutschland* (Frankfurt am Main, 1991), p. 130.

98. Barkai, *From Boycott to Annihilation*, p. 4.

99. Konrad H. Jarausch, "Jewish Lawyers in Germany, 1848–1938: The Disintegration of a Profession," *LBIY* 36 (1991): 181–82.

100. Comité des Délégations Juives, *Das Schwarzbuch*, pp. 195–96.

101. *Akten der Reichskanzlei*, part 1, vol. 1, p. 324. ("Hier müsse eine umfassende Aufklärung einsetzen.")

102. Walk, *Das Sonderrecht*, p. 17; Albrecht Götz von Olenhusen, "Die 'Nichtarischen' Studenten an den deutschen Hochschulen: Zur nationalsozialistischen Rassenpolitik 1933–1945," *VfZ* 14, no. 2 (1966): 177ff.

103. Ibid., p. 180.

104. For this point see Kurt Pätzold, *Faschismus, Rassenwahn, Judenverfolgung: Eine Studie zur politischen Strategie und Taktik des faschistischen deutschen Imperialismus (1933–1935)* ([East]Berlin, 1975), p. 105.

105. For the case of Karl Berthold (name changed) and the appended documentation, see Hans Mommsen, "Die Geschichte des Chemnitzer Kanzleigehilfen K.B.," in Detlev Peukert und Jürgen Reulecke, eds., *Die Reihen fast geschlossen: Beiträge zur Geschichte des Alltags unterm Nationalsozialismus* (Wuppertal, 1981), pp. 337ff. In present-day terminology, a *Versorgungsamt* is "a social benefits office for state employees." Here I shall refer merely to the "social benefits office."

106. Ibid., p. 348.

107. Ibid., p. 350.

108. Ibid., p. 350.

109. Ibid., p. 351.

110. Lammers to Hess, 6.6.1933, Parteikanzlei der NSDAP, microfiche 10129934, IfZ.

111. For the details of this case, see Jeremy Noakes, "The Development of Nazi Policy Towards the German-Jewish 'Mischlinge' 1933–1945," *LBIY* 34 (1989): 316–17.

112. Volker Dahm, "Anfänge und Ideologie der Reichskulturkammer," *VfZ* 34, no. 1 (1986): 78. See also Alan Edward Steinweis, *Art, Ideology and Economics in Nazi Germany: The Reich Chamber of Culture and the Regulation of the Culture Professions in Nazi Germany* (Chapel Hill, N.C., 1988), pp. 322ff.

113. For details on this issue, see in particular Herbert Freeden, "Das Ende der jüdischen Presse in Nazideutschland," *Bulletin des Leo Baeck Instituts* 65 (1983): 4–5.

114. James, "Die Deutsche Bank," p. 337.

115. Mommsen, *Beamtentum*, p. 49.

116. Quoted in Peter Pulzer, *The Rise of Political Anti-Semitism in Germany and Austria* (Cambridge, Mass., 1988), p. 112.

117. Donald M. McKale, "From Weimar to Nazism: Abteilung III of the German Foreign Office and the Support of Antisemitism, 1931–1935," *LBIY* 32 (1987): pp. 297ff. On the attitude of the Wilhelmstrasse regarding the "Jewish Question" during the early phase of the regime, see also Christoper R. Browning, *The Final Solution and the German Foreign Office* (New York, 1978).

118. Noakes and Pridham, *Nazism 1919–1945*, vol. 2, pp. 526–27.

119. A first summary of this document was published in Hebrew in *Ha'aretz* by the Israeli historian Shaul Esh on April 1, 1963; it was interpreted as a master plan for the whole Nazi anti-Jewish program. For the English translation, with comments, see Uwe Dietrich Adam, "An Overall Plan for Anti-Jewish Legislation in the Third Reich?" *Yad Vashem Studies* 11 (1976): 33–55.

120. Ibid., p. 40.

121. First published in Michaelis and Schraepler, *Ursachen*, pp. 393–95. Translated by Dieter Kuntz in Benjamin C. Sax and Dieter Kuntz, eds., *Inside Hitler's Germany: A Documentary History of Life in the Third Reich* (Lexington, Ky., 1992), pp. 401–3. The translation has been very slightly revised. For another translation see *Documents on German Foreign Policy, Series C*, pp. 253–5.

122. *Akten der Reichskanzlei: Die Regierung Hitler*, part 1, vol. 1, pp. 391–92.

123. Walk, *Das Sonderrecht*, p. 8.

124. Ibid., p. 9.

125. Ibid., p. 10.

126. Ibid., p. 13.

127. Ibid., p. 14.

128. Ibid., p. 15.

129. Ibid., p. 16. (For example, one would not be allowed to say "*A* for Abraham.")

130. Ibid., p. 19.

131. Ibid., p. 21.

132. Ibid., p. 23.

133. Ibid., p. 25.

134. Kommission zur Erforschung der Geschichte der Frankfurter Juden, ed., *Dokumente zur Geschichte der Frankfurter Juden 1933–1945* (Frankfurt am Main, 1963), p. 95.

135. *Chronik der Stadt Stuttgart 1933–1945* (Stuttgart, 1982), p. 21.

136. Ibid., p. 22.

137. Ibid., p. 22.

138. Ibid., p. 25.

139. Ibid., p. 26.

140. Ibid., p. 27.

141. William Sheridan Allen, *The Nazi Seizure of Power: The Experience of a Single German Town 1930–1935* (London, 1965), pp. 209–10.

142. Ibid., p. 212.

143. Ibid., p. 213.

144. Deborah Dwork, *Children with a Star: Jewish Youth in Nazi Europe* (New Haven, Conn., 1991), p. 22.

145. Richarz, *Jüdisches Leben in Deutschland*, p. 232.

146. Götz Aly and Karl-Heinz Roth, *Die restlose Erfassung: Volkszählen, Identifizieren, Aussondern im Nationalsozialismus* (Berlin, 1984), p. 55.

147. Robert N. Proctor, *Racial Hygiene: Medicine under the Nazis* (Cambridge, Mass., 1988), p. 95.

148. Jeremy Noakes, "Nazism and Eugenics: The Background to the Nazi Sterilization Law of 14 July 1933," in R. J. Bullen, H. Pogge von Strandmann, and A. B. Polonsky, eds., *Ideas into Politics. Aspects of European History 1880–1950* (London, 1984), pp. 83–84.

149. For the economic impact of the Great Depression on psychiatric care, see Michael Burleigh, *Death and Deliverance: "Euthanasia" in Germany 1900–1945* (Cambridge, England, 1994), pp. 33ff.

150. Ibid., pp. 84–85.

151. Gisela Bock, *Zwangssterilisation im Nationalsozialismus: Studien zur Rassenpolitik und Frauenpolitik* (Opladen, 1986), pp. 49–51, 55–56.

152. Noakes, "Nazism and Eugenics," p. 85.

153. Ibid., p. 86.

154. Hans-Walter Schmuhl, "Reformpsychiatrie und Massenmord," in Michael Prinz und Rainer Zitelmann, eds., *Nationalsozialismus und Modernisierung* (Darmstadt, 1991), p. 249.

155. Noakes, "Nazism and Eugenics," p. 87.

156. Schmuhl, "Reformpsychiatrie und Massenmord," p. 250.

Chapter 2 Consenting Elites, Threatened Elites

1. Eberhard Röhm and Jörg Thierfelder, *Juden-Christen-Deutsche*, vol. 1, *1933–1935* (Stuttgart, 1990), pp. 120ff.

2. Klaus Scholder, *Die Kirchen und das Dritte Reich*, vol. 1, *Vorgeschichte und Zeit der Illusionen 1918–1934* (Frankfurt am Main, 1977), pp. 338ff.

3. Ibid.

4. Wolfgang Gerlach, *Als die Zeugen schwiegen: Bekennende Kirche und die Juden* (Berlin, 1987), p. 42.

5. *Akten deutscher Bischöfe über die Lage der Kirche 1933–45*, vol. 1: *1933–1934*, ed. Bernhard Stasiewski (Mainz, 1968), pp. 42n, 43n.

6. Ernst Christian Helmreich, *The German Churches Under Hitler: Background, Struggle and Epilogue* (Detroit, 1979), pp. 276–77. For the German original see *Akten deutscher Bischöfe*, vol. 2, p. 54n.

7. Ernst Klee, *"Die SA Jesu Christi": Die Kirche im Banne Hitlers* (Frankfurt am Main, 1989), p. 30.

8. For the quotations see Helmreich, *The German Churches*, pp. 276–77.

9. Klaus Scholder, "Judaism and Christianity in the Ideology and Politics of National Socialism," in Otto Dov Kulka and Paul Mendes-Flohr, eds., *Judaism and Christianity Under the Impact of National Socialism 1919–1945* (Jerusalem, 1987), pp. 191ff.

10. Quoted in Doris L. Bergen, *Twisted Cross: The German Christian Movement in the Third Reich* (Chapel Hill, N.C., 1996), p. 23. On November 13, 1933, the leader of the Berlin district of German Christians, one Dr. Krause, declared at a meeting of the movement at the Sportpalast: "What belongs to it [the new Christianity] is the liberation from all that is un-German in the ritual and faith, the liberation from the Old Testament with its Jewish retribution morals and its stories of cattle dealers and pimps. . . . In the German Volk Church there is no place for people of foreign blood, either at the pulpit or below the pulpit. All expressions of a foreign spirit which have penetrated it . . . must be expelled from the German Volk Church." Ulrich Thürauf, ed., *Schulthess Europäischer Geschichtskalender* 74 (1933), p. 244.

11. Quoted in Paul R. Mendes-Flohr, "Ambivalent Dialogue: Jewish-Christian Theological Encounter in the Weimar Republic," in Kulka and Mendes-Flohr, *Judaism and Christianity*, p. 121.

12. Uriel Tal, "Law and Theology: on the Status of German Jewry at the outset of the Third Reich," in *Political Theology and the Third Reich* (Tel Aviv, 1989), p. 16. The English version of this text appeared as a brochure published by Tel Aviv University, 1982.

13. For the intense theological debates raised by the introduction of the "Aryan paragraph," see ibid.

14. Scholder, *Die Kirchen und das Dritte Reich*, pp. 612ff.

15. Robert Michael, "Theological Myth, German Anti-Semitism and the Holocaust: The Case of Martin Niemöller," *Holocaust and Genocide Studies* 2 (1987): 112. (The title "Propositions on the Aryan Question" should be considered a euphemism, in the same way as "Aryan paragraph" in fact meant "Jewish paragraph.")

16. Gerlach, *Als die Zeugen schwiegen*, p. 87.

17. Michael, "Theological Myth," p. 113.

18. Ibid.

19. Quoted in Uriel Tal, "On Structures of Political Theology and Myth in Germany prior to the Holocaust," in Yehuda Bauer and Nathan Rotenstreich, eds., *The Holocaust as Historical Experience* (New York, 1981), p. 55.

20. Richard Gutteridge, *Open Thy Mouth for the Dumb! The German Evangelical Church and the Jews 1879–1950* (Oxford, 1976), p. 122.

21. Günther van Norden, "Die Barmen Theologische Erklärung und die 'Judenfrage'," in Ursula Büttner et al., eds., *Das Unrechtsregime*, vol. 1, *Ideologie— Herrschaftssystem—Wirkung in Europa* (Hamburg, 1986), pp. 315ff.

22. Guenter Lewy, *The Catholic Church and Nazi Germany* (New York, 1964), p. 17.

23. Ibid., p. 271.

24. *Akten deutscher Bischöfe*, vol. 1, pp. 100–102.

25. Klee to Foreign Ministry, September 12, 1933, *Documents on German Foreign Policy, Series C*, pp. 793–94.

26. Scholder, *Die Kirchen und das Dritte Reich*, p. 660.

27. His Eminence Cardinal Faulhaber, *Judaism, Christianity and Germany: Advent Sermons Preached in St. Michael's, Munich, in 1933* (London, 1934), pp. 5–6. Faulhaber's argument reflects a long-standing Christian polemic tradition regarding the Talmud. See in particular Amos Funkenstein, "Changes in Christian Anti-Jewish Polemics in the Twelfth Century," in *Perceptions of Jewish History* (Berkeley, Calif., 1993), pp. 172–201 and particularly 189–96.

28. Helmreich, *The German Churches Under Hitler*, p. 262.

29. Heinz Boberach, ed., *Berichte des SD und der Gestapo über Kirchen und Kirchenvolk in Deutschland 1934–1944* (Mainz, 1971), p. 7. Although as a rule church dignitaries avoided comments regarding the contemporary aspects of the Jewish issue, some local Catholic newpapers drew the attention of their readers to the brutal treatment of the Jews. For example, on May 23, 1933, the Catholic *Bamberger Volksblatt* explicitly mentioned the death in Dachau of the young local court clerk, Willy Aron, who was Jewish. For the significance of the case, see Norbert Frei, *Nationalsozialistische Eroberung der Provinzpresse: Gleichschaltung, Selbstanpassung und Resistenz in Bayern* (Stuttgart, 1980), pp. 273–75.

30. Quoted in Walter Hofer, ed., *Der Nationalsozialismus: Dokumente 1933–1945* (Frankfurt am Main, 1957), p. 130.

31. Quoted in Konrad Kwiet and Helmut Eschwege, *Selbstbehauptung und Widerstand: Deutsche Juden im Kampf um Existenz und Menschenwürde 1933–1945* (Hamburg, 1984), p. 221.

32. Gerhard Sauder, ed., *Die Bücherverbrennung* (Munich, 1983), pp. 50–52. The sixteen named were: Bonn (Berlin), Cohn (Breslau), Dehn (Halle), Feiler (Königsberg), Heller (Frankfurt am Main), Horkheimer (Frankfurt am Main), Kantorowicz (Frankfurt), Kantorowicz (Kiel), Kelsen (Cologne), Lederer (Berlin), Löwe (Frankfurt am Main), Löwenstein (Bonn), Mannheim (Frankfurt am Main), Mark (Breslau), Tillich (Frankfurt am Main), Sinzheimer (Frankfurt am Main).

33. Doron Niederland, "The Emigration of Jewish Academics and Professionals from Germany in the First Years of Nazi Rule," *LBIY* 33 (1988): 291. The numbers vary considerably from one field to another and from university to university. In biology, for example, the scientists defined as Jews or married to Jewish spouses who were dismissed between 1933 and 1939 (including the universities of Vienna and Prague) numbered approximately 9 percent of the entire faculty in the field (30 out of 337). See Ute Deichmann, *Biologen unter Hitler: Vertreibung, Karrieren, Forschung* (Frankfurt am Main, 1992), p. 34.

34. Alan D. Beyerchen, *Scientists Under Hitler. Politics and the Physics Community in the Third Reich* (New Haven, Conn., 1977), pp. 22, 15–22. For parts of the text regarding the universities, see Saul Friedländer, "The Demise of the German Mandarins: The German University and the Jews 1933–1939," in Christian Jansen et al., eds., *Von der Aufgabe der Freiheit: Politische Verantwortung und bürgerliche Gesellschaft im 19. und 20. Jahrhundert* (Berlin, 1995), pp. 63ff.

35. Helmut Heiber, *Universität unterm Hakenkreuz*, part 2, *Die Kapitulation der Hohen Schulen: Das Jahr 1933 und seine Themen*, vol. 1 (Munich, 1992), p. 26.

36. Uwe Dietrich Adam, *Hochschule und Nationalsozialismus: Die Universität Tübingen im Dritten Reich* (Tübingen, 1977), p. 36.

37. All the details of the Freiburg case are taken from Edward Seidler, "Die Medizinische Fakultät zwischen 1926 und 1948," in Eckhard John, Bernd Martin,

Marc Mück, and Hugo Ott, eds., *Die Freiburger Universität in der Zeit des Nationalsozialismus* (Freiburg/Würzburg, 1991), pp. 76–77.

38. Arno Weckbecker, *Die Judenverfolgung in Heidelberg 1933–1945* (Heidelberg, 1985), p. 150. During the same period five "Aryan" teachers had been dismissed on political grounds.

39. Benno Müller-Hill, *Murderous Science: Elimination by Scientific Selection of Jews, Gypsies and Others, Germany 1933–1945* (Oxford, 1988), p. 24.

40. Paul Weindling, *Health, Race and German Politics Between National Unification and Nazism, 1870–1945* (Cambridge, England, 1989), p. 495.

41. Chernow, *The Warburgs*, pp. 540–41.

42. Müller-Hill, *Murderous Science*, p. 27.

43. Karen Schönwälder, *Historiker und Politik: Geschichtswissenschaft im Nationalsozialismus* (Frankfurt am Main, 1992), pp. 29ff., 33.

44. For a more detailed presentation and analysis of the indifference of German university professors regarding the fate of their Jewish colleagues, as well as for the outright expressions of hostility of some of them, see Friedländer, "The Demise of the German Mandarins," mainly pp. 70ff. For the attitude of the famous economic historian Werner Sombart, see ibid., p. 73, as well as Friedrich Lenger, *Werner Sombart 1863–1941: Eine Biographie* (Munich, 1994), p. 359; for a good analysis of Sombart's anti-Jewish intellectual position, see mainly Jeffrey Herf, *Reactionary Modernism: Technology, Culture and Politics in Weimar and the Third Reich* (Cambridge, England, 1993), pp. 130ff.

45. Kurt Pätzold, *Verfolgung, Vertreibung, Vernichtung: Dokumente des faschistischen Antisemitismus 1933 bis 1942* (Frankfurt am Main, 1984), p. 53.

46. Geoffrey J. Giles, "Professor und Partei: Der Hamburger Lehrkörper und der Nationalsozialismus," in Eckart Krause, Ludwig Huber, Holger Fischer, eds., *Hochschulalltag im Dritten Reich: Die Hamburger Universität 1933–1945* (Berlin, 1991), p. 115.

47. Christian Jansen, *Professoren und Politik: Politisches Denken und Handeln der Heidelberger Hochschullehrer 1914–1935* (Göttingen, 1992), pp. 289ff.

48. Claudia Schorcht, *Philosophie an den Bayerischen Universitäten 1933–1945* (Erlangen, 1990), pp. 159ff. It is possible that at that early stage some other collective declarations in favor of Jewish colleagues were planned that were never concretely made. According to Otto Hahn, Max Planck dissuaded him from organizing such a petition with the argument that it would only trigger a much stronger counterdeclaration. See J. L. Heilbron, *The Dilemmas of an Upright Man: Max Planck as Spokesman for German Science* (Berkeley, Calif., 1986), p. 150.

49. We know of this intervention from Max Planck's own account after the war. According to Planck, Hitler declared that he had nothing against the Jews and was only anti-Communist; then he supposedly flew into a rage. See Heilbron, *The Dilemmas of an Upright Man*, p. 153. Such postwar reports are hard to substantiate.

50. Hugo Ott, *Martin Heidegger: Unterwegs zu seiner Biographie* (Frankfurt am Main, 1988), pp. 198–200.

51. Rüdiger Safranski, *Ein Meister aus Deutschland: Heidegger und seine Zeit* (Munich, 1994), p. 299.

52. Hugo Ott, *Laubhüttenfest 1940: Warum Therese Löwy einsam sterben musste* (Freiburg, 1994), p. 113.

53. Ibid.

54. Elzbieta Ettinger, *Hannah Arendt/Martin Heidegger* (New Haven, Conn. 1995),

pp. 35–36. Heidegger's letters are paraphrased as Ettinger had not received permission to quote them directly.

55. Thomas Sheehan, "Heidegger and the Nazis," *New York Review of Books*, June 16, 1988, p. 40.

56. Ibid.

57. Ibid. See also Ott, *Laubhüttenfest 1940*, p. 183.

58. Safranski, *Ein Meister aus Deutschland*, p. 302.

59. Ibid., p. 300.

60. Heinrich Meier, *Carl Schmitt, Leo Strauss und "Der Begriff des Politischen": Dialog unter Abwesenden* (Stuttgart, 1988), p. 137.

61. Ibid., pp. 14–15.

62. For all details in this section, see Bernd Rüthers, *Carl Schmitt im Dritten Reich: Wissenschaft als Zeitgeist-Bestärkung?* (Munich, 1990), pp. 31–34.

63. Wolfgang Heuer, *Hannah Arendt* (Reinbek/Hamburg, 1987), p. 29.

64. Kommission . . . , *Dokumente zur Geschichte der Frankfurter Juden*, pp. 99–100.

65. Donald L. Niewyk, *The Jews in Weimar Germany* (Baton Rouge, La., 1980), p. 67. Michael Kater's evaluation is somewhat clearer-cut: "The number of converts to Nazism, among the professors, often motivated by antisemitism, was growing, especially in 1932, and even if most of them chose to remain outside the party, the evidence suggests that in their heart of hearts they had switched their allegiance to Hitler." Michael Kater, *The Nazi Party: A Social Profile of Members and Leaders 1919–1945* (Oxford, 1983), p. 69. Academic Judeophobia during the empire and even more during the Weimar Republic is too well documented to need much further proof. Yet some notorious incidents can be read in contrary ways. In 1924 the Jewish Nobel laureate in chemistry and professor at the University of Munich, Richard Willstätter, resigned in protest against the decision of the dean and a majority of the faculty not to appoint the geochemist Viktor Goldschmidt on obviously antisemitic grounds. Yet, conversely, a great number of faculty members and students attempted for weeks to persuade Willstätter to take back his resignation, to no avail. Regarding the issue as such and Willstätter's resignation, see Fritz Stern, *Dreams and Delusions* (New York, 1987), pp. 46–47, and John V. H. Dippel, *Bound upon a Wheel of Fire: Why So Many German Jews Made the Tragic Decision to Remain in Nazi Germany* (New York, 1996), pp. 25–27.

66. Michael H. Kater, *Studentenschaft und Rechtsradikalismus in Deutschland 1918–1933* (Hamburg, 1975), pp. 145–46.

67. Geoffrey J. Giles, *Students and National Socialism in Germany* (Princeton, N.J., 1985), p. 17.

68. Ulrich Herbert, "'Generation der Sachlichkeit': Die völkische Studenten-bewegung der frühen zwanziger Jahre in Deutschland," in Frank Bajohr et al., eds., *Zivilisation und Barbarei: Die widersprüchlichen Potentiale der Moderne* (Hamburg, 1991), pp. 115ff. For the establishment and the growth of the NSDSB until 1933 see also Michael Grüttner, *Studenten im Dritten Reich* (Paderborn, 1995), pp. 19–61.

69. On November 12, 1930, the *Berliner Tageblatt* reported that around five hundred Nazi students had launched attacks against prorepublic and Jewish students on the Berlin University campus: "During the assault, a Social Democratic student was wounded and had to get medical assistance . . . a Jewish female student was attacked by the Nazis, thrown to the ground and trampled upon. . . . The group yelled in turn 'Germany awake!' and 'Out with Jews'!" Kater, *Studentenschaft und Rechtsradikalismus*, p. 155.

70. Ibid., p. 157.

71. Kater, *The Nazi Party*, p. 184.

72. Rudolf Schottlaender, "Antisemitische Hochschulpolitik: Zur Lage an der Technischen Hochschule Berlin 1933/34," in Reinhard Rürup, ed., *Wissenschaft und Gesellschaft: Beiträge zur Geschichte der Technischen Universität Berlin 1878–1979*, vol. 1 (Berlin, 1979), p. 447.

73. Ibid.

74. Ibid., p. 448.

75. Sauder, *Die Bücherverbrennung*, p. 89. In Berlin, the publication of these theses led to an immediate conflict between students and the rector of the university, Eduard Kohlrausch, who had ordered the removal of the notices and posters from university grounds; the students countered by announcing his resignation. Giles, *Students and National Socialism*, p. 131.

76. George L. Mosse, "Die Bildungsbürger verbrennen ihre eigenen Bücher" in Horst Denkler und Eberhard Lämmert, eds., "Das war ein Vorspiel nur . . ." *Berliner Colloquium zur Literaturpolitik im "Dritten Reich"* (Berlin, 1985), p. 35.

77. Ibid., p. 42.

78. Professional schools group leader Hildburghausen to Minister Wächtler, Thuringia Education Ministry, Weimar, 6.5.1933, Nationalsozialistischer Deutscher Studentenbund (NSDStB), microfilm MA–228, IfZ.

79. District leader mid-Germany to Prime Minister Manfred von Killinger, Dresden, 12.8.1933, NSDStB, microfilm MA 228, IfZ.

80. Gerhard Gräfe to Georg Plötner, main office for political education, Berlin 16.5.1933, ibid. It seems that, on the other hand, Arthur Schnitzler's famous "Jewish" novel, *Der Weg ins Freie (The Road into the Open)*, was spared, probably because it was interpreted as carrying a Zionist message.

81. Victor Klemperer, *Ich will Zeugnis ablegen bis zum letzten: Tagebücher 1933–1945*, vol. 1, *1933–1941* (Berlin, 1995), pp. 31–43.

82. Jacob Boas, "German-Jewish Internal Politics under Hitler 1933–1938," *LBIY* 29 (1984): 3.

83. An earlier federative association of the organizations of Jewish communities (*Reichsarbeitsgemeinschaft der deutschen Landesverbände jüdischer Gemeinden*) had been established as early as January 1932; it represented the Jews of Germany during the first months of the Nazi regime before being replaced by the *Reichsvertretung*.

84. Leo Baeck has remained to this day a target of sharp criticism for what has appeared to some as subservience to and even cooperation with the Nazis. Hannah Arendt referred to him as the "Führer" of German Jewry. See Hannah Arendt, *Eichmann in Jerusalem: A Report on the Banality of Evil* (New York, 1963), p. 105. Raul Hilberg has kept to his initially harsh evaluation; for him Baeck was both pompous and pathetic all along. Raul Hilberg, *Perpetrators, Victims, Bystanders: The Jewish Catastrophe 1933–1945* (New York, 1992), p. 108.

85. Paul Sauer, "Otto Hirsch (1885–1941), Director of the Reichsvertretung," *LBIY* 32 (1987): 357.

86. For these quotations see Abraham Margalioth, "The Problem of the Rescue of German Jewry During the Years 1933–1939: The Reasons for the Delay in Their Emigration from the Third Reich," in Yisrael Guttman and Ephraim Zuroff, eds., *Rescue Attempts During the Holocaust* (Jerusalem, 1977), pp. 249ff.

87. Abraham Margalioth, *Between Rescue and Annihilation: Studies in the History of German Jewry 1932–1938* (Jerusalem, 1990), p. 5 (in Hebrew). See also Francis R. Nicosia, "Revisionist Zionism in Germany (II): Georg Kareski and the Staatszionistische Organization, 1933–1938," *LBIY* 32 ([London] 1987) 231ff.

88. Yehuda Bauer, *My Brother's Keeper: A History of the American Jewish Joint Distribution Committee 1929–1939* (Philadelphia, 1974), p. 111.

89. The Free Association for the Interests of Orthodox Jewry to the Reich Chancellor, Frankfurt, October 4, 1933, *Akten der Reichskanzlei*, vol. 2, *12/9/33–27/8/34*, pp. 884ff.

90. See Werner Rosenstock, "Exodus 1933–1939: A Survey of Jewish Emigration from Germany," *LBIY* 1 ([London] 1956): 377, and particularly Herbert A. Strauss, "Jewish Emigration from Germany: Nazi Policies and Jewish Responses (I)," *LBIY* 25: ([London] 1980): 326.

91. Ibid., p. 379.

92. Klaus Mann, *The Turning Point: Thirty-five Years in This Century* (1942; reprint, New York, 1985), p. 270.

93. Barkai, *From Boycott to Annihilation*, pp. 99ff.

94. Hans Mommsen, "Der nationalsozialistische Staat und die Judenverfolgung vor 1938," *VfZ* 1 (1962): 71–72.

95. For a detailed account of the Haavarah negotiations and agreement, see Francis R. Nicosia, *The Third Reich and the Palestine Question* (London, 1985), pp. 29ff., and in particular p. 46. See also Nicosia's updated article on the major themes of his book: "Ein nützlicher Feind: Zionismus im nationalsozialistischen Deutschland 1933–1939," *VfZ* 37, no. 3 (1989): 367ff.

96. Nicosia, "Ein nützlicher Feind," p. 383.

97. Tom Segev, *The Seventh Million: The Israelis and the Holocaust* (New York, 1993), p. 30.

98. Nicosia, *The Third Reich*, p. 42.

99. Ibid.

100. Nicosia, "Ein nützlicher Feind," p. 378. It is in the same spirit that Robert Weltsch, editor of the Zionist *Jüdische Rundschau* and possibly the best-known German Jewish journalist after 1933, wrote one of his most famous columns on April 4, 1933: "Trägt ihn mit Stolz, den gelben Fleck" (Wear the yellow badge with pride). In this memorably titled article, Weltsch argued that Nazism was offering a historic opportunity to reassert Jewish national identity. The Jews would regain the respect they had lost in assimilating, and they would launch their own national revival, as the Germans had. The Jews owed a debt of gratitude to the Nazis: Hitler had shown them the path to the recovery of their own identity. The column aroused immense enthusiasm among German Jews, Zionists and non-Zionists alike.

101. Segev, *The Seventh Million*, p. 19.

102. Ibid., p. 18.

103. Margalioth, "The Problem of the Rescue," p. 94.

104. Ibid., p. 95.

105. Quoted in Jost Hermand, "'Bürger zweier Welten?' Arnold Zweigs Einstellung zur deutschen Kultur," in Julius Schoeps, ed., *Juden als Träger bürgerlicher Kultur in Deutschland* (Bonn, 1989), p. 81.

106. Quoted in Robert Weltsch, "Vorbemerkung zur zweiten Ausgabe" (1959) in Siegmund Kaznelson, ed., *Juden im Deutschen Kulturbereich: Ein Sammelwerk* (Berlin, 1962), pp. xvff.

107. The number is taken from Eike Geisel's essay on the history of the *Kulturbund*, "Premiere und Pogrom," in Eike Geisel and Heinrich M. Broder, eds., *Premiere und Pogrom: Der Jüdische Kulturbund 1933–1941* (Berlin, 1992), p. 9.

108. Steinweis, "Hans Hinkel," p. 215.

109. Geisel, "Premiere und Pogrom," pp. 10ff.

110. Ibid., p. 12.

111. Dahm, "Anfänge und Ideologie," p. 114.

112. Ibid., p. 115. For details about the forbidding of Schiller and Goethe, see Jacob Boas, "Germany or Diaspora? German Jewry's Shifting Perceptions in the Nazi Era (1933–1938)" *LBIY* 27 (1982): 115 n 32. In his personal memoir about the epoch, Jakob Ball-Kaduri suggests that at the outset Hinkel was keen on developing the *Kulturbund* because of his ambivalent relation to Jewish matters and because the growth of the *Kulturbund* meant the growth of the domain that he was in charge of. Ball-Kaduri, *Das Leben der Juden in Deutschland im Jahre 1933*, p. 151. Such "ambivalence" appears rather as mere ambition on the part of a dedicated anti-Semitic activist.

113. Levi, *Music in the Third Reich*, pp. 51–52.

114. Ibid., pp. 33, 247.

115. Wulf, *Theater und Film im Dritten Reich*, p. 102.

116. Dahm, "Anfänge und Ideologie," p. 104.

117. Kurt Duwell, "Jewish Cultural Centers in Nazi Germany: Expectations and Accomplishments," in Jehuda Reinharz and Walter Schatzberg, eds., *The Jewish Response to German Culture: From the Enlightenment to the Second World War* (Hanover, N.H., 1985), p. 298.

118. Sir Horace Rumbold to Sir John Simon, May 11, 1933, *Documents on British Foreign Policy 1919–1939*, Second Series, vol. 5: *1933*, London, 1956, pp. 233–5.

119. The Consul General at Berlin to the Secretary of State, November 1, 1933, *Foreign Relations of the United States, 1933*, vol. 2 (Washington, D.C., 1948), p. 362. (italics added.)

120. *Akten der Reichskanzlei: Die Regierung Hitler*, part 1, vol. 1, p. 631.

121. Schacht's position was solely motivated by immediate economic goals. Otherwise he favored the "limitation of Jewish influence" in German economic life, and on several occasions he did not hesitate to make blatantly anti-Semitic speeches. In other words, Schacht fully expressed the conservative brand of anti–Semitism, and when faced with the regime's ever more radical anti-Jewish measures, he toed the line like all the Nazis' conservative allies. See in particular Albert Fischer, *Hjalmar Schacht und Deutschlands "Judenfrage,"* (Bonn, 1995), mainly pp. 126ff. One of Schacht's most outspoken anti-Semitic speeches was his "Luther speech" of November 8, 1933. The journalist and socialite Bella Fromm, who was Jewish, was in the audience and commented in her diary: "The intimate friend of Berlin Jewish society [Schacht] did not skip a single one of Martin Luther's manifold anti-Semitic remarks. . . . Certainly Jew-baiting is a legal affair since February 1, 1933. But there is no excuse for Schacht's infamy. Schacht has not always been prosperous. Everything he has he owes to friends who are no National Socialists. Bella Fromm, *Blood and Banquets: A Berlin Social Diary* (London, 1943; reprint, New York, 1990), p. 136.

122. *Akten der Reichskanzlei: Die Regierung Hitler*, part 1, vol. 1, p. 675.

123. Ibid., p. 677.

124. Noakes, "The Development of Nazi Policy Towards the German-Jewish 'Mischlinge,'" p. 303.

125. Reichsstatthalterkonferenz, 28.9.1933, *Akten der Reichskanzlei: Die Regierung Hitler*, Part 1, vol. 2, p. 865.

126. Adolf Hitler, "Die deutsche Kunst als stolzeste Verteidigung des deutschen Volkes," *Nationalsozialistische Monatshefte* 4, no. 34 (Oct. 1933), p. 437.

127. Quoted in Wolfgang Michalka, ed., *Das Dritte Reich,* vol. 1 (Munich, 1985), p. 137.

128. The "religious" dimension of Nazism, in terms both of its beliefs and its rituals, had already been noted by numerous contemporary observers; some blatant uses of Christian liturgy drew protests, mainly from the Catholic Church. The concept of "political religion" in its application to Nazism (and often to Communism as well), as a sacralization of politics and a politicization of religious themes and frameworks, was first systematically presented in Eric Voegelin, *Die politischen Religionen* (Stockholm, 1939). After the war the theme was taken up in Norman Cohn, *The Pursuit of the Millennium: Revolutionary Messianism in Medieval and Reformation Europe and Its Bearing on Modern Totalitarian Movements,* 2nd ed. (New York, 1961). The political-religious dimension of Nazi ideological themes and rituals was also analyzed in Klaus Vondung, *Magie und Manipulation: Ideologischer Kult und politische Religion des Nationalsozialismus* (Göttingen, 1971). During the seventies Uriel Tal further developed the analysis of Nazism as a political religion, mainly in his article "On Structures of Political Ideology and Myth in Germany Prior to the Holocaust," in Yehuda Bauer and Nathan Rotenstreich, eds., *The Holocaust as Historical Experience* (New York, 1981). Tal's interpretation appears as a guiding theme in Leni Yahil's *The Holocaust: The Fate of European Jewry* (New York, 1990). See also the conclusion to Saul Friedländer, "From Anti-Semitism to Extermination: A Historical Study of Nazi Policies Toward the Jews," *Yad Vashem Studies* 16 (1984).

129. Noakes and Pridham, *Nazism,* vol. 1, p.13.

Chapter 3 Redemptive Anti-Semitism

1. Lamar Cecil, *Albert Ballin: Business and Politics in Imperial Germany 1888–1918* (Princeton, N.J., 1967), p. 347. Cecil does not decide whether the overdose of sleeping pills was intentional or not. At the end of his novel *A Princess in Berlin,* Arthur R. G. Solmssen appends the (untitled) afterword: "On August 31, 1935, the Board of Directors of the Hamburg-Amerika Line announced that henceforth the SS *Albert Ballin* would carry the name SS *Hansa.*" Arthur R. G. Solmssen, *A Princess in Berlin* (Harmondsworth, England, 1980). I am grateful to Sue Llewellyn for this information.

2. Werner T. Angress, "The German Army's 'Judenzählung' of 1916: Genesis—Consequences—Significance," *LBIY* 23 ([London] 1978): 117ff. See also Egmont Zechlin, *Die deutsche Politik und die Juden im Ersten Weltkrieg* (Göttingen, 1969), pp. 528ff.

3. Angress, "The German Army's Judenzählung," p. 117.

4. Werner Jochmann, "Die Ausbreitung des Antisemitismus," in Werner E. Mosse, ed., *Deutsches Judentum in Krieg und Revolution 1916–1923* (Tübingen, 1971), p. 421.

5. Ibid., p. 423.

6. Saul Friedländer, "Political Transformations During the War and their Effect on the Jewish Question," in Herbert A. Strauss, ed., *Hostages of Modernization: Studies on Modern Anti-Semitism 1870–1933/39: Germany—Great Britain—France* (Berlin, 1993), p. 152.

7. Adolf Hitler, *Mein Kampf* (London, 1974), p. 193.

8. Friedländer, "Political Transformations," p. 152.

9. Zechlin, *Die deutsche Politik,* p. 525.

10. Ibid., in particular note 42.

11. Chernow, *The Warburgs,* p. 172.

12. Jochmann, "Die Ausbreitung des Antisemitismus," p. 427.

13. Ibid.

14. Ibid., p. 426. Jochmann quotes the classic study by the Jewish statistician and demographer Franz Oppenheimer, *Die Judenstatistik des Preussischen Kriegsministeriums* (Munich, 1922).

15. Ernst Simon, *Unser Kriegserlebnis (1919)*, quoted in Zechlin, *Die deutsche Politik*, p. 533.

16. Rathenau to Schwaner, August 4, 1916, quoted in Jochmann, "Die Ausbreitung des Antisemitismus," p. 427.

17. See especially Werner T. Angress, "The Impact of the Judenwahlen of 1912 on the Jewish Question: A Synthesis," *LBIY* 28 (1983):367ff.

18. For the shift of the Jewish vote, its dynamics and political significance, see ibid., p. 373ff., as well as Marjorie Lamberti, *Jewish Activism in Imperial Germany: The Struggle for Civil Equality* (New Haven, Conn., 1978), and Jacob Toury's classic study, *Die politischen Orientierungen der Juden in Deutschland: Von Jena bis Weimar* (Tübingen, 1966).

19. Angress, "Impact of the Judenwahlen of 1912," p. 381.

20. Ibid., p. 390.

21. Uwe Lohalm, *Völkischer Radikalismus: Die Geschichte des deutsch-völkischen Schutz-und-Trutz-Bundes 1919–1923* (Hamburg, 1970), p. 30.

22. Daniel Frymann, *Das Kaiserbuch: Politische Wahrheiten und Notwendigkeiten*, 7th ed. (Leipzig, 1925), pp. 69ff.

23. For the distinction between the traditional and the new trends in German nationalism after 1912, see Thomas Nipperdey, *Deutsche Geschichte 1866–1918*, vol. 2, *Machtstaat vor der Demokratie* (Munich, 1992), pp. 606ff. For the Kaiser's sometimes rabid anti-Jewish outbursts, see John C. G. Röhl's "Das beste wäre Gas!" *Die Zeit*, Nov. 25, 1994.

24. Roger Chickering, *We Men Who Feel Most German: A Cultural Study of the Pan-German League, 1886–1914* (Boston, 1984), p. 287.

25. Angress, "Impact of the Judenwhalen of 1912," p. 396.

26. In 1925 66.8 percent of all German Jews lived in the major cities, with Frankfurt and Berlin first and second in Jewish population. In 1871 36,326 Jews lived in the Greater Berlin area, accounting for 3.9 percent of a population of 931,984. In 1925 the official census for the same area indicated 172,672 Jews, or 4.3 percent of a general population of 4,024,165 (in Frankfurt that year, the Jewish population represented 6.3 percent). The number of Jews in Berlin was, in fact, probably higher than indicated by the official census, since many Jews did not register with Jewish communal organizations (the basis for the census), and a number of East European Jews were not registered anywhere at all. According to some estimates, as many as 200,000 Jews, or approximately 5 percent of the general population, were living in Greater Berlin in the immediate postwar period. Gabriel Alexander, "Die Entwicklung der jüdischen Bevölkerung in Berlin zwischen 1871 und 1945," *Tel Aviver Jahrbuch für Deutsche Geschichte*, vol. 20 (Tel Aviv, 1991), pp. 287ff., and particularly pp. 292ff. Such urban concentration was enhanced by the high visibility of East European Jews in the major German cities.

Jews from the East had long been present in Germany and Austria, arriving in particular after the late-eighteenth-century partitions of Poland and the annexations of Polish territory by both Prussia and Austria. A hundred years later, from 1881 on, a decisive change took place, with the beginning of a series of major pogroms against Jewish communities in the western provinces of czarist Russia. A mass exodus of Jews—most of them heading to the United States—from Russian-Polish territory began. Of the 2,750,000 Jews who left Eastern Europe for overseas between 1881 and 1914, a large proportion passed through Germany, mostly in the direction of the northern seaports Bremen and Hamburg, with a small number remaining in the country. For a detailed account see Shalom Adler-Rudel, *Ostjuden in Deutschland 1880–1940* (Tübingen, 1959). At the same time a more substantial number of Galician and Romanian Jews settled in Austria, especially in Vienna.

In 1900 7 percent of the Jews in Germany were *Ostjuden*, the percentage of East European Jews growing to 19.1 by 1925 and 19.8 by 1933. Ibid., p. 165. Moreover, their concentration in the large cities progressed at a rate faster than that of German Jewry's overall urbanization. In 1925 Eastern Jews represented 25.4 percent of Berlin's Jewish population, 27 percent of Munich's, 60 percent of Dresden's, and 80.7 percent of Leipzig's. Ibid.

27. See mainly Werner E. Mosse, "Die Juden in Wirtschaft und Gesellschaft," in Werner E. Mosse, ed., *Juden im Wilhelminischen Deutschland 1890–1914* (Tübingen, 1976), pp. 69ff., 75ff.

28. Werner E. Mosse, *Jews in the German Economy: The German-Jewish Economic Elite 1820–1935* (Oxford, 1987), p. 396.

29. Ibid., pp. 398, 400.

30. Ibid., pp. 323ff. (particularly p. 329).

31. Moritz Goldstein, "Deutsch-jüdischer Parnass," *Kunstwart* 25, no. 11 (Mar. 1912): 283.

32. Ibid., pp. 291–92.

33. Ibid., p. 293.

34. Ibid., p. 294.

35. Ferdinand Avenarius, "Aussprachen mit Juden," *Kunstwart* 25, no. 22 (Aug. 1912): 225.

36. These details and the quotations are taken from Ralph Max Engelman's Ph. D. dissertation, "Dietrich Eckart and the Genesis of Nazism" (Ann Arbor, Mich.: University Microfilms, 1971), pp. 31–32.

37. Ibid.

38. Ibid., p.32.

39. Maximilian Harden's *Die Zukunft* was "Jewish," and so was Siegfried Jacobsohn's *Schaubühne* (later *Weltbühne*). Otto Brahm's *Freie Bühne für modernes Leben*, succeeded by the *Neue Rundschau*, was "Jewish," as were the leading cultural critics of the major daily papers, Fritz Engel, Alfred Kehr, Max Osborn, and Oskar Bies. Engelman, ibid. Soon Kurt Tucholsky would become the most visible—and the most hated—journalist-author of Jewish origin of the Weimar years. Siegfried Breslauer would be associate editor of the *Berliner Lokalanzeiger*, Emil Faktor editor in chief of the *Berliner Börsen-Courier*, Norbert Falk cultural affairs editor of the *B.Z. am Mittag*, Joseph Wiener-Braunsberg editor of *Ulk*, the satirical supplement of the *Berliner Tageblatt*, and many more. Bernd Soesemann, "Liberaler Journalismus in der Kultur der Weimarer Republik," in Julius H. Schoeps, ed., *Juden als Träger bürgerlicher Kultur in Deutschland* (Bonn, 1989), p. 245.

40. Engelman, "Dietrich Eckart," p. 33.

41. Ibid.

42. Bernard Michel, *Banques et banquiers en Autriche au début du XXe Siècle* (Paris, 1976), p. 312.

43. Robert S. Wistrich, *The Jews of Vienna in the Age of Franz Josef* (Oxford, 1989), p. 170. The extraordinary role of the Jews in Viennese culture at the turn of the century

has been systematically documented in Steven Beller, *Vienna and the Jews, 1867–1938: A Cultural History* (Cambridge, England, 1989).

44. For the historical background of emancipation, see Jacob Katz, *Out of the Ghetto: The Social Background of Jewish Emancipation 1770–1870* (New York, 1978).

45. Shulamit Volkov, "Die Verbürgerlichung der Juden in Deutschland als Paradigma," in *Jüdisches Leben und Antisemitismus im 19. und 20. Jahrhundert* (Munich, 1990), pp. 112ff.

46. See in particular George L. Mosse, "Jewish Emancipation: Between *Bildung* and Respectability," in Jehuda Reinharz and Walter Schatzberg, eds., *The Jewish Response to German Culture: From the Enlightenment to the Second World War* (Hanover, N.H., 1985), pp. 1ff.

47. Michael A. Meyer, *The Origins of the Modern Jew: Jewish Identity and European Culture in Germany 1749–1824* (Detroit, 1967).

48. David Sorkin, *The Transformation of German Jewry 1780–1840* (New York, 1987).

49. Fritz Stern, *Gold and Iron: Bismarck, Bleichröder and the Building of the German Empire* (New York, 1977), p. 461.

50. Nipperdey, *Machtstaat vor der Demokratie*, p. 289.

51. Ibid., p. 290.

52. Hannah Arendt, *The Origins of Totalitarianism* (1951; reprint, New York, 1973), pp. 11ff.

53. For debates on these issues see in particular Israel Y. Yuval, "Vengeance and Damnation, Blood and Defamation: From Jewish Martyrdom to Blood Libel Accusations," *Zion* 58, no. 1 (1993): 33ff., and *Zion* 59, no. 2–3 (1994) (Hebrew).

54. Uriel Tal, *Christians and Jews in Germany: Religion, Politics and Ideology in the Second Reich, 1870–1914* (Ithaca, N.Y.,1975), pp. 96–98.

55. Ibid., pp. 209–10.

56. Jacob Katz, *From Prejudice to Destruction: Anti-Semitism 1700–1933* (Cambridge, Mass., 1980), p. 319.

57. Amos Funkenstein, "Anti-Jewish Propaganda: Pagan, Christian and Modern," *Jerusalem Quarterly* 19 (1981): 67.

58. Richard Hofstadter, *The Paranoid Style in American Politics and Other Essays* (Chicago, 1979), p. 29.

59. Jacob Katz, *Jews and Freemasons in Europe 1723–1939*, (Cambridge, England, 1970), particularly pp. 148ff.

60. Such distinctions have been implicit in some of the historical work published in the 1960s on the special course of German history during the nineteenth century; these theses have been recently reformulated and systematized by political sociologists. See in particular Pierre Birnbaum, "Nationalismes: Comparaison France-Allemagne," in *La France aux Français: Histoire des haines nationalistes* (Paris, 1993), pp. 300ff.

61. For the comparative part of the argument, see mainly Reinhard Rürup, *Emanzipation und Antisemitismus: Studien zur "Judenfrage" der bürgerlichen Gesellschaft* (Göttingen, 1975), pp. 17–18.

62. For a clear summary of German modernization and its impact, see Volker R. Berghahn, *Modern Germany: Society, Economy, and Politics in the Twentieth Century* (New York, 1987). For the *völkisch* reactions to this evolution, see Georges L. Mosse, *The Crisis of German Ideology: Intellectual Origins of the Third Reich* (New York, 1964); and Fritz Stern, *The Politics of Cultural Despair* (Berkeley, Calif., 1961).

63. The argument for the definition of this new anti-Semitic current as "revolution-

ary anti-Semitism" has been made in Paul Lawrence Rose, *Revolutionary Anti-Semitism in Germany from Kant to Wagner* (Princeton, N.J., 1990). See in particular Rose's argument about Wagner, pp. 358ff., as well as in idem, *Wagner: Race and Revolution* (New Haven, Conn., 1992).

64. See Robert W. Gutman, *Richard Wagner: The Man, His Mind and His Music* (New York, 1968), mainly pp. 389–441; Hartmut Zelinsky, *Richard Wagner: Ein deutsches Thema 1876–1976*, 3rd ed. (Vienna, 1983); Rose, *Wagner*, mainly pp. 135–70.

65. *Richard Wagner's Prose Works*, vol. 3 (London, 1894; reprint, New York, 1966), p. 100.

66. Cosima Wagner, *Die Tagebücher*, vol. 4 [1881–83], (Munich, 1982), p. 734.

67. Gustav Mahler remarked that Mime's music parodied bodily characteristics that were supposedly Jewish. For a study of the anti-Jewish imagery in Wagner's musical oeuvre, see Marc A. Weiner, *Richard Wagner and the Anti-Semitic Imagination* (Lincoln, Neb., 1995). For the Mahler remark, see ibid., p. 28.

68. Cosima Wagner, *Die Tagebücher*, p. 852.

69. Winfried Schüler, *Der Bayreuther Kreis von seiner Entstehung bis zum Ausgang der Wilhelminischen Ära* (Münster, 1971), p. 256.

70. Houston Stewart Chamberlain, *Foundations of the Nineteenth Century*, vol. 1 (1st English ed., 1910; reprint, New York, 1968) p. 578.

71. Geoffrey Field, *Evangelist of Race: The Germanic Vision of Houston Stewart Chamberlain* (New York, 1981), p. 225.

72. Ibid., p. 326.

73. Ibid.

74. On Hitler's visit to Chamberlain, see ibid., p. 436.

75. Some historians have emphasized the similarities of the reactions to the war all over Europe. See mainly Jay Winter, *Sites of Memory, Sites of Mourning: The Great War in European Cultural History* (Cambridge, England, 1995); others have pointed to the differences: the rise of an antiwar sentiment in France, that of a genocidal mood in Germany. See Bartov, *Murder in Our Midst*, mainly chap. 2. But an immense literature recognizes the apocalyptic postwar mood as such.

76. Nesta H. Webster, *World Revolution: The Plot Against Civilization* (London, 1921), p. 293.

77. Thomas Mann, *Tagebücher 1918–1921*, ed. Peter de Mendelssohn (Frankfurt am Main, 1979), p. 223.

78. The details that follow are taken from Peter Nettl, *Rosa Luxemburg*, vol. 2 (Oxford, 1966), pp. 772ff.

79. Friedländer, "Political Transformations," p. 159.

80. Among the twenty-seven members of the government of the Bavarian Republic of the Councils, eight of the most influential were of Jewish origin: Eugen Levine-Nissen, Towia Axelrod, Frida Rubiner (alias Friedjung), Ernst Toller, Erich Mühsam, Gustav Landauer, Ernst Niekisch, Arnold Wadler. Hans-Helmuth Knütter, *Die Juden und die Deutsche Linke in der Weimarer Republik, 1918–1933* (Düsseldorf, 1971), p. 118.

81. Reginald H. Phelps, "'Before Hitler Came': Thule Society and Germanen-orden," *Journal of Modern History* 35 (1963), pp. 253–54.

82. Jacques Benoist-Méchin, *Histoire de l'armee allemande*, vol. 2 (Paris, 1964), p. 216. Other Jewish left-wing politicians provoked no less negative reactions. On November 8, 1918, for instance, just after the break of relations between Germany and Russia, the Jewish Soviet ambassador in Berlin, Adolf Yoffe, about to leave Germany, transferred large sums of money to the Jewish Independent Socialist deputy Oskar

Cohn, who had become undersecretary in the Ministry of Justice. The money was meant to further revolutionary propaganda and for the acquisition of weapons. The facts soon became known and were widely discussed in the press. For the details of this transaction and of the debate in the press see Knütter, *Die Juden und die Deutsche Linke*, p. 70. Possibly even more violent was the reaction of the nationalist camp to the fact that a Jewish member of the National Assembly, Georg Gothein, became chairman of the Investigation Committee on the causes of the war and, together with Oskar Cohn and Hugo Sinzheimer, was in charge of the investigation of Hindenburg and of Ludendorff. See Friedländer, "Political Transformations," pp. 158–61, and mainly Barbara Suchy, "The Verein zum Abwehr des Antisemitismus (II): From the First World War to Its Dissolution in 1933," *LBIY* 30 (1985): 78–79.

83. Quoted in Nathaniel Katzburg, *Hungary and the Jews: Policy and Legislation 1920–1943* (Ramat-Gan, 1981), p. 35.

84. On the revolutionary events and on the leaders of the Hungarian revolution, see in particular Rudolf L. Tökés, *Béla Kun and the Hungarian Soviet Republic* (New York, 1967).

85. Two French novelists, the brothers Jérôme and Jean Tharaud, chronicled the Béla Kun regime in Hungary. Their historical fantasy appeared in 1921 and was translated into English in 1924, from the 64th French edition. Almost all of Béla Kun's revolutionary companions were Jews. Cf. Jérôme and Jean Tharaud, *When Israel Is King* (New York, 1924).

86. Isaac Deutscher, *The Non-Jewish Jew and Other Essays* (London, 1968).

87. Hamburger and Pulzer quote two sets of statistics about the Jewish vote in Weimar Germany: According to a contemporary observer, in 1924, 42 percent of the Jews cast their ballots for the SPD, 40 percent for the DDP, 8 percent for the KDP, 5 percent for the DVP, and 2 percent for the Wirtschaftspartei; according to Arnold Paucker's inquiry of 1972, the division was the following: 64 percent DDP, 28 percent SPD, 4 percent DVP, 4 percent KPD. SeeHamburger and Pulzer, "Jews as Voters in the Weimar Republic," p. 48. The main point is that in both counts more than 80 percent of Jewish voters opted for progressive liberals or for the moderate left.

88. On Jewish participation in the political life of the German Republic in its early phase, see in particular Werner T. Angress, "Juden im politischen Leben der Revolutionszeit," in Mosse, *Deutsches Judentum in Krieg und Revolution*; idem, "Revolution und Demokratie: Jüdische Politiker in Berlin 1918/19," in Reinhard Rürup, ed., *Jüdische Geschichte in Berlin: Essays und Studien* (Berlin, 1995). On Rathenau see Ernst Schulin, *Walter Rathenau: Repräsentant, Kritiker und Opfer seiner Zeit* (Göttingen, 1979).

89. Ibid., p. 137.

90. Rathenau's assassins claimed further that by sponsoring the fulfillment policy demanded by the Allies the Jewish minister was intent on the perdition of Germany, that he aimed at the Bolshevization of the country, that he was married to the sister of the Jewish Bolshevik leader Karl Radek, and so on. The anti-Jewish motivation of Rathenau's murderers is unquestionable. What remains unclear, though, is whether— beyond their hatred for the Jew Rathenau—his killers were instruments in the hands of ultra-right-wing groups that aimed to exploit his murder to destabilize the entire republican system. On this issue see Martin Sabrow, *Der Rathenaumord: Rekonstruktion einer Verschwörung gegen die Republik von Weimar* (Munich, 1994), mainly pp. 114ff.

91. For a detailed reconstruction of the origins and spreading of the *Protocols* see Norman Cohn, *Warrant for Genocide: The Myth of the Jewish World Conspiracy and the Protocols of the Elders of Zion* (London, 1967).

92. The anti–Napoleon III pamphlet was entitled "Dialogue aux enfers entre

Montesquieu et Machiavel," and composed in Brussels in 1864 by a French liberal, Maurice Joly; the novel *Biarritz*, written in 1868 by the German Hermann Gödsche, alias John Ratcliff, described the secret meeting of the heads of the Tribes of Israel in a Prague cemetery to plot Jewish domination of the world.

93. See Cohn, *Warrant for Genocide*, p. 138. For new details and nuances regarding the historical context of the *Protocols*, see Richard S. Levy's Introduction to Binjamin W. Segel, *A Lie and a Libel: The History of the Protocols of the Elders of Zion*, trans. and ed. Richard S. Levy (Lincoln, Neb., 1995). Segel's study was originally pubished in Berlin in 1926.

94. *The Protocols and the World Revolution including a Translation and Analysis of the "Protocols of the Meetings of the Zionist Men of Wisdom"* (Boston, 1920), p. 144.

95. Ibid., pp. 144–48 (the passage quoted is on pp. 147–48).

96. Quoted in Georg Franz-Willing, *Die Hitler-Bewegung*, vol. 1, *Der Ursprung 1919–1922* (Hamburg, 1962), p. 150.

97. Anything relating to the psychological, intellectual, and ideological development of "Hitler before Hitler" and, therefore, to the origins of his anti-Semitic obsession is entirely hypothetical. Were the ministrations—and particularly his morphine injections during the terminal illness of Hitler's mother—of the Jewish physician Eduard Bloch at the source of the future dictator's identification of the Jew with mortal penetration of the motherly body of the nation and the race? Did the theories of the pan-German history teacher, Leopold Pötsch, at the Realschule in Linz have any intellectual impact? Undoubtedly, early elements of Hitler's worldview stem from his sojourn in Vienna from 1908 to 1913; there he must have been influenced by Georg von Schönerer's and Karl Lüger's political campaigns. But how much further can we rely on his own declarations about this period or on the so-called recollections of his companions at the time, August Kubizek and Reinhold Hanisch?

For excellent accounts of Hitler's life before 1918 see in particular Alan Bullock, *Hitler: A Study in Tyranny* (London, 1952); Joachim C. Fest, *Hitler* (New York, 1974); as well as useful corrections regarding this period in Anton Joachimsthaler, *Korrektur einer Biographie: Adolf Hitler, 1908–1920* (Munich, 1989). For a systematic correlation between any indices of Hitler's early anti-Semitism and his later anti-Jewish worldview and policies, see Gerald Fleming, *Hitler and the "Final Solution"* (Berkeley, Calif., 1984).

98. Adolf Hitler, *Sämtliche Aufzeichnungen*, ed. Eberhard Jäckel and Axel Kuhn (Stuttgart, 1980), p. 128.

99. For the first complete publication of the text of the speech, with a detailed critical commentary, see Reginald H. Phelps, "Hitlers 'Grundlegende' Rede über den Antisemitismus," *VfZ* 16, no. 4 (1968): 390ff.

100. Hitler, *Sämtliche Aufzeichnungen*, p. 199.

101. Ibid., p. 202.

102. Eberhard Jäckel, *Hitler's Worldview: A Blueprint for Power* (Cambridge, Mass., 1981), pp. 52ff.

103. Adolf Hitler, *Hitler's Secret Conversations 1941–1944*, ed. Hugh R. Trevor-Roper (New York, 1972), p. 178.

104. Shaul Esh, "Eine neue literarische Quelle Hitlers? Eine methodologische Überlegung," *Geschichte und Unterricht*, 15 (1964), pp. 487ff.; Margarete Plewnia, *Auf dem Weg zu Hitler: Der "völkische" Publizist Dietrich Eckart* (Bremen, 1970), pp. 108–9.

105. Ernst Nolte, "Eine frühe Quelle zu Hitlers Antisemitismus," *Historische Zeitschrift* 192 (1961), particularly 604ff.

106. Esh, "Eine neue literarische Quelle Hitlers?"

107. Engelman, "Dietrich Eckart," p. 236.

108. Dietrich Eckart, *Der Bolschewismus von Moses bis Lenin. Zwiegespräch zwischen Adolf Hitler und mir* (Munich, n.d. [1924]), p. 49.

109. Hitler, *Mein Kampf*, p. 65.

110. Ibid., p. 679.

111. The most thorough presentation of Hitler's ideology as a coherent intellectual system is to be found in Jäckel, *Hitler's Worldview*; for the direct relation between the worldview and Nazi policy see in particular Eberhard Jäckel, *Hitler in History* (Hanover, N. H., 1984). This ("intentionalist") position stands in opposition to the "functionalist" approach, which dismisses the systematic aspect of Hitler's ideology and marginalizes or completely negates any direct causal relation between Hitler's ideology and the policies of the Nazi regime. The most consistent exponent of the extreme functionalist position has been Hans Mommsen. With regard to Hitler's anti-Jewish policies, see in particular Hans Mommsen, "The Realization of the Unthinkable." For an excellent evaluation of these different approaches see Ian Kershaw, *The Nazi Dictatorship: Problems and Perspectives of Interpretation*, 3rd ed. (London, 1993), mainly chaps. 4 and 5); specifically with regard to anti-Jewish policies see an evaluation of both positions in Friedländer, "From Anti-Semitism to Extermination."

112. Among the many attempts to explain Hitler's personality and particularly his anti-Jewish obsession in terms of psychopathology, mainly by using psychoanalytic concepts, see in particular Rudolph Binion, *Hitler Among the Germans* (New York, 1976); Robert G. L. Waite, *The Psychopathic God: A Biography of Adolf Hitler* (New York, 1977). See also the wartime analysis published some thirty years later: Walter C. Langer, *The Mind of Adolf Hitler: The Secret Wartime Report* (New York, 1972). The problems raised by psychobiographical inquiries have been debated at length; for an evaluation of some of the issues see Saul Friedländer, *History and Psychoanalysis: An Inquiry into the Possibilities and Limits of Psychohistory* (New York, 1978).

113. Adolf Hitler, *Hitler's Secret Book* (New York, 1961).

114. Adolf Hitler, *Reden, Schriften, Anordnungen: Februar 1925 bis Januar 1933*, vol. 1, *Die Wiedergründung der NSDAP: Februar 1925–Juni 1926*, ed. Clemens Vollnhals (Munich, 1992), p. 208.

115. Ibid., p. 421.

116. Ibid., p. 195. "Even when he [the Jew] writes the truth, the truth is only meant as a way of lying. . . . A Jewish joke is known on that account: Two Jews are sitting together on a train. . . . One asks the other: So, Stern, where are you going? Why do you want to know? Well, I would like to know it—I am going to Posemuckel! It is not true, you are not going to Posemuckel. Yes, I am going to Posemuckel. So you are really going to Posemuckel and are also saying that you are going to Posemuckel—what a liar you are!" Hitler seems to have liked this joke so much that two years later he used it in another speech. See Adolf Hitler, *Reden, Schriften, Anordnungen: Februar 1925 bis Januar 1933*, vol. 2, *Vom Weimarer Parteitag bis zur Reichstagswahl Juli 1926-Mai 1927*, ed. Bärbel Dusik (Munich, 1992), p. 584.

117. Hitler, *Reden, Schriften*, vol. 1, p. 297.

118. Ibid., vol. 2, pp. 105–6.

119. The still missing volumes will cover the period June 1931–January 1933.

120. Hitler, *Reden, Schriften*, vol. 2, part 2, *August 1927–May 1928*, pp. 699ff.

121. Ibid., vol. 3, *Zwischen den Reichstagswahlen Juli 1928–September 1930*, ed. Bärbel Dusik and Klaus A. Lankheit, part 1: *Juli 1928–Februar 1929* (Munich, 1994), p. 43.

122. Ibid., vol. 4, *Von der Reichstagswahl bis zur Reichspräsidentenwahl, Oktober 1930–März 1932*, part 1, *Oktober 1930–Juni 1931*, ed. Constantin Goschler (Munich, 1994), pp. 421–30.

123. Ibid., pp. 22-23.

124. Article of Jan. 11, 1930 (*Illustrierter Beobachter*). This article and previous texts in the same vein are quoted in Rainer Zitelmann, *Hitler: Selbstverständnis eines Revolutionärs* (Stuttgart, 1990), pp. 476ff.

125. Martin Broszat, *Hitler and the Collapse of Weimar Germany* (New York, 1987), p. 25. In his private conversations Hitler showed no restraint in his anti-Jewish fury. A telling illustration is to be found in the notes covering the years 1929–1932 and written down in 1946 by Otto Wagener, interim chief of staff of the SA and then head of the economic division of the party. Wagener remained a true believer even after the war, and thus it would have been in his interest to tone down Hitler's remarks about the "Jewish question." As they are—toned down or not—Wagener's recollections reflect the same themes and the same unbounded hatred that we know from Hitler's earlier speeches and texts. For Wagener's text see the critical edition of his notes published by Henry A. Turner, *Otto Wagener, Hitler aus nächster Nähe: Aufzeichnungen eines Vertrauten 1929–1932* (Frankfurt am Main), 1978. For the anti-Jewish tirades see in particular pp. 144ff. and 172ff.

126. For the inner core of the Nazi leadership, anti-Semitism was an essential part of their worldview from very early on. This early anti-Semitism was particularly extreme in the case of Rosenberg, Streicher, Ley, Hess, and Darré. Himmler and Goebbels also became anti-Semites before joining the Nazi Party. (The notable exceptions were Göring and the brothers Strasser.) On this issue I do not share Michael Marrus's evaluation regarding the absence of anti-Semitism among party leaders before 1925. See Michael Marrus, *The Holocaust in History* (Hanover, N.H.,1987), pp. 11–12. for a discussion of the apocalyptic dimension of the anti-Jewish creed among the Nazi elite, see Erich Goldhagen, "Weltanschauung und Endlösung: Zum Antisemitismus der nationalsozialistischen Führungsschicht," *VfZ* 24, no 4 (1976): 379ff. The marginal importance of anti-Semitism among the SA has been well documented by Theodor Abel. See the reworking and reinterpretation of Abel's questionnaires in Peter Merkl, *Political Violence Under the Swastika: 581 Early Nazis* (Princeton, N.J., 1975). The same cannot be said, however, of the middle-class future members of the SD, who often belonged to extreme-right-wing anti-Semitic organizations from the early postwar years onward. See Herbert, *Best. Biographische Studien.*

127. Russel Lemmons, *Goebbels and "Der Angriff,"* (Lexington, Ky., 1994), particularly pp. 111ff.

128. Ralf Georg Reuth, *Goebbels* (Munich, 1990), p. 200.

129. Ibid.

130. In 1932 the Nazis launched a vicious anti-Semitic attack against the DNVP candidate for the presidency, Theodor Duesterberg (one of the two leaders of the right-wing veterans' organization, the Stahlhelm), harping on the Jewish origins of his grandfather, a physician who had converted to Protestantism in 1818. For this entire episode see Volker R. Berghahn, *Der Stahlhelm: Bund der Frontsoldaten 1918–1935* (Düsseldorf, 1966), pp. 239ff.

131. Roland Flade, *Die Würzburger Juden: Ihre Geschichte vom Mittelalter bis zur Gegenwart* (Würzburg, 1987), p. 149.

132. Trude Maurer, *Ostjuden in Deutschland 1918–1933* (Hamburg, 1986), p. 346.

133. Ibid., p. 329 ff.

134. Michael Brenner, *The Renaissance of Jewish Culture in Weimar Germany* (New Haven, Conn., 1996).

135. Heinrich-August Winkler, *Weimar 1918–1933: Die Geschichte der ersten deutschen Demokratie* (Munich, 1993), p. 180.

136. Henri Béraud, "Ce que j'ai vu à Berlin," *Le Journal*, Oct. 1926. Quoted in Frédéric Monier, "Les Obsessions d'Henri Béraud," *Vingtième Siècle: Revue d'Histoire* (Oct.–Dec. 1993): 67.

137. On this whole affair see Erich Eyck, *Geschichte der Weimarer Republik*, vol. 1 (Erlenbach, 1962), pp. 433ff. (For some reason Eyck refers only to Julius Barmat.)

138. Ibid., vol. 2, pp. 316ff. See also Winkler, *Weimar 1918–1933*, p. 356.

139. Ibid. For the Barmat and Sklarek scandals see also Maurer, *Ostjuden in Deutschland*, pp. 141ff.

140. See Maurer, "Die Juden in der Weimarer Republik," in Dirk Blasius and Dan Diner, eds., *Zerbrochene Geschichte: Leben und Selbstverständnis der Juden in Deutschland* (Frankfurt am Main, 1991), p. 110.

141. Knütter, *Die Juden und die Deutsche Linke*, pp. 174ff.

142. For an analysis of the "Jewish problem" in the DDP see Bruce B. Frye, "The German Democratic Party and the 'Jewish Problem' in the Weimar Republic," *LBIY* 21 ([London] 1976), pp. 143ff.

143. Winkler, *Weimar 1918–1933*, p. 69.

144. Frye, "The German Democratic Party," pp. 145–47.

145. Berghahn, *Modern Germany*, p. 284.

146. Larry E. Jones, *German Liberalism and the Dissolution of the Weimar Party System 1918–1933* (Chapel Hill, N.C., 1988).

147. Peter Gay, *Weimar Culture: The Outsider as Insider* (New York, 1968).

148. Peter Gay, *Freud, Jews and Other Germans: Masters and Victims in Modernist Culture* (New York, 1978).

149. The same minimization of the Jewish factor appears in Carl Schorske's otherwise magnificent study of fin-de-siècle Vienna. Carl E. Schorske, *Fin-de-Siècle Vienna: Politics and Culture* (New York, 1980). For criticism on this issue see Steven Beller, *Vienna and the Jews*, pp. 5ff.

150. Istvan Deak, *Weimar Germany's Left-Wing Intellectuals: A Political History of the Weltbühne and Its Circle* (Berkeley, Calif., 1968), p. 28.

151. Peter Jelavich, *Munich and Theatrical Modernism: Politics, Playwriting and Performance, 1890–1914* (Cambridge, Mass., 1985), pp. 301ff.

152. Ibid., p. 302.

153. Ibid., p. 304.

154. Deak, *Weimar Germany's Left-Wing Intellectuals*, p. 28.

155. Quoted in Anton Kaes, ed., *Weimarer Republik: Manifeste und Dokumente zur deutschen Literatur, 1918–1933* (Stuttgart, 1983), pp. 537–39.

156. Ibid., p. 539.

157. Jakob Wassermann, *Deutscher und Jude: Reden und Schriften 1904–1933* (Heidelberg, 1984), p. 156.

158. Kaes, *Weimarer Republik*, p. 539.

159. Marion Kaplan, "Sisterhood Under Siege: Feminism and Anti-Semitism in Germany, 1904–1938, in Renate Bridenthal, Atina Grossmann, and Marion Kaplan, eds., *When Biology Became Destiny: Women in Weimar and Nazi Germany* (New York, 1984), pp. 186–87.

160. Niewyk, *The Jews in Weimar Germany*, p. 80.

161. Fest, *Hitler*, p. 355. On the unfolding of these events, see also Winkler, *Weimar 1918–1933*, pp. 508ff.

162. Ibid., p. 513.

163. Ibid., pp. 513–14.

164. Broszat, *Hitler and the Collapse of Weimar Germany*, p. 126.

Chapter 4   The New Ghetto

1. Martin Broszat and Elke Fröhlich, *Alltag und Widerstand: Bayern im Nationalsozialismus* (Munich, 1987), p. 434. All the details about Obermayer are taken from Broszat and Fröhlich's presentation of the case.

2. Ibid., pp. 450–52, 456ff.

3. Ibid., p. 437.

4. Ibid., pp. 443ff.

5. Quoted in Ian Kershaw, *The "Hitler Myth": Image and Reality in the Third Reich* (Oxford, 1987), p. 71.

6. Martin Broszat, *The Hitler State: The Foundation and the Development of the Internal Structure of the Third Reich* (London, 1981), p. 349.

7. Ibid., p. 350.

8. Ian Kershaw, "'Working Towards the Führer': Reflections on the Nature of the Hitler Dictatorship," *Contemporary European History* 2, no. 2 (1993): 116.

9. Bankier, "Hitler and the Policy-Making Process," p. 9.

10. Ibid.

11. Walk, *Das Sonderrecht*, p. 117.

12. Ibid., p. 153.

13. Lilli Zapf, *Die Tübinger Juden*, 3rd. ed. (Tübingen, 1981), p. 150.

14. Paul Sauer, ed., *Dokumente über die Verfolgung der jüdischen Bürger in Baden-Württemberg durch das Nationalsozialistische Regime 1933–1945*, vol. 1 (Stuttgart, 1966), p. 50.

15. Walk, *Das Sonderrecht*, p. 72. The Association of Jewish Frontline Soldiers had unsuccessfully turned to Hindenburg to have this exclusion rescinded. For the full text of the March 23, 1934, petition, see Ulrich Dunker, *Der Reichsbund jüdischer Frontsoldaten, 1919–1938*, (Düsseldorf, 1977), pp. 200ff.

16. See, for instance, the petition from the chairman of the Association of National German Jews, Max Naumann, addressed to Hitler on March 23, 1935, and the declaration of the Association of Jewish Frontline Soldiers of the same date in Michaelis and Schraepler, *Ursachen*, vol. 11, pp. 159–62.

17. Walk, *Das Sonderrecht*, p. 115.

18. Ibid., p. 122 (ordinance of July 25, 1935). On various aspects of the problem of the *Mischlinge* see mainly Noakes, "Wohin gehören die 'Judenmischlinge'?," pp. 69ff.

19. Communication T3/Att. Group to Adjutant's Office of Chief of the Army Command, 22.5.1934, Reichswehrministerium, Chef der Heeresleitung, microfilm MA–260, IfZ, Munich.

20. Steinweis, *Art, Ideology and Economics in Nazi Germany*, pp. 108ff.

21. Ibid., p. 111. In fact, a few Jews still remained members of the various chambers, and it was only in 1939 that the exclusion became total. Ibid.

22. Ludwig Holländer, *Deutsch-jüdische Probleme der Gegenwart: eine Auseinandersetzung über die Grundfragen des Zentralvereins deutscher Staatsbürger jüdischen Glaubens*, Berlin, 1929, p. 18. Quoted in R. L. Pierson, *German Jewish Identity in the Weimar Republic* (Ann Arbor, Mich.: University Microfilms, 1972), p. 63.

23. Kurt Loewenstein, "Die innerjüdische Reaktion auf die Krise der deutschen Demokratie" in Mosse, *Entscheidungsjahr 1932*, p. 386.

24. George L. Mosse, "The Influence of the Völkisch Idea on German Jewry," in *Germans and Jews: The Right, the Left, and the Search for a "Third Force" in Pre-Nazi Germany* (New York, 1970), pp. 77ff.

25. R. L. Pierson is quoting from an essay by Wilhelm Hanauer, "Die Mischehe," *Jüdisches Jahrbuch für Gross Berlin*, 1929, p. 37.

26. Noakes, "Wohin gehören die 'Judenmischlinge'?," p. 70.

27. Proctor, *Racial Hygiene*, p. 151.

28. Ibid.

29. Ibid., pp. 78–79.

30. Ibid., p. 79.

31. Ingo Müller, *Hitler's Justice: The Courts of the Third Reich* (Cambridge, Mass., 1991), p. 91.

32. Ibid.

33. Ibid., p. 92.

34. Ibid., p. 93.

35. Ibid., p. 94.

36. Ibid., p. 95.

37. *Chronik der Stadt Stuttgart*, p. 225.

38. Robert Thévoz, Hans Branig, and Cécile Löwenthal-Hensel, eds., *Pommern 1934/1935 im Spiegel von Gestapo-Lageberichten und Sachakten*, vol. 2, *Quellen* (Cologne, 1974), p. 118.

39. Werner T. Angress, "Die 'Judenfrage' im Spiegel amtlicher Berichte 1935," in Ursula Büttner et al., *Das Unrechtsregime*, vol. 2, p. 34.

40. For an excellent discussion of various anti-Semitic fantasies regarding the Jewish body, see Sander L. Gilman, *The Jew's Body* (New York, 1991).

41. Quoted in J. M. Ritchie, *German Literature under National Socialism* (London, 1983), p. 100.

42. Quoted in Ulrich Knipping, *Die Geschichte der Juden in Dortmund während der Zeit des Dritten Reiches* (Dortmund, 1977), p. 50.

43. Roland Müller, *Stuttgart zur Zeit des Nationalsozialismus* (Stuttgart, 1988), pp. 292–93, 296.

44. Quoted in Noakes and Pridham,, *Nazism 1919–1945*, vol. 2, p. 531. (Translation slightly revised.)

45. For the cable to Müller see Röhm and Thierfelder, *Juden-Christen-Deutsche*, vol. 1, p. 268.

46. Sauer, *Dokumente über die Verfolgung*, vol. 1, p. 62.

47. Ibid., p. 63.

48. The Minister of Justice to the Reich Chancellor, 20.5.1935, Max Kreuzberger Research Papers, AR 7183, Box 8, Folder 9, LBI, New York.

49. Lammers to Minister of Justice, 7.6.1935, ibid.

50. *Akten der Parteikanzlei der NSDAP* (abstracts), part 2, vol. 3, p. 107.

51. See Anton Doll, ed., *Nationalsozialismus im Alltag: Quellen zur Geschichte der NS-Herrschaft im Gebiet des Landes Rheinland-Pfalz* (Speyer (Landesarchiv), 1983), p. 139. Translated by Dieter Kuntz in Sax and Kuntz, *Inside Hitler's Germany:* pp. 410–11. (Translation slightly revised.)

52. Steven M. Lowenstein, "The Struggle for Survival of Rural Jews in Germany 1933–1938: The Case of Bezirksamt Weissenburg, Mittelfranken," in Arnold Paucker, ed., *The Jews in Nazi Germany 1933–1943* (Tübingen, 1986), p. 116.

53. Ibid., p. 117.

54. Ibid., p. 123.

55. Ibid., p. 121. For the opposition between the economic interests of the peasants and the pressure of party radicals regarding the activities of Jewish cattle dealers in Bavaria see Falk Wiesemann, "Juden auf dem Lande: die wirtschaftliche Ausgrenzung der jüdischen Viehhändler in Bayern," in Peukert and Reulecke, *Die Reihen fast geschlossen,*, pp. 381ff.

56. Thévoz, Branig, and Löwenthal-Hensel, *Pommern 1934/1935*, vol. 2, p. 103.

57. Klemperer, *Ich will Zeugnis ablegen*, vol. 1, p. 110.

58. Angress, "Die 'Judenfrage'," in Büttner, *Das Unrechtsregime*, vol. 2, p. 25.

59. Chief of the SD Main Office to NSDAP district court III/B, 11.10.35, SD Hauptamt, microfilm MA–554, IfZ, Munich.

60. Thévoz, Branig, and Löwenthal-Hensel, *Pommern 1934/1935*, vol. 2, p. 93.

61. Ibid., p. 118.

62. Baltic Sea Resort Management, Binz, 17.5.38, SD Main Office, microfilm MA–554, IfZ, Munich.

63. Thomas Klein, ed., *Der Regierungsbezirk Kassel 1933–1936: Die Berichte des Regierungspräsidenten und der Landräte* (Darmstadt, 1985), vol. 1, p. 72. Sometimes, mainly in small towns and villages, the reactions of some Germans were determined both by economic advantage and by the habit of buying from the Jews, who were a long-standing part of the life of the community. According to the report of a *Blockleiter* from a small town near Trier (September 20, 1935), the mayor continued his practice of buying meat from Jews. When confronted by the *Blockleiter*, he answered: "One should not be so filled with hatred; the small Jews are no Jews." See Franz Josef Heyen, *Nationalsozialismus im Alltag: Quellen zur Geschichte des Nationalsozialismus vornehmlich im Raum Mainz-Koblenz-Trier* (Boppard am Rhein, 1967), p. 138.

64. Angress, "Die 'Judenfrage,'" p. 29.

65. Commander of SD main region Rhine to SS Gruppenführer Heissmeyer, Koblenz, 3.4.1935, SD-Oberabschnitt Rhein, microfilm MA–392, IfZ, Munich. The contradictory reports about the economic relations between Germans and Jews show that the situation varied from place to place and that, in any case, manifold relations were maintained. As early as 1934, small-town shopkeepers may have displayed their commitment to Nazism by refusing to serve Jewish customers; yet, according to a September 1934 entry in Bella Fromm's diary, the real attitudes were often different: "I talked to shopkeepers and people at gas stations and inns. In many cases, their strictly National Socialist attitude was obviously a measure of precaution. Many Jewish people told me: 'although we can't enter their shops, Aryan shopkeepers give us what we need after business hours.'" Fromm, *Blood and Banquets*, p. 183.

66. Herbert Freeden, "Das Ende der jüdischen Presse in Nazideutschland," *Bulletin des Leo Baeck Instituts* 65 (1983): 6.

67. Heydrich, Gestapa to all State police local offices, 25.2.1935, Ortspolizeibehörde Göttingen, microfilm MA 172, IfZ, Munich.

68. The Mayor of Lörrach to all municipality employees and workers, 7.6.1935, Unterlagen betr. Entrechtung der Juden in Baden 1933–1940, ED 303, IfZ, Munich.

69. Robert Weltsch, "A Goebbels Speech and a Goebbels Letter," *LBIY* 10 (1965), p. 281.

70. Ibid., pp. 282–83.

71. Ibid., p. 285.

72. Ibid. (Misprinted as "West-end" in Weltsch's text.)

73. Franz Schonauer, "Zu Hans Dieter Schäfer: 'Bücherverbrennung, staatsfreie Sphäre und Scheinkultur'" in Denkler and Lämmert, eds., "*Das war ein Vorspiel nur...*," p. 131.

74. NSDAP Reichsleitung, Office of Technology, Circular 3/35, 25 Jan. 1935, Himmler Archives, Berlin Document Center, Microfilm no. 270, Roll 2 (LBI, New York, microfilm, 133g).

75. Heiber, *Universität unterm Hakenkreuz, Der Professor im Dritten Reich: Bilder aus der akademischen Provinz* (Munich, 1991), pp. 216–17. See also Beyerchen, *Scientists under Hitler*, pp. 67–68. The commemoration of Haber's death was an unusual act of courage. It was made easier, in this instance, because Haber had converted to Protestantism and had been an ultranationalist until 1933, and mainly by the fact that he had made singular contributions to science, to the German chemical industry, and to the German war effort by discovering the synthesis of ammonia (allowing for the mass production of fertilizers—but also of explosives), and also by inventing and launching the use of chlorine gas, the first poison gas used in combat during World War I. Paradoxically, though solemnly commemorated one year after his death, Haber was isolated and ostracized at the time of his resignation. and when he left Germany for England (and then Switzerland), he had abandoned much of his German nationalist stance and was actually planning to move to Palestine. On these various issues, see Stern, *Dreams and Delusions*, pp. 46ff and mainly pp. 51ff, as well as the recent massive—and at times problematic—study by Dietrich Stolzenberg, *Fritz Haber: Chemiker, Nobelpreisträger, Deutscher, Jude* (Weinheim, 1994). For a review of Stolzenberg's biography, see M. F. Perutz, "The Cabinet of Dr. Haber," *New York Review of Books*, June 20, 1996.

76. Reuth, *Goebbels* (Munich, 1990), p. 322.

77. Ibid., p. 323.

78. Rosenberg to Goebbels, 30 August 1934, Doc. CXLII–246 in Michel Mazor, *Le Phénomène Nazi: Documents Nazis Commentés* (Paris, 1957), pp. 166ff. See also Wulf, *Theater und Film*, p. 104. The debates in the plastic arts and in literature followed a similar pattern. At the outset expressionism and modern trends more generally in both domains were protected by Goebbels against Rosenberg. But the Rosenberg line, which was Hitler's position, won. In the plastic arts, the notorious turning point took place in 1937, when orthodoxy was presented at the Grosse Deutsche Kunstausstellung, and heresy was pilloried at the exhibition of *"Entartete Kunst"* (degenerate art). In literature some debates continued until the late 1930s. For literature see Dieter Schäfer, "Die nichtfaschistische Literatur der 'jungen Generation,'" in Horst Denkler and Karl Prumm, eds., *Die deutsche Literatur im Dritten Reich* (Stuttgart, 1976), pp. 464–65.

79. Levi, *Music in the Third Reich*, p. 74. Yet, Paul Graener's *Friedemann Bach* and Georg Vollerthun's *Der Freikorporal* were performed despite the fact that the libretti had been written by Jewish playwright Rudolf Lothar. Ibid., p, 75.

80. Michael H. Kater, *Different Drummers: Jazz in the Culture of Nazi Germany* (New York, 1992), p. 43.

81. Ibid.

82. Kampfbund für Deutsche Kultur/local group Greater Munich to Reich Association "Deutsche Bühne," Berlin, 16.8.1933, Rosenberg Akten, microfilm MA–697, IfZ, Munich.

83. Reich Association "Deutsche Bühne" to Kampfbund..., 23.8.1933, ibid.

84. Kampfbund... Northern Bavaria/Franconia to "Deutsche Bühne"... 2.12.33, ibid.; "Deutsche Bühne" to Kampfbund..., 5.12.33, ibid.

85. Pätzold, *Verfolgung, Vertreibung, Vernichtung*, pp. 77–78.

86. Levi, *Music in the Third Reich*, p. 76.

87. Ibid., pp. 67. Also Erik Levi, "Music and National-Socialism: The Politicization of Criticism, Composition and Performance," in Brandon Taylor and Wilfried van der Will, eds., *The Nazification of Art: Art, Design, Music, Architecture and Film in the Third Reich* (Winchester, 1990), pp. 167–71. On September 1, 1936, the *Reichskulturkammer* published an alphabetical list of mainly Jewish composers whose works were not allowed under any circumstances: Abraham, Paul; Achron, Josef; Alwin, Karl; Antheil, George; Barmas, Issay; Becker, Conrad; Benatzky, Ralph; Benjamin, Arthur; Bereny, Henry; Berg, Alban; and so on. Himmler Archives, Berlin Document Center, microfilm No. 269, Roll 1 (LBI, New York, microfilm 133f).

88. Memorandum II 112, 27.11.1936, SD-Hauptamt, microfilm MA–554, IfZ, Munich.

89. Memorandum II 112, 3.1.1938, ibid. The search for Jews and Jewish influence in music and theater illustrates but one aspect of the general identification drive in every possible cultural domain, including *völkisch* authors of the nineteenth and twentieth centuries, even those considered as belonging to the intellectual-ideological background of Nazism. For example, Ernst Häckel's "Monist League" came under scrutiny, as did the racial theorist Ludwig Woltmann. On both cases see Paul Weindling, "Mustergau Thüringen: Rassenhygiene zwischen Ideologie und Machtpolitik," in Norbert Frei, ed., *Medizin und Gesundheitspolitik in der NS-Zeit* (Munich, 1991), pp. 93, 93n.

90. Minister of the Interior to Reich Chancellor, 19.7.1935, Max Kreuzberger Research Papers, AR 7183, Box 8, LBI, New York.

91. Lammers to Frick, 31.7.35, ibid.

92. Frick to Gürtner, 14.8.35, NSDAP, Parteikanzlei, microfiche 024638ff., IfZ, Munich.

93. Pfundtner to Reich Office for Ancestry Research, 14.8.1935, ibid., microfiche 024642, ibid.

94. Steinweis, "Hans Hinkel," p. 213.

95. Heydrich to all Gestapo stations, 19.8.1935, Aktenstücke zur Judenverfolgung, Ortspolizeibehörde Göttingen, MA–172, IfZ, Munich.

96. Steinweis, "Hans Hinkel," p. 215.

97. Heydrich to all State Police Offices, 13.8.1935, Aktenstücke zur Judenverfolgung, Ortspolizeibehörde Göttingen, MA–172 IfZ, Munich.

98. Adam, *Judenpolitik*, p. 115.

99. Report of the Police Directorate Munich, April/Mai 1935 (Geheimes Staatsarchiv, Munich, MA 104990), Fa 427/2, IfZ, Munich, pp. 24ff.

100. Jochen Klepper, *Unter dem Schatten deiner Flügel: Aus den Tagebüchern der Jahre 1932–1942* (Stuttgart, 1983), p. 269.

101. Ibid., p. 270.

102. Ibid., pp. 282–83.

103. SOPADE, *Deutschland-Berichte* 2 (1935): 803.

104. Ibid., p. 804.

105. Ibid., 921.

106. Marlis Steinert, *Hitlers Krieg und die Deutschen: Stimmung und Haltung der deutschen Bevölkerung im Zweiten Weltkrieg* (Düsseldorf, 1970), p. 57.

107. Ian Kershaw may have overstressed the negative reactions of the population to the violence against the Jews in his "The Persecution of the Jews and German Popular Opinion in the Third Reich," *LBIY* 26 (1981).

108. Fischer, *Hjalmar Schacht*, pp. 154-55.

109. For the list of participants see Otto Dov Kulka, "Die Nürnberger Rassengesetze und die deutsche Bevölkerung," *VfZ* 32 (1984): 616.

110. Ibid.

111. Parts of the protocols of this meeting presented at the Nuremberg trial as NG-4067 are quoted in Michalka, *Das Dritte Reich*, vol. 1, p. 155. For additional material to the published German text see Otto Dov Kulka, "Die Nürnberger Rassengesetze," pp. 615ff. See also *Documents on German Foreign Policy, Series C*, vol. 4 (Washington, D.C., 1962), pp. 568ff.

112. For a comparison of different versions of Wagner's suggestions, see Peter Longerich, *Hitlers Stellvertreter*(Munich, 1992), pp. 212-13.

113. Michalka, *Das Dritte Reich*, p. 155.

114. Ibid.; Kulka, "Die Nürnberger Rassengesetze," p. 617.

115. Ibid.

116. Wildt, *Die Judenpolitik des SD*, pp. 23-24, 70-78.

117. Quoted in Ephraim Maron, "The Press Policy of The Third Reich on the Jewish Question" (Ph. D. diss., Tel Aviv University, 1992), pp. 81n-82n.

118. Adolf Hitler, *Speeches and Proclamations 1932-1945*, ed. Max Domarus, trans. Chris Wilcox and Mary Fran Gilbert, vol. 2, *The Chronicle of a Dictatorship, 1935-1938*, (Wauconda, Ill., 1992), p. 702. [Hitler, *Reden und Proklamationen, 1932-1945: Kommentiert von einem deutschen Zeitgenossen*, ed. Max Domarus, vol. 1, *Triumph (1932-1938)* (Würzburg, 1962), p. 534.]

119. For Hitler's speech see Hitler, *Speeches*, ed. Max Domarus (English), vol. 2, pp. 706-7.

120. Walk, *Das Sonderrecht*, p. 127.

121. Ibid.

122. Noakes and Pridham, *Nazism 1919-1945*, vol. 2, p 463.

123. Hitler, *Speeches*, vol. 2, p. 708.

124. Ibid.

125. Ibid., p. 731.

126. Helmut Heiber, ed., *Goebbels-Reden*, vol. 1, *1932-1939* (Düsseldorf, 1971), p. 246.

127. Manuscript notes taken by Fritz Wiedemann, Institut für Zeitgeschichte, Munich. Quoted in Helmut Krausnick, "Judenverfolgung," in Hans Buchheim et al., *Anatomie des SS-Staates*, vol. 2 (Munich, 1967), p. 269.

128. "An dieser Stelle erklärteer noch, dass er in dem Falle eines Krieges auf allen Fronten, bereit zu allen Konsequenzen, sei." Philippe Burrin, *Hitler and the Jews: The Genesis of the Holocaust* (New York, 1994), pp. 48-49. In his book Burrin emphasizes the significance of Hitler's threat in case of a war "on all fronts"—that is, in a situation similar to that of 1914-18. For Germany World War II became such a war, as the Russian campaign did not result in a rapid German victory. The relation between this situation and Hitler's decision to exterminate the Jews will be discussed in volume 2 of *Nazi Germany and the Jews: The Years of Extermination*.

Chapter 5   The Spirit of Laws

1. Minister of Education ..., decree, 13.9.1935, Reichsministerium für Wissenschaft u. Erziehung, microfilm MA-103/1, IfZ, Munich.

2. Klemperer, *Ich will Zeugnis ablegen*, vol. 1, p. 195.

3. Neliba, *Wilhelm Frick*, pp. 198ff.

4. For these various details see David Bankier, *The Germans and the Final Solution: Public Opinion under Nazism* (Oxford, 1992), pp. 43–44.

5. Neliba, *Wilhelm Frick*, pp. 200ff.

6. Goebbels, *Tagebücher*, part 1, vol. 2, p. 488.

7. For Lösener's attitude see Bankier, *The Germans and the Final Solution*, p. 43.

8. Robert L. Koehl, *The Black Corps: The Structure and Power Struggles of the Nazi SS* (Madison, Wis., 1983), p. 102.

9. Bernhard Lösener, "Als Rassereferent im Reichsministerium des Innern," *VfZ* 3 (1961): 264ff.

10. Ibid., pp. 273–75.

11. Ibid., p. 276.

12. Ibid., p. 281.

13. For a thorough analysis see Noakes, "Wohin gehören die 'Judenmischlinge'?" pp. 74ff.

14. Ibid.

15. Walk, *Das Sonderrecht*, p. 139.

16. Ibid.

17. Noakes, "Wohin gehören die 'Judenmischlinge'?" pp. 85–86.

18. Walk, *Das Sonderrecht*, p. 139.

19. Führer's Deputy Circular No 228/35, 2.12.1935, Stellvertreter des Führers (Anordnungen . . .), Db 15.02, IfZ, Munich.

20. Friedlander and Milton, *Archives of the Holocaust*, vol. 20, *Bundesarchiv of the Federal Republic of Germany, Koblenz and Freiburg*, pp. 28–30.

21. Führer's Deputy Circular 2.12.1935, Stellvertreter des Führers (Anordnungen . . .), 1935, Db 15.02, IfZ, Munich.

22. Herbert A. Strauss, "Jewish Emigration from Germany: Nazi Policies and Jewish Responses (I)," *LBIY* 25 ([London] 1980): 317. Werner Cohn, in his 1988 study on "non-Aryan" Christians, also offers a thorough statistical analysis. He estimates the population of partial Jews at 228,000 in 1933, which could roughly correspond to the 200,000 estimate for 1935. See Werner Cohn, "Bearers of a Common Fate? The 'Non-Aryan' Christian 'Fate Comrades' of the *Paulus-Bund*, 1933–1939," *LBIY* 33 (1988): 350ff. H. W. Friedmann, of the *Paulus-Bund*, also evaluated the number of "non-Jewish non-Aryans" at 200,000, which according to him was considered much too low by the Racial Policy Office of the Party. See *Akten deutscher Bischöfe*, vol. 2, *1934–1935*, p. 133.

23. Dr. E. R——x, "Die nichtjüdischen Nichtarier in Deutschland," *CV Zeitung* 20, no. 1 (Beiblatt): 16 May 1935. I am grateful to Sharon Gillerman for drawing my attention to this article.

24. Bernhard Lösener and Friedrich U. Knost, *Die Nürnberger Gesetze* (Berlin, 1936), pp. 17–18.

25. Wilhelm Stuckart and Hans Globke, *Kommentare zur deutschen Rassengesetzgebung*, vol. 1 (Munich, 1936).

26. Ibid., pp. 65–66.

27. The example given by Stuckart and Globke was obviously meant as the most extreme illustration of the principle that lay at the very basis of the Nuremberg Laws. Yet the manifest absurdity of determining race by religion must have been troublesome

enough to induce the ministerial bureaucracy to issue at least one clarification. On November 26, 1935, a circular was issued by the Ministry of the Interior: "In assessing whether a person is Jewish or not, it is basically not the fact of belonging to the Jewish religious community that is decisive, but that of belonging to the Jewish race. However, in order to avoid difficulties in dealing with [individual] cases, it has been expressly decided that a grandparent who has belonged to the Jewish religion unquestionably belongs to the Jewish race; counter-evidence is not permitted." Quoted in Noakes, "Wohin gehören die 'Judenmischlinge'?" p. 84.

28. Stuckart and Globke, *Kommentare,* p. 5.

29. Burleigh and Wippermann, *The Racial State,* p. 49.

30. Ibid.

31. Adjutantur des Führers, microfilm MA–287, IfZ, Munich.

32. Monthly Report, 8.12.1937, *Die Kirchliche Lage in Bayern nach den Regierungspräsidentenberichten 1933–1943,* vol. 2, *Regierungsbezirk Ober- und Mittelfranken,* ed. Helmut Witeschek (Mainz, 1967), p. 254.

33. Karl Haushofer, the founder of German geopolitics, was Hess's teacher at the University of Munich, and by way of Hess he influenced parts of *Mein Kampf* regarding international affairs and world strategy; although himself a declared anti-Semite, Haushofer was married to a "half-Jewish" woman, Martha Mayer-Doss. From 1934 to 1938 Karl's son, Albrecht, was employed by the foreign affairs agency "Office Ribbentrop"—*Amt Ribbentrop.* For Karl's and Albrecht's attitudes to Judaism and the Jews, and for their personal situation in this respect, see Hans-Adolf Jacobsen, *Karl Haushofer: Leben und Werke,* 2 vols. (Boppard, 1979); and Ursula Laak-Michael, *Albrecht Haushofer und der Nationalsozialismus* (Stuttgart, 1974); for an overall interpretation see Dan Diner, "Grundbuch des Planeten: Zur Geopolitik Karl Haushofers," in Dan Diner, *Weltordnungen: Über Geschichte und Wirkung von Recht und Macht* (Frankfurt am Main, 1993), pp. 131 ff.

34. *Akten der Parteikanzlei der NSDAP,* microfiches, 30100219–30100223, IfZ, Munich.

35. See Shlomo Aronson, *Reinhard Heydrich und die Frühgeschichte von Gestapo und SD* (Stuttgart, 1971), pp. 11–12; Werner Maser, *Adolf Hitler: Legende, Mythos, Wirklichkeit* (Munich, 1971), pp. 11ff.

36. *Akten der Parteikanzlei der NSDAP* (abstracts), part 1, vol. 2, p. 226.

37. Lothar Gruchmann, "Blutschutzgesetz und Justiz: Zu Entstehung und Auswirkung des Nürnberger Gesetzes vom 15 September 1935," *VfZ* 3 (1983): 419.

38. *Akten der Parteikanzlei der NSDAP* (abstracts), part 1, vol. 1, p. 55. On various levels, German racial laws and racial discrimination continued to be a source of difficulties in the relations between the Reich and numerous countries. Thus, according to a 1936 report from the German legation in Bangkok, discriminatory measures were applied to "colored" passengers (Japanese, Chinese, and Siamese, among others) on German ships in the Far East. The Ministry of Transportation in Berlin requested German shipping companies to be aware of the negative consequences of such measures. Ibid., p. 178. During the same year the Wilhelmstrasse had to assuage the worries of Egyptian authorities: There were no obstacles to the marriage of a German non-Jewish woman with an Egyptian non-Jewish man; as for the difficulties regarding the marriage of non-Jewish German men with non-Jewish foreign women, they were of a general nature and in no way discriminated against Egyptians. Ibid., part 2, vol. 3, p. 108. All in all, various states in the Middle East felt targeted by German legislation regarding non-Aryans, despite all efforts of the Foreign Ministry in Berlin. Ibid., p. 109. Turkey was placated by a German declaration that the Turks were of "related racial stock," but the ruling as far as other Middle Eastern nations were concerned was not clear at all. Ibid., p. 104.

39. Ibid., part 1, vol. 2, p 168.

40. For a strong affirmation of the primacy of the wider biological vision, and for the victimization of women that it implied, see, in particular, Bock, *Zwangssterilisation im Nationalsozialismus;* regarding the 1935 laws see particularly pp. 100–103. In her more recent writings Gisela Bock has formulated positions closer to those presented here. See Gisela Bock, "Krankenmord, Judenmord und nationalsozialistische Rassenpolitik," in Frank Bajohr et al., eds., *Zivilisation und Barbarei: die widersprüchlichen Potentiale der Moderne* (Hamburg, 1991), pp. 285ff. and particularly pp. 301–3. Throughout the twelve years of the Nazi regime, a number of university research institutes were bolstering the racial policies with so-called scientific data: The Kaiser Wilhelm Institute of Biology in Berlin; the Institute of Anthropology and Ethnography at Breslau University; the Institute of Hereditary Biology and Racial Hygiene at Frankfurt University; the Racial-Biological Institutes at Königsberg and Hamburg Universities; the Thüringian Center for Racial Questions, linked to Jena University; and the Research Institute in Hereditary Biology at Alt-Rhese in Mecklenburg. Klaus Drobisch et al., *Juden unterm Hakenkreuz: Verfolgung und Ausrottung der deutschen Juden 1933–1945* (Frankfurt am Main, 1973), pp. 162–63.

41. For the beginning of this story see chapter 1, pp. 33ff.

42. Mommsen, "Die Geschichte," p. 352.

43. Ibid., pp. 353–57.

44. For the inquiry and the quotes see Noakes, "The Development of Nazi Policy," pp. 299ff.

45. Ibid., pp. 300–301.

46. Ursula Büttner, "The Persecution of Christian-Jewish Families in the Third Reich," *LBIY* 34 (1989): 277–78.

47. Adam, *Hochschule und Nationalsozialismus*, p. 117.

48. Cohn, "Bearers of a Common Fate?" pp. 360–61.

49. Müller, *Hitler's Justice*, pp. 99–100.

50. Ibid., pp. 100–101.

51. Ibid., pp. 101–2.

52. I am using the title of Klaus Theweleit's study, *Male Fantasies*, 2 vols. (Minneapolis, Minn., 1987–89).

53. Noam and Kropat, *Juden vor Gericht*, pp. 125–27.

54. Müller, *Hitler's Justice*, p. 102–3.

55. *Akten der Parteikanzlei*, microfiche No. 031575, IfZ, Munich.

56. See Götz von Olenhusen, "Die 'Nichtarischen' Studenten," note 52, and also Michael H. Kater, "Everyday Anti-Semitism in Prewar Nazi Germany: The Popular Bases," *Yad Vashem Studies* 16 (1984): 150.

57. Adolf Diamant, *Gestapo Frankfurt am Main* (Frankfurt am Main, 1988), p. 91.

58. Robert Gellately, *The Gestapo and German Society: Enforcing the Racial Policy 1933–1945* (Oxford, 1990).

59. Ibid., p. 164.

60. Ibid., pp. 163–64.

61. Robert Gellately, "The Gestapo and German Society: Political Denunciations in the Gestapo Case Files," *Journal of Modern History* 60, no. 4 (December 1988): 672–74. According to Sarah Gordon, some evidence to the contrary notwithstanding, although in the thirties some *Rassenschänder* were first held in ordinary prisons (*Gefängnisse*), whereas the Jewish *Rassenschänder* were sent to the much harder forced-labor establishments (*Zuchthäuser*), the fate of both categories of prisoners was ulti-

mately the same. Sarah Gordon, *Hitler, Germans and the "Jewish Question,"* (Princeton, N.J., 1984), pp. 238ff.

62. Ministry of Justice, the Spokesman to all Justice press offices, 11.3.1936, Reichsjustizministerium, Fa 195/1936, IfZ, Munich.

63. Bankier, *The Germans and the Final Solution*, p. 77.

64. Ibid., p. 78.

65. Ibid., pp. 78–79.

66. Ibid., p. 79.

67. Richard Gutteridge, "German Protestantism and the Jews in the Third Reich," in Kulka and Mendes-Flohr, *Judaism and Christianity under the Impact of National Socialism*, p. 237. See also Gutteridge, *Open Thy Mouth for the Dumb!* pp. 153ff. and particularly pp. 156–58.

68. Bankier, *The Germans and the Final Solution*, p. 80.

69. Kulka, "Die Nürnberger Rassengesetze," p. 602–3.

70. For this interpretation of the long-term impact of the laws on the population, see Drobisch, *Juden unterm Hakenkreuz*, p. 160.

71. Gestapa [the Gestapa was the central office of the Gestapo, in Berlin] to State police offices, 3.12.1935, Ortspolizeibehörde Göttingen, microfilm MA–172, IfZ, Munich.

72. Gestapa to Central Association of German Jews (CV) June 1, 1934, ibid.; State police Hannover, 16.8.1934, ibid.

73. Gestapa to all State police offices, 24.11.35, ibid.

74. Gestapa to all State police offices, 4.4.1936, ibid.

75. Gerlach, *Als die Zeugen schwiegen*, p. 166.

76. For this case see Friedlander and Milton, *Archives of the Holocaust*, vol. 11, *Berlin Document Center*, ed. Henry Friedlander and Sybil Milton (New York, 1992), part 1, pp. 210–22.

77. *Akten der Parteikanzlei* (abstracts), part 1, vol. 1, p. 121.

78. Friedlander and Milton, *Archives of the Holocaust*, vol. 11, part 1, pp. 210–22.

79. Abraham Margalioth, "The Reaction of the Jewish Public in Germany to the Nuremberg Laws," *Yad Vashem Studies* 12 (1977): 76.

80. Bankier, "Jewish Society Through Nazi Eyes 1933–1936," pp. 113–14.

81. Margarete T. Edelheim-Mühsam, "Die Haltung der jüdischen Presse gegenüber der nationalsozialistischen Bedrohung," in Robert Weltsch, ed., *Deutsches Judentum: Aufstieg und Krise* (Stuttgart, 1963), p. 375.

82. Some Gestapo reports, such as the one emanating from Koblenz and dealing with October 1935, reported greater pessimism among the Jews and an urge to emigrate, also to Palestine. According to this report, the Jews did not believe in the possibility of staying in Germany and envisioned that "within approximately ten years the last Jew would have left Germany." Heyen, *Nationalsozialismus im Alltag*, pp. 138–39.

83. Boas, "German-Jewish Internal Politics," p. 3.

84. Ibid., p. 4, n. 4.

85. Edelheim-Mühsam, "Die Haltung der jüdischen Presse," pp. 376–77.

86. Claudia Koonz, *Mothers in the Fatherland: Women, the Family and Nazi Politics* (New York, 1987), p. 358.

87. Dawidowicz, *The War Against the Jews*, p. 178.

88. William L. Shirer, *Berlin Diary: The Journal of a Foreign Correspondent 1934–1941* (New York, 1941; reprint, New York, 1988), p. 36.

89. Quoted in Lowenstein, "The Struggle for Survival of Rural Jews," p. 120.

90. Yoav Gelber, "The Zionist Leadership's Response to the Nuremberg Laws," *Studies on the Holocaust Period* 6 (Haifa, 1988) (Hebrew).

91. Chernow, *The Warburgs*, pp. 436ff.

92. *Akten der Parteikanzlei der NSDAP* (abstracts), part 1, vol. 2, p. 208.

93. Chernow, *The Warburgs*, pp. 436ff.

94. Charlotte Beradt, *Das Dritte Reich des Traums* (Frankfurt am Main, 1981), p. 98.

95. Ibid.

96. Ibid., p. 104.

97. Feuchtwanger and Zweig, *Briefwechsel*, vol. 1, p. 97.

98. C. G. Jung, "Civilization in Transition," in *Collected Works*, vol. 10 (New York, 1964), p. 166. This text is but one of many more or less identical statements made by Jung during the years 1933 to 1936 at least. The controversy concerning Jung's attitude toward National Socialism has continued since the end of the war. The mildest appraisal of the issue by a historian not belonging to either camp is that of Geoffrey Cocks: "It is by no means clear that the personal philosophical beliefs and attitudes behind Jung's dubious, naive and often objectionable statements during the Nazi era about 'Aryans' and Jews motivated his actions with regard to psychotherapists in Germany. The statements themselves reveal a destructive ambivalence and prejudice that may have served Nazi persecution of the Jews. But Jung conceded much more to the Nazis by his words than by his actions." *Psychotherapy in the Third Reich: The Göring Institute* (New York, 1985), p. 132. Cocks's evaluation would have to be thoroughly examined; nonetheless, given the circumstances, Jung's attitude seems repellent enough.

99. Ernst L. Freud, ed., *The Letters of Sigmund Freud and Arnold Zweig* (New York, 1970), p. 110.

100. Kurt Tucholsky, *Politische Briefe* (Reinbek/Hamburg, 1969), pp. 117–23.

Chapter 6 Crusade and Card Index

1. Goebbels, *Tagebücher*, part 1, vol. 3, p. 55.

2. Ibid., p. 351.

3. The primacy of the anti-Bolshevik crusade has been argued by Arno J. Mayer. As will be seen, the speeches of 1936–37 explicitly indicate that the Jews were considered as the enemy behind the Bolshevik threat. For Mayer's argument see his *Why Did the Heavens Not Darken? The Final Solution in History* (New York, 1988).

4. Lipstadt, *Beyond Belief*, p. 80.

5. Arad, "The American Jewish Leadership's Response," pp. 418–19.

6. Arnd Krüger, *Die Olympischen Spiele 1936 und die Weltmeinung* (Berlin, 1972), pp. 128–31. On June 13, 1936, notwithstanding a jump of five feet three inches (equaling the German women's record) during the training period, athlete Gretel Bergmann received a letter from the German Olympic Committee that read in part: "Looking back on your recent performances, you could not possibly have expected to be chosen for the team." In the spring of 1996, eighty-two-year-old Margaret Bergmann Lambert, a U.S. citizen who lives in New York, accepted the invitation of the German Olympic Committee to be its guest of honor at the Centennial Games in Atlanta. Ira Berkow, "An Olympic Invitation Comes 60 Years Late," *New York Times*, June 18, 1996, pp. A1, B12.

7. Eliahu Ben-Elissar, *La Diplomatie du IIIe Reich et les Juifs, 1933–1939* (Paris, 1969), p. 179.

8. Ibid., p. 173.

9. Goebbels, *Tagebücher*, part 1, vol. 2, p. 630.

10. *Ibid.*, p. 655.

11. Walk, *Das Sonderrecht*, p. 153.

12. Hitler, *Speeches and Proclamations*, pp. 750–51.

13. Goebbels, *Tagebücher*, part 1, vol. 2, p. 718.

14. Heinrich Himmler, *Die Schutzstaffel als antibolschewistische Kampforganisation* (Munich, 1936), p. 30.

15. Noakes and Pridham, *Nazism 1919–1945*, vol. 2, p. 281.

16. *Akten der Parteikanzlei* (abstracts), part 1, vol. 2, p. 249.

17. *Der Parteitag der Ehre: Vom 8 bis 14 September 1936* (Munich, 1936), p. 101.

18. Hitler, *Reden und Proklamationen*, p. 638.

19. *Der Parteitag der Ehre*, p. 294. In his Reichstag speech of January 30, 1937, Hitler had already broached the theme of the Judeo-Bolshevik revolutionary action attempting to penetrate Germany. Hitler, *Reden und Proklamationen*, p. 671.

20. Klee, *Die "SA Jesu Christi,"* p. 127.

21. *Der Parteitag der Arbeit vom 6 bis 13 September 1937: Offizieller Bericht über den Verlauf des Reichsparteitages mit sämtlichen Kongressreden* (Munich, 1938), p. 157. Alfred Rosenberg's contribution was unusual, even by Nazi standards. In his speech he described in gory detail the murderous rule of the Jews in the Soviet Union. He then produced a book "published in New York," entitled *Now and Forever*, a "dialogue" between the Jewish writer Samuel Roth and the purportedly Zionist politician Israel Zangwill, with an introduction by Zangwill; the book was dedicated to the "president of the Jewish university in Jerusalem." *Der Parteitag der Arbeit*, pp. 102–3. The texts mentioned by Rosenberg, who did not hesitate to quote chapter and verse, make the *Protocols of the Elders of Zion* seem like a harmless lullaby. In reality, as becomes clear even from the two-part article devoted to Roth's book in the *NS Monatshefte* of January and February 1938, the book is based on a fictitious dialogue between Roth and Zangwill, mainly about anti-Semitism and the difficulties of political Zionism. See Georg Leibbrandt, "Juden über das Judentum," *Nationalsozialistische Monatshefte* 94, 95 (January, February, 1938). No occasion was missed in the Rosenberg-Goebbels feud. In Rosenberg's August 25, 1937, letter informing Goebbels that he, Rosenberg, would speak first at the rally, the master of ideology enjoyed a parting barb, closing with the following remark: "Finally, I would like to draw your attention to a small mistake. The quotation defining the Jew as the visible demon of the decay of humanity comes not from Mommsen but from Richard Wagner." Rosenberg to Goebbels, 25.8.1937, Rosenberg files, microfilm MA–596, IfZ, Munich.

22. Hitler, *Speeches and Proclamations*, p. 938; German original in vol. 2, p. 728.

23. Ibid., p. 939.

24. Ibid., p. 940.

25. Ibid., p. 941.

26. Ibid.

27. The Führer's Deputy, Directive, 19.4.1937, NSDAP Parteikanzlei (Anordnungen . . .), Db 15.02, IfZ, Munich.

28. Goebbels, *Tagebücher*, part 1, vol. 3, p. 21.

29. See the various studies in Hans-Erich Volkmann, ed., *Das Russlandbild im Dritten Reich* (Cologne, 1994). For Heydrich's statement see Gerhart Hass, "Zum Russlandbild der SS," Ibid., p. 209.

30. See Michael Burleigh, *Germany Turns Eastwards: A Study of Ostforschung in the Third Reich* (Cambridge, 1988), p. 146.

31. Peter-Heinz Seraphim, *Das Judentum im osteuropäischen Raum* (Essen, 1938), p. 266.

32. Ibid., p. 262.

33. Ibid., p. 267.

34. Commander of main region Rhine to SS-Gruppenführer Heissmeyer, 3.4.35 ("Lagebericht Juden," 30 Lenzing [from the old German form of "springtime," *der Lenz*] 1935), Sicherheitsdienst des Reichsführers SS, SD Oberabschnitt Rhein, microfilm MA–392, IfZ, Munich.

35. Helmut Krausnick and Hildegard von Kotze, eds., *Es spricht der Führer: Sieben exemplarische Hitler-Reden* (Gütersloh, 1966), pp. 147–48.

36. State police station Hildesheim to county prefects, mayors . . . 28.10.1935, Ortspolizeibehörde Göttingen, microfilm MA–172, IfZ, Munich.

37. Ibid., 23.10.1935.

38. Gutteridge, "German Protestantism," p. 238. See also Gutteridge, *Open Thy Mouth for the Dumb!* pp. 158ff.

39. Gutteridge, "German Protestantism," p. 238.

40. Gutteridge, *Open thy Mouth for the Dumb!* pp. 159–60.

41. Schönwälder, *Historiker und Politik*, pp. 86–87.

42. Helmut Heiber, *Walter Frank und sein Reichsinstitut für Geschichte des neuen Deutschlands* (Stuttgart, 1966), pp. 279–80.

43. Karl Alexander von Müller, "Zum Geleit," *Historische Zeitschrift* 153, no. 1 (1936): 4–5.

44. Heiber, *Walter Frank*, p. 295.

45. Ibid.

46. *Historische Zeitschrift* 153, no. 2 (1936): 336ff. Sometimes reviews of Jewish publications that could appear hostile and damning for the Nazi reader could have been understood as praise from a non-Nazi perspective. One of the strangest examples is the review published in 1936 in the *NS Monatshefte* by Joachim Mrugowsky (later of criminal notoriety for euthanasia) on letters of fallen Jewish soldiers. Mrugowsky compared these letters with those of fallen German soldiers and came to the conclusion that the absolute racial incompatibility of the two groups was clearly revealed in the main ideals expressed by each group. Whereas the German ideal was the race, the *Volk*, and the struggle for the right to live, the Jewish letters idealized equality, humanity, and world peace. Joachim Mrugowsky, "Jüdisches und deutsches Soldatentum: Ein Beitrag zur Rassenseelenforschung," *Nationalsozialistische Monatshefte* 76 (July 1936): 638.

47. For a detailed presentation of Frank's and Grau's activities regarding the "Jewish Question" see Heiber, *Walter Frank*, mainly pp. 403–78.

48. *DAZ*, 20 Nov. 1936, Nationalsozialismus/1936, *Miscellanea*, LBI, New York.

49. Heiber, *Walter Frank*, pp. 444ff.

50. See *Das Judentum in der Rechtswissenschaft*, vol. 1, Die Rechtswissenschaft im Kampf gegen den jüdischen Geist (Berlin, 1937), pp. 14ff, 28ff. See also Bernd Rüthers, *Carl Schmitt im Dritten Reich: Wissenschaft als Zeitgeist-Bestärkung?* (Munich, 1990), pp. 81ff, 95ff.

51. Ibid., pp. 97ff.

52. Ibid.

53. Ibid., p. 30.

54. Carl Schmitt, *Der Leviathan in der Staatslehre des Thomas Hobbes* (Hamburg, 1938), p. 18. For the translation see Susan Shell, "Taking Evil Seriously: Schmitt's Concept of the Political and Strauss's True Politics," in Kenneth L. Deutsch and Walter Nigorski, eds., *Leo Strauss: Political Philosopher and Jewish Thinker* (Lanham, Md., 1994), p. 183, n. 22. I am grateful to Eugene R. Sheppard for having drawn my attention to this text. All in all Schmitt's anti-Semitism appears to have run deeper than mere opportunism, and his political and ideological commitment between 1933 and 1945 cannot, it seems, be equated with mere "card-carrying," as his defenders would have it. See, for example, Dan Diner, "Constitutional Theory and 'State of Emergency' in Weimar Republic: The Case of Carl Schmitt," *Tel Aviver Jahrbuch für Deutsche Geschichte* 17 (1988): 305.

55. For a good overview of the impact of Nazi ideology on German scientific research, see the essays in H. Mehrtens and S. Richter, eds., *Naturwissenschaft, Technik und NS-Ideologie* (Frankfurt, 1980). For a very thorough survey of the development of biology in Nazi Germany, see Deichmann, *Biologen unter Hitler*

56. On this issue see Cocks, *Psychotherapy in the Third Reich*, p. 7.

57. Beyerchen, *Scientists under Hitler*, pp. 156ff.

58. See Hans Buchheim, "Die SS—Das Herrschaftsinstrument," in Hans Buchheim et al., *Anatomie des SS-Staates*, 2 vols., Olten, 1965, vol. 1, pp. 55 ff; in particular George C. Browder, *Foundations of the Nazi Police State. The Formation of Sipo and SD*, Lexington, 1990.

59. Browder, *Foundations of the Nazi Police State*, p. 231.

60. Buchheim, "Die SS," p. 54.

61. All the details about Aus den Ruthen's brides are taken from William L. Combs, *The Voice of the SS: A History of the SS Journal Das Schwarze Korps*, vol. 1 (Ann Arbor, Mich.: University Microfilms, 1985), pp. 29-30.

62. Heinrich Himmler, "Reden, 1936-1939," F 37/3, IfZ, Munich.

63. Helmut Heiber, ed., *Reichsführer! . . . Briefe an und von Himmler* (Stuttgart, 1968), p. 44. In his answer the researcher, SS-Haupsturmführer Dr. K. Mayer, mentioned that, although no Jewish ancestry was found, Mathilde von Kemnitz had no fewer than nine theologians among her forefathers which, for him, offered the explanation. To which Walther Darré remarked: "I have three Reformators among my ancestors. Does it make me unacceptable to the SS?" Ibid., p. 45, n. 3.

64. Ibid., p. 52, as well as pp. 64, 66, 75, 231, 245.

65. See in Friedlander and Milton, *Archives of the Holocaust*, vol. 11, part 2, pp. 124-25.

66. Heiber, *Reichsführer*, p. 50.

67. "Warum wird über das Judentum geschult?" *SS-Leitheft* 3, no. 2, 22 Apr. 1936.

68. Ibid., quoted in Josef Ackermann, *Heinrich Himmler als Ideologe* (Göttingen, 1970), p. 159.

69. For the entire case see Friedlander and Milton, *Archives of the Holocaust*, vol. 11, part 2, , pp. 55ff.

70. For the reorganization of the SD, see Wildt, *Die Judenpolitik des SD*, pp. 25, 73ff. I wish to express my thanks to Dr. Wildt and to Dr. Norbert Frei for allowing me to have access to this study and its appended documents before publication.

71. For the administrative structure of the SD in 1936-37, see Hebert, *Best*, p. 578,

Drobisch, "Die Judenreferate," pp. 239–40. For Wisliceny's indication see Hans Safrian, *Die Eichmann-Männer* (Vienna, 1993), p. 26.

72. For the development and organization of the Gestapo see Johannes Tuchel and Reinhold Schattenfroh, *Zentrale des Terrors* (Berlin, 1987).

73. Herbert, *Best*, p. 187.

74. About the topics discussed at that meeting see Wildt, *Die Judenpolitik des SD*, pp. 45ff.

75. II.112 to II.11, 15.6.37, Sicherheitsdienst des Reichsführers SS, SD Hauptamt, Abt. II 112, microfilm MA–554, IfZ, Munich. It is hard to tell on the basis of what concrete "evidence" the SD spun such fantastic links.

76. Ibid.

77. Drobisch, "Die Judenreferate," p. 242.

78. Ibid., as well as Götz Aly and Karl-Heinz Roth, *Die restlose Erfassung: Volkszählen, Identifizieren, Aussondern im Nationalsozialismus* (Berlin, 1984), pp. 77–79.

79. Wildt, *Die Judenpolitik des SD*, p. 134.

80. Shlomo Aronson, *Heydrich und die Anfänge des SD und der Gestapo (1931–1935)* (Berlin, 1967), p. 275.

81. Reinhard Heydrich, *Wandlungen unseres Kampfes* (Munich, 1935).

82. Wildt, *Die Judenpolitik des SD*, p. 33.

83. Ibid., p. 66–67. Excerpts of this document were previously published in Susanne Heim, "Deutschland muss ihnen ein Land ohne Zukunft sein: Die Zwangsemigration der Juden 1933 bis 1938," in *Beiträge zur Nationalsozialistischen Gesundheits- und Sozialpolitik*, vol. 11, *Arbeitsmigration und Flucht* (Berlin, 1993).

84. Safrian, *Die Eichmann-Männer*, p. 28.

85. SD main region Rhine to SD-Commander, SD main region Fulda-Werra, 18.9.37, Himmler Archives, Berlin Document Center, microfilm No. 270, roll 2 (LBI, NY, microfilm 133g).

86. The Commander, concentration camp Columbia, to Inspector of the concentration camps, SS-Gruppenführer Eicke, 28.1.1936, SS-Standort Berlin, microfilm MA–333, IfZ, Munich.

87. The Commander of the SS Death Head units to Chief of the SS Personnel Office, 30.1.1936, ibid.

88. The Chief of the SS Personnel Office to Standortführer-SS, Berlin, 4.2.1936, ibid.

89. Martin Broszat, "Nationalsozialistische Konzentrationslager" in Buchheim et al., *Anatomie des SS-Staates*, p. 75.

90. Ibid., pp. 173–74.

91. Ibid., pp. 78–79.

92. Ibid., p. 81.

93. Ibid., pp. 81–82.

94. Burleigh and Wippermann, *The Racial State*, p. 116.

95. Ibid., pp. 119–20. For details about Ritter's research, see in particular Michael Zimmermann, *Verfolgt, vertrieben, vernichtet: Die nationalsozialistische Vernichtungspolitik gegen Sinti und Roma* (Essen, 1989), pp. 25ff.

96. For details on the Gypsy camps see in particular Sybil Milton, "Vorstufe zur Vernichtung: Die Zigeunerlager nach 1933," *VfZ* 43, no. 1 (1995): 121ff.

97. Burleigh and Wippermann, *The Racial State*, p. 191.

98. Ibid., p. 196.

99. Ibid., p. 197.

100. Broszat and Fröhlich, *Alltag und Widerstand*, p. 466.

101. Ibid., pp. 450ff.

102. Ibid., p. 461.

103. Ibid., p. 463.

104. Ibid., pp. 475–76.

105. For the most complete investigation of this subject, see Reiner Pommerin, *Sterilisierung der "Rheinlandbastarde": Das Schicksal einer farbigen deutschen Minderheit 1918–1937* (Düsseldorf, 1979).

106. Ibid., pp. 44ff.

107. *Dokumente des Verbrechens: Aus Akten des Dritten Reiches 1933–1945*, vol. 2 (Berlin, 1993), pp. 83ff.

108. Ibid., pp. 122ff. Also Pommerin, *Sterilisierung der "Rheinlandbastarde,"* pp. 71ff.

109. Burleigh and Wippermann, *The Racial State*, p. 130.

110. Führer's Deputy (the chief of staff) to all Gauleiters, 30. März 1936, Stellvertreter des Führers (Anordnungen . . .), 1936, Db 15.02, IfZ, Munich.

111. For the sterilization policy see the already mentioned studies by Bock, Proctor, Schmuhl, and others as well as Henry Friedlander, *The Origins of Nazi Genocide: From Euthanasia to the Final Solution* (Chapel Hill, N.C., 1995), pp. 23ff.

112. Burleigh, *Death and Deliverance*, p. 43.

113. Ibid., p. 187.

114. Burleigh and Wippermann, *The Racial State*, p. 154.

115. Ibid.

116. Ernst Klee, *"Euthanasie" im NS-Staat: Die Vernichtung "lebensunwerten Lebens"* (Frankfurt am Main, 1985), p. 62. According to Hans-Walter Schmuhl, some psychiatric patients were killed between 1933 and 1939 on the basis of local initiatives. Hans-Walter Schmuhl, *Rassenhygiene, Nationalsozialismus, Euthanasie* (Göttingen, 1987), p. 180.

117. Ibid., p. 61. The attitudes of these pastors should not hide the fact that from the outset, even the sterilization policy encountered mostly silent but nonetheless tangible opposition from the wider population, particularly in Catholic areas. See Dirk Blasius, "Psychiatrischer Alltag im Nationalsozialismus," in Peukert and Reulecke, *Die Reihen fast geschlossen*, pp. 373–74.

118. Klee, *"Euthanasie" im NS-Staat*, p. 67.

119. Burleigh and Wippermann, *The Racial State*, p. 142; Burleigh, *Death and Deliverance*, p. 93–96; Friedlander, *The Origins of Nazi Genocide*, p. 39.

120. Martin Höllen, "Episkopat und T4," in Götz Aly, ed., *Aktion T4 1939–1945: Die "Euthanasie"-Zentrale in der Tiergartenstrasse 4* (Berlin, 1987), pp. 84–85; Gitta Sereny, *Into That Darkness: From Mercy Killing to Mass Murder* (New York, 1974), pp. 64ff.

121. Höllen, "Episkopat und T4"; Sereny, *Into That Darkness*, pp. 67–68.

122. Ibid., pp. 68–69. Burleigh has doubts about the reliability of Hartl's testimony, but does not question the existence of Mayer's memorandum. Cf. Burleigh, *Death and Deliverance*, p. 175.

Chapter 7   Paris, Warsaw, Berlin—and Vienna

1. See especially Pierre Birnbaum, *Le peuple et les gros: Histoire d'un mythe* (Paris, 1979).

2. Georges Bernanos, *La grande peur des bien-pensants*, in *Essais et ecrits de combat* (Paris, 1971), p. 329.

3. Ibid., p. 350.

4. Louis-Ferdinand Céline, *Bagatelles pour un Massacre* (Paris, 1937); André Gide, "Les Juifs, Céline et Maritain," *Nouvelle Revue Française* (Apr. 1, 1938). Within months Céline's *Bagatelles* was translated into German under the title *Judenverschwörung in Frankreich* (Jewish conspiracy in France) and received rave reviews from Streicher's *Stürmer* and from the SS weekly *Das Schwarze Korps* as well as an array of provincial papers. Albrecht Betz, "Céline im Dritten Reich," in Hans Manfred Bock et al., eds., *Entre Locarno et Vichy: Les relations culturelles franco-allemandes dans les années 1930* (Paris, 1993), vol. 1, p. 720.

5. Jean Giraudoux, *Pleins pouvoirs* (Paris, 1939).

6. Quoted in Zeev Sternhell, *Neither Right nor Left: Fascist Ideology in France* (Berkeley, Calif., 1986), p. 265. For the evolution of Bergery's attitude on the Jewish issue (he himself was probably part Jewish), see a highly nuanced analysis in Philippe Burrin, *La Dérive fasciste: Doriot, Déat, Bergery 1933-1945* (Paris, 1986), pp. 237ff.

7. Ezra Mendelsohn, *The Jews of East Central Europe Between the World Wars* (Bloomington, Ind., 1983), p. 1.

8. Béla Vago, *The Shadow of the Swastika: The Rise of Fascism and Anti-Semitism in the Danube Basin, 1936-1939* (London, 1975), pp. 15-16.

9. Friedlander and Milton, *Archives of the Holocaust*, vol. 8, *American Jewish Archives Cincinnati: The Papers of the World Jewish Congress 1939-1945*, ed. Abraham L. Peck (New York, 1990), p. 21. (The translation from the Polish has been kept as it was.)

10. Ibid., p. 20.

11. Mendelsohn, *The Jews of East Central Europe*, pp. 23-24.

12. Ibid., p. 27. Mendelsohn uses the statistics compiled by the foremost historian of Polish Jewry between the wars, Rafael Mahler, whose standard work, *Yehudei Polin bein Shtei Milhamot ha-Olam* [The Jews of Poland between the two world wars], was published in Tel Aviv in 1968.

13. Ibid., pp. 29-30.

14. Joseph Marcus, *Social and Political History of the Jews in Poland 1919-1939* (Berlin, 1983), p. 362.

15. S. Andreski, "Poland," in S. J. Woolf, ed., *European Fascism* (London, 1968), pp. 178ff.

16. Ibid., pp. 362-63.

17. Mendelsohn, *The Jews of East Central Europe*, p. 74.

18. These percentages are quoted in Leslie Buell, *Poland: Key to Europe* (New York, 1939), p. 303, and reproduced in Götz Aly and Susanne Heim, *Vordenker der Vernichtung: Auschwitz und die deutschen Pläne für eine neue europäische Ordnung* (Hamburg, 1991), p. 86.

19. Mendelsohn, *The Jews of East Central Europe*, p. 75.

20. Ibid., p. 71.

21. Ibid., p. 73.

22. For the details of the plan mentioned here see Leni Yahil, "Madagascar—Phantom of a Solution for the Jewish Question," in Bela Vago and George L. Mosse, eds., *Jews and Non-Jews in Eastern Europe* (New York, 1974), pp. 315ff. For an account of Polish efforts to obtain the support of the League of Nations and that of foreign countries for the immigration of Jews to their colonies (Madagascar) or to Palestine see Pawel Korzec, *Juifs en Pologne: La question juive pendant l'entre-deux guerres* (Paris 1980), pp. 250ff.

23. Rolf Vogel, *Ein Stempel hat gefehlt: Dokumente zur Emigration deutscher Juden* (Munich, 1977), pp. 170-71.

24. Yahil, "Madagascar," p. 321.

25. Jacques Adler, *Face à la persecution: Les organisations juives à Paris de 1940 à 1944* (Paris, 1985), p. 25.

26. Ibid., pp. 26–27.

27. It is extremely difficult to evaluate the exact number of foreign Jews living in France in the late thirties, as a result of the reemigration of some of the immigrants. Approximately 55,000 Jews entered France between 1933 and the beginning of the war. Michael R. Marrus and Robert O. Paxton, *Vichy France and the Jews* (New York, 1981), p. 36.

28. Michael R. Marrus, "Vichy Before Vichy: Antisemitic Currents in France During the 1930's," *Wiener Library Bulletin* 33 (1980): 16.

29. Vicki Caron, "Loyalties in Conflict: French Jewry and the Refugee Crisis, 1933–1935," *LBIY* 36 (1991): 320.

30. Ibid.

31. Ibid., p. 326.

32. Marrus and Paxton, *Vichy France and the Jews*, pp. 54ff.

33. Marrus, "Vichy Before Vichy," pp. 17–18.

34. Some historians see a distinct regression of French anti-Semitism between the end of the Dreyfus Affair and the mid-thirties, others—with whom I tend to agree—perceive persistent strands of anti-Jewish attitudes, mainly in the cultural field, even throughout the "quieter" years. For the first interpretation, see Paula Hyman, *From Dreyfus to Vichy: The Remaking of French Jewry, 1906–1939* (New York, 1979); for the second see Léon Poliakov, *Histoire de l'Antisémitisme*, vol. 4, *L'Europe Suicidaire 1870–1933* (Paris, 1977), pp. 281ff.

35. Jean Lacouture, *Léon Blum* (Paris, 1977), p. 305.

36. See mainly David H. Weinberg, *A Community on Trial: The Jews of Paris in the 1930s* (Chicago, 1977), pp. 78ff.

37. Ibid., pp. 114–16.

38. Eugen Weber, *Action Française: Royalism and Reaction in Twentieth-Century France* (Stanford, Calif., 1962), p. 363.

39. There were three Jewish ministers in the first cabinet (Blum, Cécile Leon-Brunschwicg, and Jules Moch) and four in the second (Blum, Moch, L. O. Frossard, and Pierre Mendès-France). Stephen A. Schuker, "Origins of the 'Jewish Problem' in the Later Third Republic," in Frances Malino and Bernard Wasserstein, eds., *The Jews in Modern France* (Hanover, N.H., 1985), pp. 156–57.

40. Robert Soucy, *French Fascism: The Second Wave, 1933–1939* (New Haven, Conn., 1995), pp. 55, 278–79. According to Soucy, Doriot himself dismissed anti-Semitism at least until 1937. In 1936 his party received financial support from three Jewish-owned banks (Rothschild, Worms, and Lazard), and among his closest collaborators there was a Jew, Alexander Abremski, and the partly Jewish Bertrand de Jouvenel. Abremski was killed in a automobile accident in 1938; in that same year Doriot changed his position on the Jewish issue.

41. Michel Laval, *Brasillach ou la trahison du clerc* (Paris, 1992), pp. 75–76. See also Pierre-Marie Dioudonnat, *Je suis partout, 1930–1944* (Paris, 1973).

42. Rita Thalmann, "Du Cercle de Sohlberg au Comité France-Allemagne: une évolution ambigüe de la coopération franco-allemande," in Bock, *Entre Locarno et Vichy*, vol. 1, pp. 67ff.

43. Reinhard Bollmus, *Das Amt Rosenberg und seine Gegner: Zum Machtkampf im nationalsozialistischen Herrschaftssystem* (Stuttgart, 1970), pp. 121ff.

44. For the protocol of the meeting as established by Wilhelm Stuckart, see Hans Mommsen and Susanne Willems, eds., *Herrschaftsalltag im Dritten Reich: Studien und Texte* (Düsseldorf, 1988), pp. 445ff. For Stuckart's remark see ibid., p. 446.

45. Ibid., p. 448.

46. Ibid., p. 457.

47. Fridolf Kudlien, *Ärtzte im Nationalsozialismus* (Cologne, 1985), p. 76.

48. *Akten der Reichskanzlei*, vol. 5 (24 Jan. 1935–5 Feb. 1938), serial number 859, IfZ, Munich.

49. Friedlander and Milton, *Archives of the Holocaust*, vol. 20, pp. 85–87, and *Akten der Parteikanzlei der NSDAP* (abstracts), part 1, vol. 1, p. 245. In a meeting with Hitler on December 3, 1937, it was decided that "within weeks" the minister of the interior would submit to the chief of the Reich Chancellery the draft of a Law for the exclusion of Jewish physicians from medical practice. Ibid., p. 97.

50. Ibid.

51. Reich Minister of Education . . . , 25.11.1936, Reichsministerium für Wissenschaft . . . , microfilm MA 103/1, IfZ, Munich.

52. Ibid., 19.4.1937.

53. *Akten der Parteikanzlei*, microfiches 016639–40, IfZ, Munich.

54. *Akten der Parteikanzlei der NSDAP* (abstracts), part 1, vol. 2, p. 262.

55. Ibid. The reason for Hitler's decision can be tentatively surmised on the basis of the issues raised by the minister of education himself. Moreover, when it appeared, on September 10, 1935, that a similar law about Jewish schooling would be enforced from the beginning of the school year 1936, Cardinal Bertram sent a protest to Minister of Education Rust on precisely the issue of the converted Jewish pupils. See *Akten deutscher Bischöfe,* vol. 3, *1935–1936,* p. 57.

56. Regarding the general situation of Jewish students in Nazi Germany see Götz von Olenhusen, "Die 'nichtarischen' Studenten" and Grüttner, *Studenten im Dritten Reich*, pp. 212ff. For details on the doctorates issue, see also Friedländer, "The Demise of the German Mandarins," pp. 75ff.

57. Wilhelm Grau to State Secretary Kunisch, Reich Ministry of Education . . . , 18.2.1936, Reichsministerium für Wissenschaft u. Erziehung, microfilm MA 103/1 IfZ, Munich.

58. Reich Minister of Education, 28.4.1936, ibid.

59. The Dean, Philosophy Faculty of the Friedrich Wilhelm University, 29.2.1936, ibid.

60. The Führer's Deputy to the Reich Minister of the Interior, 15.10.1936, ibid.

61. Reich Minister of Education . . . , 15.4.1937, ibid.

62. Dean Weinhandel, Philosophical Faculty, Kiel, to Reich Minister of Education, 21.4.1937, ibid. The issue of Heller's dissertation, one of the elements that triggered the revision process in regard to doctoral degrees for Jews, had a delayed aftermath. Heller defended his dissertation on July 5, 1934, and was awarded summa cum laude. Soon after, Dr. Heller left for Tel Aviv, where he was informed, on November 23, 1935, by the dean's office in Berlin that his diploma would be sent to him on receipt of 4.25 DM to cover postage. But, instead of the diploma, Heller received the following letter from Dean Bieberach on January 10, 1936:

"You claim that on October 16, 1935 [the official graduation date] you were awarded the doctoral degree by the philosophy faculty of Berlin University. I demand that you refrain from making this false statement. You will not be granted this degree in the future either, as you are unworthy of bearing a German academic title. This has been unequivocally confirmed by a verification of your dissertation. The faculty regrets that you had been allowed to accede to the doctoral examination."

In 1961 Heller wrote to Humboldt University in East Berlin to receive his doctoral diploma. The university did not answer, but the senator for education of East Berlin sent an authorization allowing Heller to use the doctoral title. With the opening of the archives of the German Democratic Republic, the reason for the university's silence in 1961 became clear: Heller's dissertation was considered to be anticommunist. In 1992, fifty-seven years after Heller had in fact been deemed worthy of the doctoral degree, two representatives of Humboldt University came to his home in Israel and presented him with his diploma. Abraham Heller, personal archives, Ramat-Gan, Israel. I am indebted to Dr. Heller, and to his daughter, Mrs. Nili Bibring, for having given me access to the documentation in this case.

63. Peter Hanke, *Zur Geschichte der Juden in München zwischen 1933 und 1945* (Munich, 1967), p. 139.

64. Ibid., pp. 139–40.

65. Kommission . . . , *Dokumente zur Geschichte der Frankfurter Juden*, p. 163.

66. Ibid., pp. 163–64.

67. Ibid., pp. 167–71.

68. Ibid., p. 172.

69. Müller, *Stuttgart*, p. 296.

70. Ibid., pp. 296–97.

71. Ibid., p. 297.

72. Dr. Hugo Schleicher, Offenburg i/B, to District Office Offenburg, 19 March, 1937, Unterlagen betr. Entrechtung der Juden in Baden 1933–1940, ED 303, IfZ, Munich.

73. The Mayor as Chairman of the Hospital Fund to District Office Offenburg, 2.4.1937, ibid. When he referred to "the obscurantists of our time," the mayor of Gengenbach was using the title of Alfred Rosenberg's anti-Catholic pamphlet *An die Dunkelmänner unserer Zeit* (To the obscurantists of our time).

74. District Office, Offenburg, to Mayor, Gengenbach, 5.4.1937, ibid.

75. *Chronik der Stadt Stuttgart*, vol. 3, p. 354.

76. Ibid., p. 368.

77. "Otto Bernheimer, 'Kunde Göring,'" in Hans Lamm, ed., *Von Juden in München* (Munich, 1959), pp. 351–52.

78. Thomas Klein, ed., *Die Lageberichte der Geheimen Staatspolizei über die Provinz Hessen-Nassau 1933–1936*, vol. 1 (Vienna, 1986), p. 515.

79. Broszat, Fröhlich, and Wiesemann, *Bayern in der NS-Zeit*, vol. 1, p. 462.

80. Ibid., p. 458.

81. Wildt, *Die Judenpolitik des SD*, pp. 40, 108. An SD quarterly report for the period January through April 1937 states that some large Jewish firms had doubled their revenues by comparison to 1933. Ibid., p. 108.

82. Hayes, "Big Business and Aryanisation," p. 260.

83. Ibid., pp. 260–61.

84. Ibid., p. 262.

85. Barkai, *From Boycott to Annihilation*, p. 108.

86. Ibid., p. 84.

87. Wildt, *Die Judenpolitik des SD*, p. 165.

88. See Wilhelm Treue, "Hitlers Denkschrift zum Vierjahresplan," *VfZ* 3 (1955).

89. *Akten der Parteikanzlei der NSDAP* (abstracts), part 1, vol. 2, p. 267.

90. Adjutantur des Führers 1934–1937, microfilm MA 13/2, IfZ, Munich.

91. The Führer's Deputy, the Chief of Staff, directive, 23.10.37, Stellvertreter des Führers (Anordnungen . . .), 1937, Db 15.02, IfZ, Munich.

92. Ben-Elissar, *La Diplomatie du IIIe Reich*, p. 191. I follow Ben-Elissar for most details on this issue.

93. Ibid., p. 194 (see English translation in *Documents on German Foreign Policy*, Series D, vol. 5, pp. 746–47).

94. Ibid., pp. 209ff. On the whole issue see Avraham Barkai, "German Interests in the Haavarah-Transfer Agreement 1933–1939," *LBIY* 35 ([London] 1990).

95. *Jüdische Rundschau*, Jan. 14, 1938, LBI, New York.

96. Hitler, *Speeches and Proclamations*, p. 1057.

97. Kwiet and Eschwege, *Selbstbehauptung und Widerstand*, p. 201.

98. Thomas Bernhard, *Heldenplatz* (Frankfurt am Main, 1988), pp. 136–37.

Chapter 8   An Austrian Model?

1. Peter Gay, *Freud: A Life for Our Time* (New York, 1988), p. 628. A minor postscript may be added to the story of this departure. As the emigration and an entry permit to France had been arranged through the intervention of the U.S. ambassador to Paris, William Bullitt (an ex-patient and devoted admirer of Freud's), an American official accompanied the Freuds from Vienna to Paris. Years later a person who knew the official wrote: "When I saw him . . . he told me about the trip and also vehemently described his personal feelings of repugnance for Freud, his friends and relatives, Jews and psychoanalysis." Quoted in Linda Donn, *Freud and Jung: Years of Friendship, Years of Loss* (New York, 1988), p. 20.

2. Tonny Moser, "Österreich," in Benz, *Dimension des Völkermords*, p. 68n.

3. F. L. Carstens, *Faschismus in Österreich: Von Schönerer zu Hitler* (Munich, 1978), p. 185.

4. Ibid., pp. 231–32.

5. Ibid., p. 233.

6. Wildt, *Die Judenpolitik des SD*, pp. 52–53.

7. Safrian, *Die Eichmann-Männer*, p. 32.

8. Götz Aly and Susanne Heim, *Vordenker der Vernichtung*, p. 33.

9. Ibid., p. 38.

10. Ibid.

11. Ibid., p. 39.

12. The Führer's Deputy to the Reich Commissary for the Reunification of Austria with the Reich, Gauleiter Party comrade Josef Bürckel, 18.7.1938, Reichskomissar für die Wiedervereinigung Österreichs mit dem Deutschen Reich, microfilm MA 145/1, IfZ, Munich.

13. Hilberg, *The Destruction of the European Jews*, p. 61.

14. The State Commissary for Private Business (Walter Rafelsberger) to Heinrich Himmler, 14.8.1939, Persönlicher Stab des Reichsführers SS, microfilm MA–290, IfZ, Munich.

15. Bruce F. Pauley, *From Prejudice to Persecution: A History of Austrian Antisemitism* (Chapel Hill, N.C., 1992), p. 289. About the confiscation of Jewish dwellings in Vienna, see mainly Gerhard Botz, *Wohnungspolitik und Judendeportation in Wien, 1938–1945* (Vienna, 1975).

16. Wildt, *Die Judenpolitik des SD*, p. 52.

17. Eichmann to Hagen, 8.5.1938, in Yitzhak Arad, Yisrael Guttman, and Abraham Margalioth, eds., *Documents on the Holocaust* (Jerusalem, 1981), pp. 93–94. There were other ways of perceiving the situation that was unfolding in former Austria. In a letter to the London *Times* of April 4, 1938, a Mr. Edwin A. Stoner wrote: "At St Anton—a village beloved by British skiers—the railway station was a blaze of color; even the station dog wore his swastika, but he looked unhappy and wagged a reluctant tail. Ninety per cent of Viennese now sport the swastika, popularly referred to as 'the safety pin.' One of the strangest sights was the vast crowd struggling to get into the British consulate in Wallnerstrasse. Many were Jews desirous of British nationality or anxious to leave a country where only Aryans are tolerated. Poor demented folk, they had little chance of success. Quoted in George Clare, *Last Waltz in Vienna: The Rise and Destruction of a Family, 1842–1942* (New York, 1981), p. 199.

18. Herbert Rosenkranz, "Austrian Jewry: Between Forced Emigration and Deportation," in Yisrael Guttman and Cynthia J. Haft, eds., *Patterns of Jewish Leadership in Nazi Europe 1933–1945* (Jerusalem, 1979), pp. 70–71. During his interrogation by Israeli police in 1960, Eichmann described how Löwenherz, just released from prison, authored the new plan for the centralization of the emigration procedures: "I gave Dr. Löwenherz paper and pencil and said: Please go back for one more night and write up a memo telling me how you would organize this whole thing, how you would run it. Object: stepped-up emigration. . . . The next day, this Dr. Löwenherz brought me his draft. I found it excellent and we immediately took action on his suggestions." Jochen von Lang, ed., *Eichmann Interrogated: Transcripts from the Archives of the Israeli Police* (New York, 1983), pp. 50–51.

19. Safrian, *Die Eichmann-Männer*, p. 41.

20. Quoted in Heinz Höhne, *The Order of the Death's Head: The Story of Hitler's SS* (New York, 1970), p. 337.

21. Ibid., p. 338.

22. On the forcible expulsion of Jews from the Reich, mainly over Germany's western borders, see Jacob Toury, "Ein Auftakt zur Endlösung: Judenaustreibungen über nichtslawische Reichsgrenzen 1933 bis 1939," in Büttner, Johe, and Voss, *Das Unrechtsregime*, vol. 2, pp. 164ff.; for Austria, pp. 169ff.

23. Memorandum of II 112/4, 2.11.38, idem.

24. Moser, "Österreich," p. 68n.

25. SD, II 112, to Racial Policy Office of the NSDAP, 3.12.38; Racial Policy Office to Chief of the SD Main Office, 14.12.38, SD Hauptamt, microfilm MA–554, IfZ, Munich.

26. Aly and Heim, *Vordenker der Vernichtung*, p. 40. In May, on Eichmann's orders, some nineteen hundred Jews with prior records of jail sentences were shipped to Dachau, spreading fear in the community and hastening the exodus. Herbert, *Best.*, p. 213.

27. Gordon J. Horwitz, *In the Shadow of Death: Living Outside the Gates of Mauthausen* (London, 1991), p. 23.

28. Ibid., p. 28.

29. Ibid., p. 29.

30. Ibid., p. 12.

31. Ibid., pp. 13–14.

32. Aly and Heim, *Vordenker der Vernichtung*, p. 36.

33. Ibid., pp. 41–42.

34. Henry L. Feingold, *Bearing Witness: How America and Its Jews Responded to the Holocaust* (Syracuse, N.Y., 1995), p. 75.

35. *Foreign Relations of the United States, 1938*, vol. 1 (Washington, D.C., 1950), pp. 740–41.

36. Shlomo Z. Katz, "Public Opinion in Western Europe and the Evian Conference of July 1938," *Yad Vashem Studies* 9 (1973): 106.

37. Ibid., 108.

38. Ibid., 111.

39. Ibid., 113.

40. Ibid., 114.

41. Heinz Boberach, ed., *Meldungen aus dem Reich: Die geheimen Lageberichte des Sicherheitsdienstes der SS 1938–1945*, vol. 2 (Herrsching, 1984), p. 23.

42. David S. Wyman, *Paper Walls: America and the Refugee Crisis 1938–1941* (New York, 1985), p. 50.

43. Ben-Elissar, *La Diplomatie*, p. 251.

44. Hitler, *Reden und Proklamationen*, vol. 2, p. 899.

45. For the situation of the Jews in Italy before 1938 and for the 1938 laws, see, among others, Meir Michaelis, *Mussolini and the Jews: German-Italian Relations and the Jewish Question in Italy 1922–1945* (London, 1978), particularly pp. 152ff; Jonathan Steinberg, *All or Nothing: The Axis and the Holocaust 1941–1943* (London, 1990), pp. 222ff; Susan Zuccotti, *The Italians and the Holocaust: Persecution, Rescue, and Survival* (New York, 1987), pp. 28ff.

46. Michaelis, *Mussolini and the Jews*, p. 191.

47. For the situation of the Jews in Hungary before 1938 and for the laws of 1938 and 1939, see, among others, Randolph L. Braham, *The Politics of Genocide: The Holocaust in Hungary*, vol. 1 (New York, 1981), particularly pp. 118ff; Nathaniel Katzburg, *Hungary and the Jews: Policy and Legislation 1920–1943* (Ramat Gan, Israel, 1981), particularly pp. 94ff; Mendelsohn, *The Jews of East Central Europe*, pp. 85ff.

48. All details are taken from Georges Passelecq and Bernard Suchecky, *L'Encyclique cachée de Pie XI: Une occasion manquée de l'Église face à l'antisémitisme* (Paris, 1995). The full text of the encyclical is published for the first time in this study. Regarding Pius XI's meeting with LaFarge and his instructions to him, see ibid., pp. 69ff.

49. Ibid., pp. 113ff.

50. Ibid., pp. 180–81.

51. Ibid., pp. 285ff.

52. Ibid., pp. 116ff., and particularly 138.

53. Ibid., pp. 139, 208.

54. Letter of State Secretary Zschintsch, 17.3.1938 (NG–1261) in Mendelsohn, *The Holocaust*, vol. 1, p. 75.

55. Michael P. Steinberg, *The Meaning of the Salzburg Festival: Austria as Theater and Ideology, 1890–1938* (Ithaca, N.Y., 1990), pp. 164ff.

56. Ibid., pp. 233ff.

57. Shirakawa, *The Devil's Music Master*, p. 221.

58. *Deutsche Allgemeine Zeitung*, Nov. 4, 1937, Nationalsozialismus/1937 (misc.), LBI, New York.

59. SOPADE, *Deutschland-Berichte* 5 (1938): 195–96. Strangely enough, in their all-encompassing propaganda effort, the Nazis did not make major use of film until the

beginning of the war. Thus, during the second half of the thirties, the only anti-Semitic productions shown in German theaters were an adaptation of a Swedish comedy, *Peterson und Bandel* (1935), a merely allusive scene in the German film *Pour le mérite* (1938), and, finally, a minor anti-Jewish film, *Robert und Bertram* (1939). Dorothea Hollstein, *"Jud Süss" und die Deutschen: Antisemitische Vorurteile im nationalsozialistischen Spielfilm* (Frankfurt am Main, 1971), pp. 38ff.

60. *Akten der Parteikanzlei der NSDAP* (abstracts), part 1, vol. 2, p. 364.

61. Minister of Justice to State Prosecutors . . ., 24.2.1938, Reichsjustizministerium, Fa 195/1938, IfZ, Munich.

62. Memorandum, II 112, 28.3.38, SD Hauptamt, microfilm No. MA–554, IfZ, Munich.

63. See Adam, *Judenpolitik*, pp. 198–99. The seeming absurdity of this measure did not escape the victims: "Now we have also to turn in our passports," noted Berlin Jewish physician Hertha Nathorff in her diary. "Jews are not allowed to have passports anymore. They are afraid that we might get across the border! But isn't that what they want? Strange logic." Wolfgang Benz, ed. *Das Tagebuch der Hertha Nathorff: Berlin–New York, Aufzeichnungen 1933 bis 1945* (Munich, 1987), p. 105.

64. For the text of the decree see Pätzold, *Verfolgung, Vertreibung, Vernichtung,* p. 155.

65. Walk, *Das Sonderrecht,* p. 237.

66. Pätzold, *Verfolgung, Vertreibung, Vernichtung,* p. 159.

67. Christiane Hoss, "Die jüdischen Patienten in rheinischen Anstalten zur Zeit des Nationalsozialismus," in Mathias Leipert, Rudolf Styrnal, Winfried Schwarzer, eds., *Verlegt nach unbekannt: Sterilisation und Euthanasie in Galkhausen 1933–1945* (Cologne, 1987), pp. 67–68.

68. I owe this information to the late Amos Funkenstein.

69. Internal memorandum of the SD, August 29, 1938, regarding letter of Streicher to Himmler, July 22, 1938, and Rosenberg to Henlein, October 15, 1938, in Mendelsohn, *The Holocaust,* vol. 4, pp. 216–17.

70. Reich leadership of the NSDAP, Office for the Fostering of German Letters to SS-Hauptsturmführer Hartl, Gestapo Vienna, 17.6.1938; SD II 112 to Reich leadership of the NSDAP, Office for the Fostering of German Letters, 17.8.1939, SD Hauptamt, microfilm MA-554, IfZ, Munich.

71. SS-Oberführer Albert to SS-Standartenführer Six, 18.1.39; SS-Standartenführer Six to SS-Oberführer Albert, 26.1.39, SD Hauptamt, microfilm MA–554, IfZ, Munich.

72. Mendelsohn, *The Holocaust,* vol. 4, p. 138.

73. Karl Winter to Rosenberg, 9.3.38, NSDAP, Hauptamt Wissenschaft, microfilm MA–205, IfZ, Munich.

74. Main Office for Science (NSDAP) to Karl Winter, 18.3.38, ibid.

75. Karl Winter to Rosenberg, 30.3.38, ibid.

76. Main Office for Science (NSDAP) to Karl Winter, 12.4.38, ibid.

77. Max Kreuzberger Research Papers, AR 7183, Box 8, Folder 9, LBI, New York.

78. Ibid.

79. Ibid.

80. Barkai, *From Boycott to Annihilation,* p. 114.

81. Walk, *Das Sonderrecht,* p. 223.

82. Ibid., p. 229. I am using the simplified translation of the law as presented in Hilberg, *The Destruction of the European Jews,* p. 82.

83. Walk, *Das Sonderrecht*, p. 232; Hilberg, *The Destruction of the European Jews*, pp. 83–84.

84. Walk, *Das Sonderrecht*, p. 234.

85. Hilberg, *The Destruction of the European Jews*, p. 84.

86. Walk, *Das Sonderrecht*, p. 234.

87. Ibid., p. 242. Seven hundred physicians were allowed to attend to the Jewish population as "caretakers of the sick" and two hundred lawyers were similarly authorized as "consultants." See Arndt and Boberach, "Deutsches Reich," p. 28. The procedure that enabled a Jewish lawyer to become a consultant—and the status of consultants—is analyzed in Lothar Gruchmann, *Justiz im Dritten Reich 1933–1949: Anpassung und Unterwerfung in der Ära Gürtner* (Munich, 1988), pp. 181ff.

88. Ibid., pp. 178–79.

89. Reich Chamber of Physicians to Ministry of Education, 3.10.38, Reichsministerium für Wissenschaft u. Erziehung, microfilm MA 103/1, IfZ, Munich.

90. Minister of Justice to Minister of Education . . . , 3.10.38, ibid.

91. Interior Minister to Minister of Education, 14.12.38, ibid.

92. Barkai, *From Boycott to Annihilation*, p. 129.

93. Hayes, "Big Business," p. 266.

94. Ibid., p. 267.

95. See in particular Hilberg, *The Destruction of the European Jews*, pp. 60–90; Genschel, *Die Verdrängung*, mainly chap. 10; Barkai, *From Boycott to Annihilation*, p. 75.

96. Hilberg, *The Destruction of the European Jews*, p. 79.

97. Barkai, *From Boycott to Annihilation*, p. 118.

98. For the details of this affair and the supporting documentary evidence, see Wolf Gruner, "Die Reichshauptstadt und die Verfolgung der Berliner Juden 1933–1945," in Reinhard Rürup, ed., *Jüdische Geschichte in Berlin: Essays und Studien* (Berlin, 1995), pp. 238, 260–61.

99. None of this was apparently mentioned by Speer in his talks with Gitta Sereny. See Gitta Sereny, *Albert Speer: His Battle with Truth* (New York, 1995).

100. It was the first time that the SD had taken the initiative of arresting a large number of German Jews and sending them to concentration camps. Herbert, *Best*, p. 213.

101. For the text of the Gestapo memorandum and its historical context, see Wolf Gruner, " 'Lesen brauchen sie nicht zu können': Die Denkschrift über die Behandlung der Juden in der Reichshauptstadt auf allen Gebieten des öffentlichen Lebens, von Mai 1938," *Jahrbuch für Antisemitismusforschung* 4 (1995): 305ff.

102. Goebbels, *Tagebücher*, part 1, vol. 3, p. 452.

103. Hugh R. Wilson to Secretary of State, June 22, 1938, in Mendelsohn, *The Holocaust*, vol. 1, pp. 139–40.

104. Fromm, *Blood and Banquets*, p. 274.

105. Undated SD report on the Evian Conference and the Berlin "Judenaktion," SD-Hauptamt, microfilm MA 557, IfZ, Munich.

106. Wildt, *Die Judenpolitik des SD*, p. 57.

107. Goebbels, *Tagebücher*, part 1, vol. 3, p. 490.

108. Sybil Milton, "Menschen zwischen Grenzen: Die Polenausweisung 1938," *Menora* (1990), pp. 189–90.

109. Carl Ludwig, *Die Flüchtlingspolitik den Schweiz in den Jahren 1933 bis 1945: Bericht an den Bundesrat zuhanden der eidgenössischen Räte*, Bern, 1957.

110. Conseil Fédéral, "Procès-verbal de la séance du 28 mars 1938," *Documents Diplomatiques Suisses*, vol. 12 (1.1.1937–31.12.1938), ed. (under the direction of Oscar Gauye) Gabriel Imboden and Daniel Bourgeois (Bern, 1994), p. 570.

111. For all these details and for relevant documents see Ludwig, *Die Flüchtlingspolitik der Schweiz*, pp. 124ff.

112. *Documents Diplomatiques Suisses*, vol. 12, p. 938n. 5.

113. Quoted in Ben-Elissar, *La Diplomatie*, p. 286.

114. Reproduced in Arad, Guttman, Margalioth, *Documents on the Holocaust*, pp. 101–2.

115. See mainly Toury, "Judenaustreibung," pp. 173ff.

116. Maier, District Office Überlingen, to the mayors of the district, 20.9.1938, Unterlagen betr. Entrechtung der Juden in Baden 1933–1940, ED 303, IfZ, Munich.

117. Milton, "Menschen zwischen Grenzen"; Trude Maurer, "Abschiebung und Attentat: Die Ausweisung der polnischen Juden und der Vorwand für die 'Kristallnacht,'" in Pehle, *Der Judenpogrom 1938*, pp. 52ff.

118. Maurer, "Abschiebung und Attentat," pp. 59–66.

119. Sauer, *Dokumente*, vol. 2, pp. 423ff.

120. For the agreement between Germany and Poland on this matter see *DGFP*, series D, vol. 5 (Washington, 1953), p. 169.

121. Arndt and Boberach, "Deutsches Reich," p. 34.

122. Michael R. Marrus, "The Strange Story of Herschel Grynszpan," *American Scholar* 57, no.1 (Winter 1987–88): 70–71.

123. Ibid., pp. 71–72.

Chapter 9   The Onslaught

1. Sauer, *Dokumente*, vol. 2, pp. 25–28.

2. Kulka, "Public Opinion in Nazi Germany and the 'Jewish Question,'" *Jerusalem Quarterly* 25 (Fall 1982): 136.

3. Georg Landauer to Martin Rosenblüth, 8 February 1938, in Friedlander and Milton, *Archives of the Holocaust*, vol. 3, *Central Zionist Archives*, ed. Francis R. Nicosia (New York, 1990), p. 57.

4. Drobisch, *Juden unterm Hakenkreuz*, pp. 159–60.

5. Hugh R. Wilson to Secretary of State, June 22, 1938, in Mendelsohn, *The Holocaust*, vol. 1, p. 144.

6. II 112 to I 111, 31.10.1938, SD-Hauptamt, microfilm MA 554, IfZ, Munich.

7. Friedlander and Milton, *Archives of the Holocaust*, vol. 20, p. 113.

8. Adam, "Wie spontan war der Pogrom?" in Pehle, *Der Judenpogrom 1938*, p. 76. Graml, *Anti-Semitism*, p. 8.

9. Friedlander and Milton, *Archives of the Holocaust*, vol. 20, p. 374.

10. "50, dann 75 Synagogen brennen: Tagebuchschreiber Goebbels über die Reichskristallnacht," *Der Spiegel*, July 13, 1992, p. 126.

11. Walter Buch to Göring, 13.2.1939, Michaelis and Schraepler, *Ursachen*, vol. 12, p. 582.

12. Dieter Obst, "Die 'Reichskristallnacht' im Spiegel westdeutscher Nachkriegsprozessakten und als Gegenstand der Strafverfolgung," *Geschichte in Wissenschaft und Unterricht* 44, no. 4 (1993): 212.

13. Goebbels, "50, dann 75 Synagogen brennen," pp. 126–28.

14. Carl Östreich, "Die letzten Stunden eines Gotteshauses" in Lamm, *Von Juden*, p. 349.

15. Graml, *Anti-Semitism*, p. 13.

16. Goebbels, "50, dann 75 Synagogen brennen," p. 128.

17. Ibid.

18. Ibid.

19. Adam, "Wie spontan war der Pogrom?" p. 89. For the orders given on November 9 and 10, see Walk, *Das Sonderrecht*, pp. 249–54.

20. *Nazi Conspiracy and Aggression* (Washington, D.C., 1946), vol. 5, doc. no. 3051–PS, pp. 799–800.

21. Michaelis and Schraepler, *Ursachen*, vol. 12, p. 584.

22. The sequence of events in Innsbruck is taken from Michael Gehler, "Murder on Command: The Anti-Jewish Pogrom in Innsbruck 9th–10th November 1938," *LBIY* 38 (1993): 119–33. The details about Eichmann's trip have been corrected.

23. Michalka, *Das Dritte Reich*, vol. 1, p. 165.

24. Heinz Lauber, *Judenpogrom "Reichskristallnacht" November 1938 in Grossdeutschland* (Gerlingen, 1981), pp. 123–24.

25. The Mayor of Ingolstadt to the Government of Upper Bavaria, Munich, 1.12.1938, Monatsberichte des Stadtrats Ingolstadt, 1929–1939 (Stadtarchiv Ingolstadt No. A XVI/142), IfZ, Fa 411.

26. Gestapo Würzburg to . . . , 6.12.38 (Himmler Archives, Berlin Document Center, microfilm No. 269, Roll 1) LBI, New York, microfilm 133f.

27. Mendelsohn, *The Holocaust*, vol. 3, p. 301.

28. Pätzold, *Verfolgung, Vertreibug, Vernichtung*, p. 221. A precise inquiry into the events in Schleswig-Holstein and in North Germany more generally indicates that the concrete murder orders were often decided upon by local middle-ranking SA officers. Thus in Kiel, SA-Stabführer Carsten Vorquardsen of the SA Group Nordmark organized a meeting with delegates from the party district, the SS, the SD, and the Gestapo in which the decision was taken that at least two of the city's Jewish businessmen, Lask and Leven, were to be put to death in reprisal for Rath's assassination. The two were severely wounded but survived. In Bremen five Jews (three men and two women) were killed by members of the SA Group Nordsee after receiving their orders from Munich from the leader of their group and mayor of Bremen, Heinrich Böhnker. See Gabriele Ferk, "Judenverfolgung in Norddeutschland," in Frank Bajohr, ed., *Norddeutschland im Nationalsozialismus* (Hamburg, 1993), pp. 291–92. It seems therefore that rather than individual initiatives of low-level SA or SS men, the murders were perpetrated after orders were given by regional SA or SS leaders, who "translated" in their own way the orders they received from Munich. The Innsbruck case confirms the same pattern.

29. Peter Loewenberg, "The Kristallnacht as a Public Degradation Ritual", *LBIY* 32 (1987): 309ff.

30. Gauye, Imboden, and Bourgeois, *Documents Diplomatiques Suisses*, p. 1020.

31. Alfons Heck, *The Burden of Hitler's Legacy* (Frederick, Colo., 1988), p. 62.

32. Some historians have nonetheless attempted to reinterpret the events of November 9 and 10 in terms of a process of chaotic radicalization in which anti-Jewish hatred as such played a minor role, once the initial orders had been given. For such an interpretation, see in particular Dieter Obst, *"Reichskristallnacht": Ursachen und Verlauf des antisemitischen Pogroms vom November 1938* (Frankfurt am Main, 1991).

33. Hitler, *Reden und Proklamationen*, pp. 971, 973ff.

34. Ulrich von Hassell, *Die Hassell Tagebücher 1938–1944*, (Berlin, 1988), p. 70.

35. Local Group Hüttenbach to district leader's office, 25.11.1938, "Hist." Ordner No. 431, Zuwachs, Fa 506/14, IfZ, Munich.

36. Local Group Hüttenbach to district leader's office, 7.2.39, ibid.

37. Michaelis and Schraepler, *Ursachen*, vol. 12, p. 581.

38. Hans Mommsen, "Reflections on the Position of Hitler and Göring in the Third Reich," in Thomas Childers and Jane Caplan, eds., *Reevaluating the Third Reich* (New York, 1993), pp. 86ff.

39. Michaelis and Schraepler, *Ursachen*, vol. 2, p. 600.

40. For a full text of the meeting see *Trial of Major War Criminals Before the International Military Tribunal* [hereafter *IMT*], vol. 28, pp. 499ff.

41. Walk, *Das Sonderrecht*, pp. 254–55.

42. Ibid., p. 254.

43. *IMT,* 28, pp. 508–9.

44. Ibid., pp. 509–10.

45. Ibid., pp. 510–11. The benches actually carried the sign FOR ARYANS ONLY (NUR FÜR ARYER). For a photograph of a bench with this sign, see Gerhard Schoenberner, *Die Judenverfolgung in Europa, 1933–1945* (Frankfurt am Main, 1982), p. 38.

46. *IMT,* 28, p. 532.

47. Ibid., pp. 533–35. Heydrich's opposition to the creation of ghettos in German cities was not new; in the September 9, 1935, memorandum sent to the participants in the conference that had been called in August by Schacht, the chief of the State Police and the SD explicitly took a stand against ghettoization of the Jews. See Wildt, *Die Judenpolitik des SD*, p. 71.

48. *IMT,* 28, pp. 536–39.

49. Ibid., pp. 538–539.

50. Freeden, "Das Ende der jüdischen Presse," p. 8.

51. Ibid., p. 9.

52. Lynn H. Nicholas, *The Rape of Europa: The Fate of Europe's Treasures in the Third Reich and the Second World War* (New York, 1994), p. 43.

53. Walk, *Das Sonderrecht*, p. 256.

54. The Minister of Education . . . to the Education Administration of the *Länder*, the Reich Commissary for the Saar, etc. . . ., 15.11.1938, Reichsministerium für Wissenschaft u. Erziehung, microfilm MA–103/1, IfZ, Munich.

55. Walk, *Das Sonderrecht*, p. 260.

56. Ibid., p. 262. For the full text of the edict see Hans-Adolf Jacobsen and Werner Jochmann, eds., *Ausgewählte Dokumente zur Geschichte des Nationalsozialismus 1933–1945* (Bielefeld, 1961), section D, pp. 2–3.

57. Walk, *Das Sonderrecht*, pp. 262, 264, 270.

58. Michaelis and Schraepler, *Ursachen*, vol. 12, pp. 614–15.

59. Walk, *Das Sonderrecht*, p. 261.

60. Criminal police Memmingen to the Mayor of Memmingen, 10.11.1938 (Himmler Archives, Document Center Berlin, microfilm No. 270, Roll 2), LBI, New York, microfilm 133g.

61. Sauer, *Dokumente*, vol. 2, pp. 47–49. Although the seizure of archives took place immediately all over the Reich, some of the local SA and police units may not have hurried to transfer them to the Gestapo. On May 5, 1939, an order was issued by SA

Headquarters in Munich to all regional and local units that Jewish archives seized during the November 1938 action had to be delivered as they were to the Gestapo. Himmler Archives, Berlin Document Center, microfilm No. 269, Roll 1 (LBI, NY 133f).

62. Susanne Heim and Götz Aly, "Staatliche Ordnung und 'Organische Lösung': Die Rede Hermann Görings 'über die Judenfrage' vom 6 Dezember 1938," *Jahrbuch für Antisemitismusforschung* 2 (1993): 387.

63. Ibid., 391–92.

64. Ibid., 384.

65. Ibid., 393ff.

66. Ibid., 387.

67. Ibid., 384.

68. Ibid., 385–86.

69. Ibid., 386.

70. Ibid., 387–88.

71. Gestapo Würzburg to office heads . . . 9.12.1938, Himmler Archives, LBI, New York, pp. 133ff.

72. Frick to Reichsstatthalter, Interior Ministers of the Länder, . . . 4.12.1938, Reichsministerium für Wissenschaft u. Erziehung, microfilm MA 103/1. IfZ, Munich.

73. Department East, memorandum, 13.12.1938, Amt Osten, microfilm MA 128/3 IfZ, Munich.

74. A summary of the meeting, uncovered in the Hamburg municipal archives, was first published in 1991. See Susanne Heim and Götz Aly, eds., *Beiträge zur Nationalsozialistischen Gesundheits– und Sozialpolitik*, vol. 9, *Bevölkerungsstruktur und Massenmord: Neue Dokumente zur deutschen Politik der Jahre 1938–1945* (Berlin, 1991), pp. 15ff.

75. For Göring's decree see Michaelis and Schraepler, *Ursachen*, vol. 12, pp. 615–16; see also Ursula Büttner, "The Persecution of Chistian-Jewish Families in the Third Reich," *LBIY* 34 (1989): 284.

76. Klepper, *Unter dem Schatten deiner Flügel*, p. 726. Quoted and translated in Büttner, "The Persecution of Christian-Jewish Families," 284.

77. Sauer, *Dokumente*, vol. 2, p. 84.

78. Walk, *Das Sonderrecht*, p. 292.

79. Alexander Kirk to Secretary of State, May 11, 1939, in Mendelsohn, *The Holocaust*, vol. 1, pp. 189–90.

80. Walk, *Das Sonderrecht*, p. 275.

81. Minister of Justice, 15.2.1939, Reichsjustizministerium, Fa 195/1939, IfZ, Munich.

82. Minister of Justice to the President of the Reich Supreme Court, . . . 7.3.1939, idem.

83. *Heinrich Himmler, Geheimreden 1933 bis 1945 und andere Ansprachen*, ed. Bradley F. Smith and Agnes F. Peterson (Berlin, 1974), pp. 37–38.

84. Walk, *Das Sonderrecht*, p. 137.

85. *Akten der Parteikanzlei der NSDAP* (abstracts), part 1, vol. 2, p. 247.

86. Schmitthenner to the Minister of Religion and Education, Karlsruhe, 10.11.1938, Reichsministerium für Wissenschaft u. Erziehung microfilm MA 103/1, IfZ, Munich.

87. Minister of Religion and Education, Karlsruhe, to the Reich Ministry of Education, 24.11.1938, idem.

88. Pätzold, *Verfolgung, Vertreibung, Vernichtung,* p. 222.

89. Sauer, *Dokumente,* vol. 1, p. 246.

90. District Leader Neustadt a. d. Aisch to Fritz Kestler, Ühlfeld, 21.11.1938 (Himmler Archives, Berlin Document Center, microfilm No. 270, Roll 2), LBI, New York, microfilm 133g.

91. Ogilvie-Forbes to Halifax, Nov. 16, 1938, *Documents on British Foreign Policy 1919–1938, Third Series,* vol. 3, *1938–39,* (London, 1950), pp. 275–76.

92. De Montbas to Bonnet, 15.11.38, *Documents Diplomatiques Français 1932–1939,* 2nd series (1936–1939), vol. 12 (3 Octobre–30 Novembre 1938) (Paris, 1978), p. 570.

93. Kulka, "Public Opinion in Nazi Germany," p. 138.

94. *Deutschlandberichte* 5 (1938): 1352ff. For the excerpt and translation see Detlev J. K. Peukert, *Inside Nazi Germany: Conformity, Opposition and Racism in Everyday Life* (New Haven, Conn., 1987), p. 59.

95. Albert Speer, *Inside the Third Reich* (London, 1970), p. 111.

96. Sereny, *Albert Speer,* p. 164.

97. Ibid.

98. Ibid., p. 165.

99. Gutteridge, *Open Thy Mouth for the Dumb!* pp. 188–89ff.

100. Report, 8.12.1938, *Die Kirchliche Lage in Bayern,* vol. 2, p. 301.

101. Saul Friedländer, *Pius XII und das Dritte Reich* (Hamburg, 1965), p. 70.

102. Helmreich, *The German Churches under Hitler,* p. 294.

103. Klaus Schwabe, Rolf Reichardt, and Reinhard Hauf, eds., *Gerhard Ritter: Ein politischer Historiker in seinen Briefen* (Boppard, 1984), p. 339.

104. Ibid., n.1.

105. Ibid., pp. 769ff.

106. Hugo Ott, "Der Freiburger Kreis," in Rudolf Lill and Michael Kissener, eds., *20. Juli 1944 in Baden und Württemberg* (Constance, 1994), p. 147; Klaus Schwabe, "Der Weg in die Opposition," in John, Martin, Mück, and Ott, *Die Freiburger Universität,* p. 201.

107. The text runs as follows: "*Um der Liebe zum eigenen Volke willen muss jedoch der Christ die Augen offen halten, ob enge Berührung oder gar Vermischung mit anderen Rassen sich nicht schädlich auswirken kann für Leib und Seele,*" in Schwabe, Reichardt, and Hauf, *Gerhard Ritter,* p. 769.

108. Schwabe, "Der Weg in die Opposition . . . ," p. 201. Whether the other members of the Freiburg Circle as well as other related opposition groups were aware of Dietze's text is not entirely clear, but, as has been shown by Christoph Dipper, Carl Goerdeler's ideas were no different; these ideas had been presented in Freiburg in late 1942. All in all Dietze expressed the themes of a conservative anti-Semitism accepted by most of the German resistance to Hitler—and by the great majority of the German academics. For the anti-Semitism of the German conservative resistance see Christoph Dipper," Der Deutsche Widerstand und die Juden," *Geschichte und Gesellschaft,* 9, no. 3 (1983): esp. pp. 367ff.

109. Bertram to Rust, 16.11.1938, *Akten deutscher Bischöfe,* vol. 4, *1936–1939,* ed. Ludwig Volk (Mainz, 1981), pp. 592–93.

110. Goebbels, *Tagebücher,* part 1, vol. 3, p. 532.

111. Mendelsohn, *The Holocaust,* vol. 3, p. 241.

112. Lipstadt, *Beyond Belief,* p. 99.

113. Martin Gilbert, "British Government Policy towards Jewish Refugees (November 1938–September 1939), *Yad Vashem Studies*, vol. 13, 1979, p. 150.

114. Wyman, *Paper Walls*, p. 221.

115. Ibid., pp. 75ff.

116. Haskel Lookstein, *Were We Our Brothers' Keepers? The Public Response of American Jews to the Holocaust, 1938–1944* (New York, 1985), p. 82.

117. For the details see, among others, Arthur Morse, *While Six Million Died: A Chronicle of American Apathy* (New York, 1968), pp. 270ff.

118. Mann, *The Letters*, p. 297.

119. Vicki Caron, "Prelude to Vichy: France and the Jewish Refugees in the Era of Appeasement," *Journal of Contemporary History* 20 (1985): 161. According to a memorandum of December 20, 1938, circulated by Sonderreferat Deutschland (Germany Department), the following countries protested against the pogrom, usually in relation to damages caused to their Jewish nationals living in Germany: Italy, Britain, the Netherlands, Hungary, Brazil, Lithuania, the USSR, Guatemala, Latvia, Finland, Poland, the United States of America. Cf. *Documents on German Foreign Policy*, Series D (1937–1945), Vol. 5, *Poland* et al. . . ., *June 1937–March 1939*, Washington/London, 1951, pp. 916–7.

120. Caron, "Prelude to Vichy," p. 163.

121. *Documents on German Foreign Policy, Series D, vol. 4, Oct. 1938–Mar. 1939 (London, 1951)*, pp. 481ff. The great majority of the French population's fear of war, and the widespread belief that the Jews were the instigators of a military confrontation with Nazi Germany, were exacerbated by the Sudeten crisis. In September 1938, anti-Jewish incidents took place in Paris and in a number of other French cities. The prevailing tension prompted Julien Weill, the Grand Rabbi of Paris, to warn his correligionists to avoid gatherings in front of synagogues during the High Holidays. Some French Jewish personalities again expressed their hostility to the foreign Jews in their midst, who supposedly were responsible for anti-German incitement. Michael R. Marrus and Robert O. Paxton, *Vichy France and the Jews* (New York, 1981), p. 40. The pogrom of November 9–10 did not change some of these attitudes and declarations. Thus, on November 19, Grand Rabbi Weill declared to the daily *Le Matin* that the Consistoire was unable to make "the least contribution" to the refugee question: the problem could be solved only on an international scale and France could not take in more refugees. Moreover, the Grand Rabbi declared, he did not want to take any initiative "that could in any way hamper the attempts presently made for a Franco-German rapprochement." On the other hand, the Comte de Paris, the pretender to the French throne, stressed in an interview of December 1938 that the French Jews were Frenchmen equal to all others and that "excluding them . . . meant weakening the country." For both quotations see Ralph Schor, *L'Antisémitisme en France pendant les années trente* (Bruxelles, 1992), pp. 215, 221.

122. Jean-Baptiste Duroselle, *La Décadence 1932–1939* (Paris, 1979), p. 385.

123. Marrus, "The Strange Story," p. 73.

124. Ibid., pp. 74–78.

125. Report of February 6, 1939, *Deutschlandberichte der Sozialdemokratischen Partei Deutschlands (SOPADE) 1934–1940*, p. 219. Translated by Dieter Kuntz in Sax and Kuntz, *Inside Hitler's Germany*, p. 422.

126. Pätzold, *Verfolgung, Vertreibung, Vernichtung,* pp. 207, 225; Toury, "Judenaustreibung," p. 180.

127. Stefan Keller, *Grüningers Fall: Geschichten von Flucht und Hilfe* (Zurich, 1993).

128. Grüninger was sentenced in 1941. He died in 1972 and was rehabilitated by the Swiss cantonal and federal authorities in the fall of 1995.

129. David Kranzler, "The Jewish Refugee Community of Shanghai, 1938–1945," *Wiener Library Bulletin* 26 (1972–73): 28ff.

130. Ibid.

131. *Documents on German Foreign Policy*, series D, vol. 5, p. 936.

132. Dalia Ofer, *Escaping the Holocaust: Illegal Immigration to the Land of Israel, 1939–1944* (New York, 1990), chapter 1; Yehuda Bauer, *Jews for Sale? Nazi-Jewish Negotiations 1933–1945* (New Haven, Conn., 1994), chap. 3.

133. Bernard Wasserstein, *Britain and the Jews of Europe, 1939–1945* (Oxford, 1988), p. 40.

134. Eichmann was replaced in Vienna by SS-Hauptsturmführer Rolf Günther and Haupsturmführer Alois Brunner.

135. George F. Kennan, *From Prague after Munich: Diplomatic Papers 1938–1940* (Princeton, N.J., 1968), p. 86.

Chapter 10   A Broken Remnant

1. National Socialism/1939 (misc.), LBI, New York. On German ships carrying mainly German passengers, Jewish emigrants were segregated as they were in the Reich. In the dining rooms, for example, their tables were set in the "Jewish corner." Benz, *Das Tagebuch der Hertha Nathorff*, p. 163.

2. Noam and Kropat, *Juden vor Gericht*, pp. 41–45.

3. Reich Minister of Justice to presidents of regional higher courts, 22.6.1939, Reichsjustizministerium/Fa 195/1939, IfZ, Munich.

4. Hitler, *Reden und Proklamationen*, part 1, vol. 2, p. 955.

5. Ibid., p. 1055.

6. Ibid., pp. 1056–58.

7. For a presentation and an analysis of these various statements by Hitler see Burrin, *Hitler and the Jews*, pp. 60–61.

8. Hitler, *Reden und Proklamationen*, p. 955.

9. Michaelis and Schraepler, *Ursachen*, vol. 12, pp. 616ff.

10. Friedländer, *L'Antisémitisme Nazi: Histoire d'une psychose collective* (Paris, 1971), p. 197.

11. Ibid., p. 198.

12. Richard Breitman, *The Architect of Genocide: Himmler and the Final Solution* (New York, 1991), p. 58.

13. Ibid., p. 59.

14. II 112 to II, 19.1.39, SD-Hauptamt, MA–554, IfZ, Munich.

15. Wildt, *Die Judenpolitik des SD*, p. 48.

16. Pätzold, *Verfolgung, Verteibung, Vernichtung*, pp. 212–13.

17. These negotiations have often been described. For excellent summaries see, among others, Ben-Elissar, *La Diplomatie*, pp. 378–415, 434–56; Bauer, *Jews for Sale?* pp. 30–44.

18. Vogel, *Ein Stempel*, p. 194.

19. Ben-Elissar, *La Diplomatie*, p. 377.

20. Strauss, "Jewish Emigration from Germany," I, p. 326.

21. The most recent computation, that of Arndt and Boberach, gives the following breakdown: 177,000 full Jews emigrated between the census of June 1933 and that of May 1939; the surplus of deaths over births until the end of 1939 was 47,500; between 15,000 and 17,000 Jews were expelled in October 1938. The authors evaluate the number of full Jews living in the Reich at the end of 1939 at approximately 190,000; it would mean that emigration between May 1939 and December 1939 amounted to approximately 30,000. Arndt and Boberach, "Deutsches Reich," p. 34.

22. II 112, 15.6.1939, SD-Hauptamt, microfilm MA 554, IfZ, Munich.

23. Ibid.

24. Gruner, "Die Reichshauptstadt," p. 239.

25. Friedlander and Milton, *Archives of the Holocaust*, vol. 3 pp. 93–94. (Translation somewhat revised.)

26. Pätzold, *Verfolgung, Vertreibung, Vernichtung*, p. 225.

27. Ibid., p. 222.

28. Arad, Gutman, and Margalioth, *Documents on the Holocaust*, p. 140.

29. Ibid., pp. 141–42.

30. Ibid., pp. 125–26. (Translation slightly revised.)

31. Himmler Archives (misc.) (Berlin Document Center, microfilm 270, Roll 2), LBI, New York, microfilm 133g.

32. Konrad Kwiet, "Forced Labor of German Jews in Nazi Germany," *LBIY* 36 (1991): 392.

33. Pätzold, *Verfolgung, Vertreibung, Vernichtung*, p. 228, see also Sauer, *Dokumente*, vol. 2, p. 77.

34. Ibid., vol. 2, p. 75.

35. Ibid.

36. Ibid., p. 76.

37. District Governor, Hildesheim, to heads of administrative regions and mayors of the district, 29.8.1939, Ortspolizeibehörde Göttingen, microfilm MA 172, IfZ, Munich.

38. Gruner, "Die Reichshauptstadt," pp. 241–42.

39. Ibid., p. 242.

40. Ibid.

41. For the prior stages of this story see chapter 1, pp. 33–34 and chapter 5, pp. 157–58. Regarding Ada Berthold's letter see Mommsen, "Die Geschichte," p. 357.

42. Ibid., p. 358.

43. Ibid., p. 361.

44. Ibid., pp. 362–63.

45. Ibid., p. 365.

46. Valentin Senger, *No. 12 Kaiserhofstrasse: The Story of an Invisible Jew in Nazi Germany*, New York, 1980.

47. District Leader's Office, Bernburg to District Leader's Office Rosenheim, 6.2.39, Himmler Archives (misc.), (Berlin Document Center, microfilm 270, Roll 2) LBI, New York, microfilm 133g.

48. Gendarmerie Station Fischbach to Labor Office Augsburg, 6.5.39, *Idem*.

49. Monthly report, 8.2.39, *Die Kirchliche Lage*, vol. 2 . . . , pp. 305–6. In March 1937 the Gestapo had seized all the copies of the new catechism published under Cardinal Bertram's responsibility by the vicar-general of Breslau. Question and answer no. 17, which quoted these words of Jesus about the Jews, were considered "a glorifica-

tion of the Jewish race." On this whole matter, see Bertram's March 20, 1937, letter of protest sent to Minister of Religious Affairs Kerrl and to the Gestapo in Berlin. Bertram's argument was that the sentence had to be read in a purely religious way: Jesus, the Savior, came from the Jewish fold. See *Akten Deutscher Bischöfe*, vol. 4, *1936–1939*, edited by Ludwig Volk (Mainz, 1981), pp. 184ff.

50. Monthly report, 7.1.1939, ibid., p. 303.

51. Hans Donald Cramer, *Das Schicksal der Goslarer Juden 1933–45* (Goslar, 1986), p. 42.

52. Klemperer, *Ich will Zeugnis ablegen*, vol. 1, p. 447.

53. An interpretation of the events assuming the widespread presence in Germany society at large, throughout the modern era, of an "eliminationist anti-Semitism," craving the physical annihilation of the Jews, is not convincing on the basis of the material presented in this study. For such an interpretation see Goldhagen, *Hitler's Willing Executioners*.

54. Frick to Reichsstatthalter, Reich Commissaries, etc . . . , 10.1939, Reichsministerium für Wissenschaft u. Erziehung, microfilm MA 103/1, IfZ, Munich.

55. Karl Schué to local group leader Dornbusch (Frankfurt am Main), 14.1.39, Max Kreuzberger Research Papers, AR 7183, Box 8, Folder 9, LBI, New York.

56. Gendarmerie Station Theilheim to the Prosecution, Land Court Schweinfurt, 12 July 1939, Würzburg Gestapo Akten 1933–1945 (St-Archiv Würzburg), Fa 168/4, IfZ, Munich.

57. Gestapo Würzburg, 20 Juli 1939, ibid.

58. Criminal police office Würzburg to Gestapo Würzburg, 20 März 1941, ibid.

59. J. S. Conway, *The Nazi Persecution of the Churches, 1933–45* (London, 1968), p. 230.

60. Helmreich, *The German Churches Under Hitler*, pp. 233–34. At the Wartburg Luther translated the New Testament into German.

61. Main Education Office of the NSDAP to Main Security Office, 17.3.39; II 112 to Main Education Office, 26.4.39, SD Hauptamt, microfilm MA 554, IfZ, Munich.

62. Klee, *"Die SA Jesu Christi,"* pp. 137–38.

63. For the correspondence among Krebs, Rust, and Hess on this matter, see Max Kreuzberger Research Papers, AR 7183, Box 8, Folder 9, LBI, New York.

64. For the 1937 list see Oliver Rathkolb, *Führertreu und Gottbegnadet: Künstlereliten im Dritten Reich* (Vienna, 1991), pp. 25ff.

65. From 1937 on Hinkel was increasingly taking charge of the demands made in the minister's name. See Reichskulturkammer files Fa 224/1, Fa 224/2, Fa 224/3, and Fa 224/4, IfZ, Munich.

66. President, Reich Music Chamber to Reich Propaganda Minister, 25.2.1939, Reichskulturkammer file Fa 224/4, IfZ, Munich.

67. List of Jewish authors (Vorläufige Zusammenstellung des Amtes Schriftumspflege bei dem Beauftragten des Führers für die Überwachung der gesamten geistigen und weltanschaulichen Schulung und Erziehung der NSDAP und der Reichsstelle für Förderung des deutschen Schrifttums [Temporary compilation at the Führer's delegate office for the supervision of the entire spiritual and ideological education of the NSDAP and the Reich office for the fostering of German letters], part VI, S-V), MA 535, IfZ, Munich. According to the U.S. National Archives, the provenance of this item is unknown.

68. Christiane Hoss, "Die Jüdischen Patienten in rheinischen Anstalten zur Zeit des

Nationalsozialismus," in Mathias Leipert, Rudolf Styrnal, Winfried Schwarzer, eds., *Verlegt nach unbekannt: Sterilization und Euthanasie in Galkhausen 1933–1945* (Cologne, 1987), p. 68.

69. Klee, *Die "SA Jesu Christi,"* p. 132.

70. Walk, *Das Sonderrecht*, p. 230.

71. Administration of the municipal hospital Offenburg, 29.12.38; municipal hospital, Singen, 5.1.39, Unterlagen betr. Entrechtung der Juden in Baden 1933–1940, ED 303, IfZ, Munich.

72. Klee, *Die "SA Jesu Christi,"* p. 132.

73. Friedlander and Milton, *Archives of the Holocaust*, vol. 20, pp. 202–3.

74. Ibid., p. 204.

75. Burleigh, *Death and Deliverance*, pp. 99ff. See also Friedlander, *The Origins of Nazi Genocide*, pp. 39ff.

76. Burleigh, *Death and Deliverance*, pp. 98, 111–12. See also Friedlander, *The Origins of Nazi Genocide*, pp. 40ff.

77. Norbert Frei, *Der Führerstaat: Nationalsozialistische Herrschaft 1933 bis 1945* (Munich, 1987), p. 86.

78. Hilmar Hoffmann, *"Und die Fahne führt uns in die Ewigkeit": Propaganda im NS-Film* (Frankfurt, 1988), p. 197.

79. Max Kreuzberger Research Papers, AR 7183, Box 8, Folder 9, LBI, New York. The play was first performed in November 1937 in London. See *The Plays of J. B. Priestley*, vol. 3 (London, 1950), pp. 69ff. (Lochner's report has been slightly revised.)

Bibliography

Unpublished Sources

Institut für Zeitgeschichte, Munich

Adjutantur des Führers, 1934–1937, MA–287, MA–13/2.
Aktenstücke zur Judenverfolgung 1933–1945, Ortspolizeibehörde Göttingen, MA–172.
Geheime Staatspolizeistelle Würzburg, Fa 168/4.
Heinrich Himmler, Reden, 1936–1939, F 37/3.
Historischer Ordner No. 431–Zuwachs, Fa 506/14.
Landeshauptstadt Düsseldorf 1933–1945, Einzelschicksale von Bürgern, die im Bereich des heutigen Stadtbezirks 3 wohnten, Ms 456.
Monatsberichte des Stadtrats Ingolstadt 1929–1939, Fa 411.
NSDAP: Aussenpolitisches Amt/Amt Osten, MA–128/3.
NSDAP: Hauptamt Wissenschaft, MA–205.
NSDAP: Parteikanzlei (microfiches).
Nationalsozialistischer Deutscher Studentenbund (NSDStB), MA–228.
Polizeipräsident München (Lageberichte/Monatsberichte) (Misc.), Fa 427/2.
Reichsführer SS–persönlicher Stab, MA–290.
Reichsführer SS–SS Standort Berlin, MA–333.
Reichsjustizministerium (1933–1939), Fa 195/1933 . . . 1939.
Reichskanzlei (24.1.1935–5.2.1938), Serial No. 859.
Reichskulturkammer, Fa 224/1–4.
Reichskommissar für die Wiedervereinigung Österreichs mit dem deutschen Reich, MA–145/1.
Reichsministerium für Wissenschaft und Erziehung, MA 103/1.
Reichswehrministerium, Chef der Heeresleitung, MA–260.
Rosenberg Akten, MA–697 and MA–596.
Sicherheitsdienst des Reichsführers SS, (Lageberichte/Monatsberichte, misc.), MA–557.
Sicherheitsdienst des Reichsführers SS, SD Hauptamt/Abt.II 112, MA–554.
Sicherheitsdienst des Reichsführers SS, SD Oberabschnitt Rhein, MA–392.
Stellvertreter des Führers (Anordnungen . . .), Db 15.02.
Unterlagen betr. Entrechtung der Juden in Baden 1933–1940, ED–303.
Item of unknown provenance, MA 535.

Leo Baeck Institute, New York

Himmler Archives (misc.), 133f, 133g.
Kulturbund (misc.).
Max Kreuzberger Research papers (misc.), AR 7183, Boxes 3–8, Folders 1–9.
National Socialism (misc.) 1933–1939.

Abraham Heller Personal Archives, Ramat Gan, Israel

Published Documentary Sources

COLLECTIONS OF DOCUMENTS

Akten deutscher Bischöfe über die Lage der Kirche 1933–1945: Vol. 1, *1933–1934*.
  Edited by Bernard Stasiewski. Mainz, 1968.
Vol. 2, *1934–1935*. Edited by Bernard Stasiewski. Mainz, 1976.
Vol. 3, *1935–1936*. Edited by Bernard Stasiewski. Mainz, 1979.
Vol. 4, *1936–1939*. Edited by Ludwig Volk. Mainz, 1981.
Akten der Parteikanzlei der NSDAP (abstracts). Part 1, vols. 1 and 2, edited by
  Helmut Heiber. Munich, 1983. Part 2, vol. 3, edited by Peter Longerich.
  Munich, 1992.
Akten der Reichskanzlei: Die Regierung Hitler 1933–1938. Edited by Karl-
  Heinz Minuth. Part 1, *1933–1934*, vol. 1, *30 January to 31 August 1933*.
  Vol. 2, *12 September 1933 to 27 August 1934*. Boppard am Rhein, 1983.
Archives of the Holocaust: An International Collection of Selected Documents, 22
  vols. Edited by Henry Friedlander and Sybil Milton. New York, 1990–93.
Ausgewählte Dokumente zur Geschichte des Nationalsozialismus 1933–1945.
  Edited by Hans-Adolf Jacobsen and Werner Jochmann. Bielefeld, 1961.
Beiträge zur Nationalsozialistischen Gesundheits- und Sozialpolitik. Edited by
  Götz Aly and Susanne Heim. Vol. 9, *Bevölkerungsstruktur und Massenmord:
  Neue Dokumente zur deutschen Politik der Jahre 1938–1945*. Berlin, 1991.
Berichte des SD und der Gestapo über Kirchen und Kirchenvolk. Edited by Heinz
  Boberach. Mainz, 1971.
*Deutschlandberichte der Sozialdemokratischen Partei Deutschlands (SOPADE)
  1934–1940*. 8 vols. Frankfurt am Main, 1980.
Documents on British Foreign Policy 1919–1939. Second series, vol. 5, *1933*.
  London, 1953. Third series, vol. 3, *1938–39*. London, 1950.
Documents Diplomatiques Français 1932–1939. First series, *1932–1935*, vol. 3,
  17 March–15 July 1933. Paris, 1967. Vol. 4, *16 July–12 November 1933*.
  Paris, 1968. Second series, *1936–1939*, vol. 12, *3 October–30 November
  1938*. Paris, 1978.
Documents Diplomatiques Suisses 1848–1945. Vol. 12, *1.1.1937–31.12.1938*.
  Bern, 1994.
Documents on German Foreign Policy. Series C, *1933–1937*. Washington, D.C.,
  1957–62. Series D, *1937–1945*, vol. 4. Washington, D.C., 1951. Vol. 5,
  Washington, D.C., 1953.
*Documents on the Holocaust: Selected Sources on the Destruction of the Jews of
  Germany, Austria, Poland, and the Soviet Union.* Edited by Yitzhak Arad,
  Yisrael Guttman, and Abraham Margalioth. Jerusalem, 1981.
Dokumente des Verbrechens: Aus Akten des Dritten Reiches 1933–1945. 3 vols.
  Berlin, 1993.
*Dokumente über die Verfolgung der jüdischen Bürger in Baden-Württemberg durch
  das Nationalsozialistische Regime 1933–1945*. 2 vols. Edited by Paul Sauer.
  Stuttgart, 1966.

Dokumente zur "Euthanasie." Edited by Ernst Klee. Frankfurt am Main, 1985.
Dokumente zur Geschichte der Frankfurter Juden 1933–1945. Edited by Kommission zur Erforschung der Geschichte der Frankfurter Juden. Frankfurt am Main, 1963.
Das Dritte Reich. 2 Vols. Edited by Wolfgang Michalka. Munich, 1985.
Foreign Relations of the United States, 1933. Vol. 2, Washington, D.C., 1948.
Foreign Relations of the United States, 1938. Vol. 1, Washington, D.C., 1950.
Herrschaftsalltag im Dritten Reich: Studien und Texte. Edited by Hans Mommsen and Susanne Willems. Düsseldorf, 1988.
The Holocaust: Selected Documents. 18 vols. Edited by John Mendelsohn. New York, 1982.
Inside Hitler's Germany: A Documentary History of Life in the Third Reich. Edited by Benjamin C. Sax and Dieter Kuntz. Lexington, Ky., 1992.
Juden vor Gericht, 1933–1945: Dokumente aus hessischen Justizakten. Edited by Ernst Noam and Wolf-Arno Kropat. Wiesbaden, 1975.
Kennan, George F. *From Prague after Munich: Diplomatic Papers 1938–1940.* Princeton, N.J., 1968.
Die Kirchliche Lage in Bayern nach den Regierungspräsidentenberichten 1933–1943. Vol. 2, *Regierungsbezirk Ober- und Mittelfranken.* Edited by Helmut Witeschek. Mainz, 1967.
Kleinstadt und Nationalsozialismus: Ausgewählte Dokumente zur Geschichte von Eutin 1918–1945. Edited by Lawrence D. Stokes. Neumünster, 1984.
Die Lageberichte der Geheimen Staatspolizei über die Provinz Hessen-Nassau 1933–1936. 2 vols. Edited by Thomas Klein. Vienna, 1986.
Meldungen aus dem Reich: Die Geheimen Lageberichte des Sicherheitsdienstes der SS 1938–1945. 17 vols. Edited by Heinz Boberach. Herrsching, 1984.
Nationalsozialismus im Alltag: Quellen zur Geschichte der NS-Herrschaft im Gebiet des Landes Rheinland-Pfalz. Edited by Anton Doll. Speyer, 1983.
Der Nationalsozialismus: Dokumente 1933–1945. Edited by Walter Hofer. Frankfurt am Main, 1957.
Nazi Conspiracy and Aggression. 10 vols. Washington, D.C., 1947.
Nazism 1919–1945: A Documentary Reader. 3 vols. Edited by Jeremy Noakes and Geoffrey Pridham. Exeter, England, 1983.
Le Phénomène Nazi: Documents Nazis Commentés. Edited by Michel Mazor. Paris, 1957.
Pommern 1934/1935 im Spiegel von Gestapo-Lageberichten und Sachakten. 2 vols. Edited by Robert Thevoz, Hans Branig, Cécile Löwenthal-Hensel. Cologne, 1974.
Der Regierungsbezirk Kassel 1933–36: Die Berichte der Regierungspräsidenten und der Landräte. 2 vols. Edited by Thomas Klein. Darmstadt, 1985.
Das Sonderrecht für die Juden im NS-Staat. Edited by Joseph Walk. Heidelberg, 1981.
Ein Stempel hat gefehlt: Dokumentation zur Emigration deutscher Juden. Edited by Rolf Vogel. Munich, 1977.

Treue, Wilhelm. "Hitlers Denkschrift zum Vierjahresplan," *VfZ* 3 (1955).

Trial of the Major War Criminals Before the International Military Tribunal. 42 vols. Nuremberg, 1948.

Ursachen und Folgen, Vom deutschen Zusammenbruch 1918 und 1945 bis zur staatlichen Neuordnung Deutschlands in der Gegenwart: Eine Urkunden- und Dokumentensammlung zur Zeitgeschichte. Edited by Herbert Michaelis and Ernst Schraepler. Vols. 9–23. Berlin, 1964–75.

Verfolgung, Vertreibung, Vernichtung: Dokumente des faschistischen Antisemitismus 1933 bis 1942. Edited by Kurt Pätzold. Frankfurt am Main, 1984.

Weimarer Republik: Manifeste und Dokumente zur deutschen Literatur, 1918–1933. Edited by Anton Kaes. Stuttgart, 1983.

Wulf, Joseph. *Die bildenden Künste im Dritten Reich: Eine Dokumentation.* Reinbek/Hamburg, 1966.

SPEECHES, DIARIES, LETTERS, AND ALL OTHER PRE–1945 LITERATURE

Atrocity Propaganda Is Based on Lies, Say the Jews of Germany Themselves [Die Greuel-Propaganda ist eine Lügenpropaganda, sagen die deutschen Juden selbst]. Jakov Trachtenberg. Berlin, 1933.

Avenarius, Ferdinand. "Ausprachen mit Juden," *Kunstwart* 25, no. 22 (Aug. 1912).

Benjamin, Walter. *The Correspondence of Walter Benjamin.* Edited by Gershom Scholem and Theodor Adorno. Chicago, 1994.

Bernanos, Georges. *La grande peur des bien-pensants.* In *Essais et ecrits de combat.* Vol. 1. Paris, 1971.

Buber, Martin. *The Letters of Martin Buber.* Edited by Nahum N. Glatzer and Paul Mendes-Flohr. New York, 1991.

Chamberlain, Houston Stewart. *Foundations of the Nineteenth Century.* London, 1910. Reprint, New York, 1968.

Comité des Délegations Juives, ed. *Das Schwarzbuch: Tatsachen und Dokumente: Die Lage der Juden in Deutschland 1933.* Paris, 1934. Reprint, Berlin, 1983.

Eckart, Dietrich. *Der Bolschewismus von Moses bis Lenin: Zwiegespräch zwischen Adolf Hitler und mir.* Munich, 1924.

Faulhaber, Michael. *Judaism, Christianity and Germany: Advent Sermons Preached in St. Michael's, Munich, in 1933.* London, 1934.

Feuchtwanger, Lion, and Arnold Zweig. *Briefwechsel 1933–1958.* 2 vols. Berlin, 1984.

Freud, Sigmund, and Arnold Zweig. *The Letters of Sigmund Freud and Arnold Zweig.* Edited by Ernst L. Freud. New York, 1970.

Fromm, Bella. *Blood and Banquets: A Berlin Social Diary.* London, 1943. Reprint, New York, 1990.

Frymann, Daniel. *Das Kaiserbuch: Politische Wahrheiten und Notwendigkeiten.* Leipzig, 1925.

Gercke, Achim. "Die Lösung der Judenfrage." *Nationalsozialistische Monatshefte* 38 (May 1933).

Gide, André. "Les Juifs, Céline et Maritain." *Nouvelle Revue Française* (Apr. 1938).

Goebbels, Joseph. *Goebbels-Reden.* Edited by Helmut Heiber. Vol. 1, *1932–1939.* Düsseldorf, 1971.

———. *Die Tagebücher von Joseph Goebbels: Sämtliche Fragmente.* Edited by Elke Fröhlich. Part 1, *1924–1941,* vol.1 27.6.1924—31.12.30. Vol. 2, *1.1.1931–31.12.1936.* Vol. 3, *1.1.1937–31.12.1939.* Munich, 1987.

———. "50, dann 75 Synagogen brennen": Tagebuchschreiber Goebbels über die 'Reichskristallnacht.'" *Der Spiegel,* July 13, 1992.

Goldstein, Moritz. "Deutsch-jüdischer Parnass." *Kunstwart* 25, no. 11 (Mar. 1912).

Göring, Hermann. "Staatliche Ordnung und 'Organische Lösung': Die Rede Hermann Görings 'über die Judenfrage' vom 6 Dezember 1938." Edited by Götz Aly and Susanne Heim. *Jahrbuch für Antisemitismusforschung* 2 (1993).

Grau, Wilhelm. "Um den jüdischen Anteil am Bolschewismus." *Historische Zeitschrift* 153, no. 2 (1936).

Hassell, Ulrich von. *Die Hassell Tagebücher 1938–1944.* Berlin, 1988.

Heydrich, Reinhard. *Wandlungen unseres Kampfes.* Munich, 1935.

Himmler, Heinrich. *Geheimreden 1933 bis 1945 und andere Ansprachen.* Edited by Bradley F. Smith and Agnes F. Peterson. Berlin, 1974.

———. *Reichsführer!... Briefe an und von Himmler.* Edited by Helmut Heiber. Stuttgart, 1968.

———. *Die Schutzstaffel als antibolschewistische Kampforganisation.* Munich, 1936.

Hitler, Adolf. "Die deutsche Kunst als stolzeste Verteidigung des deutschen Volkes." *Nationalsozialistische Monatshefte* 4, no. 34 (Oct. 1933).

———. *Hitler's Secret Book.* New York, 1961.

———. *Es spricht der Führer: Sieben exemplarische Hitler-Reden.* Edited by Helmut Krausnick and Hildegard von Kotze. Gütersloh, 1966.

———. *Hitler's Secret Conversations 1941–1944.* Edited by Hugh R. Trevor-Roper. New York, 1972.

———. "Hitlers Rede zur Eröffnung der 'Grossen Deutschen Kunstausstellung' 1937." In Peter-Klaus Schuster, ed., *Nationalsozialismus und 'Entartete Kunst': Die 'Kunststadt' München 1937.* Munich, 1987.

———. *Mein Kampf.* London, 1974.

———. *Reden, Schriften, Anordnungen, Februar 1925 bis Januar 1933:*

Vol. 1, *Die Wiedergründung der NSDAP, Februar 1925-Juni 1926.* Edited by Clemens Vollnhals. Munich, 1992.

Vol. 2, *Vom Weimarer Parteitag bis zur Reichstagswahl Juli 1926-Mai 1928,* part 1, *Juli 1926-August 1927.* Part 2, *August 1927-Mai 1928.* Edited by Bärbel Dusik. Munich, 1992.

Vol. 3, *Zwischen den Reichstagswahlen Juli 1928-September 1930,* part 1, *Juli 1928-Februar 1929.* Edited by Bärbel Dusik and Klaus A. Lankheit. Munich, 1994.

Vol. 4, *Von der Reichstagswahl bis zur Reichspräsidentenwahl, Oktober 1930-März 1932*, part 1, *Oktober 1930-Juni 1931*. Edited by Constantin Goschler. Munich, 1994.

———. *Reden und Proklamationen, 1932–1945: Kommentiert von einem deutschen Zeitgenossen*. Edited by Max Domarus. 4 vols. Munich, 1965.

———. *Hitler: Speeches and Proclamations, 1932–1945*. Edited by Max Domarus. Vol. 2, *The Chronicle of a Dictatorship, 1935–1938*. Wauconda, Ill., 1992.

———. *Sämtliche Aufzeichnungen 1905–1924*. Edited by Eberhard Jäckel and Axel Kuhn. Stuttgart, 1980.

Jung, Carl G. "Civilization in Transition." In *Collected Works*. Vol. 10. New York, 1964.

Klepper, Jochen. *Unter dem Schatten deiner Flügel: Aus den Tagebüchern der Jahre 1932–1942*. Stuttgart, 1983.

Klemperer, Victor. *Ich will Zeugnis ablegen bis zum letzten: Tagebücher 1933–1945*. 2 vols. Berlin, 1995.

Leibbrandt, Georg. "Juden über das Judentum." *Nationalsozialistische Monatshefte* 94, 95 (January–February 1938).

Lösener, Bernhard, and Friedrich U. Knost. *Die Nürnberger Gesetze*, Berlin, 1936.

Mann, Klaus. *The Turning Point: Thirty-five Years in This Century*. 1942. Reprint, New York, 1975.

———. *Mephisto*. 1936. Reprint, New York, 1977.

Mann, Thomas, *The Letters of Thomas Mann 1889–1955*. 2 vols. Selected and edited by Richard and Clara Winston. New York, 1975.

———. *Tagebücher 1918–1921*. Edited by Peter de Mendelssohn. Frankfurt am Main, 1979.

———. *Tagebücher 1933–1934*. Edited by Peter de Mendelssohn. Frankfurt am Main, 1977.

Mrugowsky, Joachim, "Jüdisches und deutsches Soldatentum: Ein Beitrag zur Rassenseelenforschung." *Nationalsozialistische Monatshefte* 76 (July 1936).

Müller, Karl Alexander von. "Zum Geleit." *Historische Zeitschrift* 153, no. 1 (1936).

Nathorff, Hertha. *Das Tagebuch der Hertha Nathorff: Berlin–New York: Aufzeichnungen 1933 bis 1945*, edited by Wolfgang Benz. Munich, 1987.

Oppenheimer, Franz. *Die Judenstatistik des Preussischen Kriegsministeriums*. Munich, 1922.

Der Parteitag der Ehre: Vom 8 bis 14 September 1936. Munich, 1936.

Der Parteitag der Arbeit vom 6 bis 13 September 1937: Offizieller Bericht über den Verlauf des Reichsparteitages mit sämtlichen Kongressreden. Munich, 1938.

Phelps, Reginald H. "Hitlers 'Grundlegende' Rede über den Antisemitismus." *Vierteljahrshefte für Zeitgeschichte (VfZ) 16*, no. 30 (1968).

Priestley, J. B., "People at Sea. [1937]" In *The Plays of J. B. Priestley*. Vol. 3. London, 1950.

The Protocols and the World Revolution including a Translation and Analysis of the "Protocols of the Meetings of the Zionist Men of Wisdom." Boston, 1920.

R———x, E., "Dic nichtjüdischen Nichtarier in Deutschland." *C.V. Zeitung* 20, no. 1 (Beiblatt), 16 May 1935.

Schmitt, Carl. *Der Leviathan in der Staatslehre des Thomas Hobbes.* Hamburg, 1938.

———. "Das Judentum in der Rechtswissenschaft" ("Presentation" and "Concluding Remarks") in *Die deutsche Rechtswissenschaft im Kampf gegen den Jüdischen Geist* (Berlin, 1937).

Schulthess Europäischer Geschichtskalender. Vol. 74, *1933.* Edited by Ulrich Thuerauf. Munich, 1934.

Seraphim, Peter-Heinz. *Das Judentum im osteuropäischen Raum.* Essen, 1938.

Shirer, William L. *Berlin Diary: The Journal of a Foreign Correspondent 1934–1941.* New York, 1941. Reprint, New York, 1988.

Stuckart, Wilhelm, and Hans Globke. *Kommentare zur deutschen Rassengesetzgebung.* Vol. 1. Munich, 1936.

Stuckart, Wilhelm, and Rolf Schiedemair. *Rassen- und Erbpflege in der Gesetzgebung des Dritten Reiches.* Leipzig, 1938.

Tharaud, Jérôme, and Jean Tharaud. *When Israel Is King.* New York, 1924.

Tucholsky, Kurt. *Politische Briefe.* Reinbek, 1969.

———. *Briefe aus dem Schweigen 1932–1935*, Reinbek, 1977.

Wagener, Otto. *Otto Wagener, Hitler aus nächster Nähe: Aufzeichnungen eines Vertrauten 1929–1932.* Edited by Henry A. Turner. Frankfurt am Main, 1978.

Wagner, Cosima. *Die Tagebücher 1869–1883.* 4 vols. Munich, 1982.

Wagner, Gerhard. *Reden und Aufrufe.* Edited by Leonardo Conti. Berlin, 1943.

Wagner, Richard. *Richard Wagner's Prose Works.* 10 vols. London, 1984. Reprint, New York, 1966.

Wassermann, Jakob. *Jakob Wassermann, Deutscher und Jude: Reden und Schriften 1904–1933.* Heidelberg, 1984.

Webster, Nesta H. *World Revolution: The Plot Against Civilization.* London, 1921.

Secondary Sources

Ackermann, Josef. *Heinrich Himmler als Ideologe.* Göttingen, 1970.

Adam, Uwe Dietrich. "An Overall Plan for Anti-Jewish Legislation in the Third Reich?" *Yad Vashem Studies* 11 (1976).

———. *Hochschule und Nationalsozialismus: Die Universität Tübingen im Dritten Reich.* Tübingen, 1977.

———. *Judenpolitik im Dritten Reich*, Düsseldorf, 1972.

———. "Wie spontan war der Pogrom?" In *Der Judenpogrom 1938: von der "Reichskristallnacht" zum Völkermord*, edited by Walter Pehle. Frankfurt am Main, 1987.

Adler, Jacques. *Face à la persécution: Les organisations juives à Paris de 1940 à 1944*. Paris, 1985.

Adler-Rudel, Shalom. *Ostjuden in Deutschland 1880–1940*. Tübingen, 1959.

Alexander, Gabriel. "Die Entwicklung der jüdischen Bevölkerung in Berlin zwischen 1871 und 1945." *Tel Aviver Jahrbuch für Deutsche Geschichte* 20 (1991).

Allen, William Sheridan. *The Nazi Seizure of Power: The Experience of a Single German Town 1930–1935*. London, 1965.

Aly, Götz, and Karl-Heinz Roth. *Die restlose Erfassung: Volkszählen, Identifizieren, Aussondern im Nationalsozialismus*. Berlin, 1984.

Aly, Götz, and Susanne Heim. *Vordenker der Vernichtung: Auschwitz und die deutschen Pläne für eine neue europäische Ordnung*. Hamburg, 1991.

Andreski, S. "Poland." In *European Fascism*, edited by S. J. Woolf. London, 1968.

Angress, Werner T. "The German Army's 'Judenzählung' of 1916: Genesis—Consequences—Significance." *LBIY* 23 ([London] 1978).

———. "The Impact of the Judenwahlen of 1912 on the Jewish Question: A Synthesis." *LBIY* 28 (1983).

———. "Juden im politischen Leben der Revolutionszeit." In *Deutsches Judentum in Krieg und Revolution 1916–1923*, edited by Werner E. Mosse. Tübingen, 1971.

———. "Die 'Judenfrage' im Spiegel amtlicher Berichte 1935." In *Das Unrechtsregime: Internationale Forschung über den Nationalsozialismus*, edited by Ursula Büttner, Werner Johe, and Angelika Voss. 2 Vols. Hamburg, 1986.

———. "Revolution und Demokratie: Jüdische Politiker in Berlin 1918/19." In *Jüdische Geschichte in Berlin: Essays und Studien*, edited by Reinhard Rürup. Berlin, 1995.

Arendt, Hannah. *The Origins of Totalitarianism*. New York, 1951.

———. *Eichmann in Jerusalem: A Report on the Banality of Evil*. New York, 1963.

Arndt, Ino, and Heinz Boberach. "Deutsches Reich." In *Dimension des Völkermords: Die Zahl der Jüdischen Opfer des Nationalsozialismus*, edited by Wolfgang Benz. Munich, 1991.

Aronson, Shlomo. *Reinhard Heydrich und die Frühgeschichte von Gestapo und SD*. Stuttgart, 1971.

Aschheim, Steven E. *Culture and Catastrophe: German and Jewish Confrontations with National Socialism and Other Crises*. New York, 1996.

Baldwin, Peter, ed., *Reworking the Past: Hitler, the Holocaust and the Historians*. Boston, 1990.

Ball-Kaduri, Kurt Jacob. *Das Leben der Juden in Deutschland im Jahre 1933: Ein Zeitbericht*. Frankfurt am Main, 1963.

Bankier, David. "The German Communist Party and Nazi Antisemitism, 1933–1938." *LBIY* 32 (1987).

———. *The Germans and the Final Solution: Public Opinion Under Nazism.* Oxford, 1992.

———. "Hitler and the Policy-Making Process on the Jewish Question." *Holocaust and Genocide Studies* 3, no. 1 (1988).

———. "Jewish Society through Nazi Eyes 1933–1936." *Holocaust and Genocide Studies* 6, no. 2 (1991).

Barkai, Avraham. *From Boycott to Annihilation: The Economic Struggle of German Jews 1933–1943.* Hanover, N.H., 1989.

———. "German Interests in the Haavara-Transfer Agreement 1933–1939. *LBIY* 35 ([London] 1990).

Bartov, Omer. *Hitler's Army: Soldiers, Nazis and War in the Third Reich* (New York, 1991).

———. *Murder in our Midst: the Holocaust, Industrial Killing, and Representation.* New York, 1996.

Bauer, Yehuda. *Jews for Sale? Nazi-Jewish Negotiations 1933–1945.* New Haven, 1994.

———. *My Brother's Keeper: A History of the American Joint Distribution Committee 1929–1939.* Philadelphia, 1974.

Bauman, Zygmunt. *Modernity and the Holocaust.* New York, 1989.

Beller, Steven. *Vienna and the Jews 1867–1938: A Cultural History.* Cambridge, England, 1989.

Ben-Elissar, Eliahu. *La Diplomatie du IIIe. Reich et les Juifs, 1933–1939.* Paris, 1969.

Bennathan, Esra. "Die demographische und wirtschaftliche Struktur der Juden." In *Entscheidungsjahr 1932: Zur Judenfrage in der Endphase der Weimarer Republik*, edited by Werner E. Mosse. Tübingen, 1965.

Benoist-Méchin, Jacques. *Histoire de l'armée allemande.* 4 vols. Paris, 1964.

Benz, Wolfgang, ed. *Das Exil der kleinen Leute: Alltagserfahrung deutscher Juden in der Emigration.* Munich, 1991.

———. *Die Juden in Deutschland 1933–1945: Leben unter nationalsozialistischer Herrschaft.* Munich, 1988.

Beradt, Charlotte. *Das Dritte Reich des Traums.* Frankfurt am Main, 1981.

Bergen, Doris L. *Twisted Cross: The German Christian Movement in the Third Reich.* Chapel Hill, N.C., 1996.

Berghahn, Volker R. *Modern Germany: Society, Economy and Politics in the Twentieth Century.* Cambridge, England, 1982.

———. *Der Stahlhelm: Bund der Frontsoldaten 1918–1935.* Düsseldorf, 1966.

Berkow, Ira. "An Olympic Invitation That Is Sixty Years Late." *New York Times,* June 18, 1996.

Bernhard, Thomas. *Heldenplatz.* Frankfurt am Main, 1988.

Bernheimer, Otto. "Kunde Göring." In *Von Juden in München*, edited by Hans Lamm. Munich, 1959.

Bessel, Richard. *Political Violence and the Rise of Nazism: The Storm Troopers in Eastern Germany 1925–1934.* New Haven, 1984.

Betz, Albrecht. "Céline im Dritten Reich." In Hans Manfred Bock et al., eds. *Entre Locarno et Vichy: les relations culturelles franco-allemandes dans les années 1930*. Paris, 1993.

Beyerchen, Alan D. *Scientists Under Hitler: Politics and the Physics Community in the Third Reich*. New Haven, Conn., 1977.

Binion, Rudolph. *Hitler Among the Germans*. New York, 1976.

Birnbaum, Pierre. "Nationalismes: La comparaison France-Allemagne." In *La France aux Français: Histoire des haines nationalistes*. Paris, 1993.

———. *Le Peuple et les gros: Histoire d'un mythe*. Paris, 1979.

Blasius, Dirk. "Zwischen Rechtsvertrauen und Rechtszerstörung: Deutsche Juden 1933–1935." In *Zerbrochene Geschichte, Leben und Selbstverständnis der Juden in Deutschland*, edited by Dirk Blasius and Dan Diner. Frankfurt am Main, 1991.

———. "Psychiatrischer Alltag im Nationalsozialismus." In *Die Reihen fast geschlossen: Beiträge zur Geschichte des Alltags unterm Nationalsozialismus*, edited by Detlev Peukert and Jürgen Reulecke. Wuppertal, 1981.

Boas, Jacob. "German-Jewish Internal Politics under Hitler 1933–1938." *LBIY* 29 ([London] 1984).

———. "Germany or Diaspora? German Jewry's Shifting Perceptions in the Nazi Era 1933–1938." *LBIY* 27 ([London] 1982).

Bock, Gisela. "Krankenmord, Judenmord und nationalsozialistische Rassenpolitik." In *Zivilisation und Barbarei: Die Widersprüchlichen Potentiale der Moderne*, edited by Frank Bajohr et al. Hamburg, 1991.

———. *Zwangssterilisation im Nationalsozialismus: Studien zur Rassenpolitik und Frauenpolitik*. Opladen, 1986.

Bollmus, Reinhard. *Das Amt Rosenberg und seine Gegner: Zum Machtkampf im nationalsozialistischen Herrschaftssystem*. Stuttgart, 1970.

Botz, Gerhard. *Wohnungspolitik und Judendeportation in Wien, 1938 bis 1945: Zur Funktion des Antisemitismus als Ersatz nationalsozialistischer Sozialpolitik*. Vienna, 1975.

Bracher, Karl-Dietrich, et al. *Die nationalsozialistische Machtergreifung*. Cologne, 1962.

Braham, Randolph, L. *The Politics of Genocide: The Holocaust in Hungary*. 2 vols. New York, 1981.

Breitman, Richard. *The Architect of Genocide: Himmler and the Final Solution*. New York, 1991.

Brenner, Michael. *The Renaissance of Jewish Culture in Weimar Germany*. New Haven, Conn., 1996.

Broszat, Martin. *Hitler and the Collapse of Weimar Germany*. New York, 1987.

———. *The Hitler State: The Foundation and Development of the Internal Structure of the Third Reich*. London, 1981.

———. "Nationalsozialistische Konzentrationslager 1933–1945." In Hans Buchheim et al., *Anatomie des SS-Staates*. 2 vols. Olten, 1965.

———. "A Plea for the Historicization of National Socialism." In *Reworking the Past: Hitler, the Holocaust and the Historians*, edited by Peter Baldwin. Boston, 1990.

Broszat, Martin, Elke Fröhlich, and Falk Wiesemann, eds. *Bayern in der NS-Zeit: Soziale Lage und politisches Verhalten der Bevölkerung im Spiegel vertraulicher Berichte.* Munich, 1977.

Broszat, Martin, and Elke Fröhlich. *Alltag und Widerstand: Bayern im Nationalsozialismus.* Munich, 1987.

Broszat, Martin, and Saul Friedländer. "A Controversy about the Historicization of National Socialism," In *Reworking the Past: Hitler, the Holocaust and the Historians,* edited by Peter Baldwin. Boston, 1990.

Browder, George C. *Foundations of the Nazi Police State: The Formation of Sipo and SD.* Lexington, Ky., 1990.

Browning, Christopher R. *The Final Solution and the German Foreign Office.* New York, 1978.

———. *Ordinary Men: Reserve Police Battalion 101 and the Final Solution in Poland.* New York, 1992.

Buchheim, Hans. "Die SS—Das Herrschaftsinstrument." In Hans Buchheim et al., *Anatomie des SS-Staates.* 2 vols. Olten, 1965.

Bullock, Alan. *Hitler: A Study in Tyranny* (London, 1952).

Burleigh, Michael R. *Germany Turns Eastwards: A Study of Ostforschung in the Third Reich.* Cambridge, England, 1988.

———. *Death and Deliverance: "Euthanasia" in Germany 1900–1945.* Cambridge, England, 1994.

Burleigh, Michael, ed. *Confronting the Nazi Past: New Debates on Modern German History* (London, 1996).

Burleigh, Michael R., and Wolfgang Wippermann. *The Racial State.* Cambridge, England, 1991.

Burrin, Philippe. *La dérive fasciste: Doriot, Déat, Bergery 1933–1945.* Paris, 1986.

———. *Hitler and the Jews: The Genesis of the Holocaust.* New York, 1994.

Büttner, Ursula. "The Persecution of Christian-Jewish Families in the Third Reich." *LBIY* 34 (1989).

Büttner, Ursula, Werner Johe, and Angelika Voss, eds. *Das Unrechtsregime: Internationale Forschung über den Nationalsozialismus.* 2 vols. Hamburg, 1986.

Caron, Vicki. "Loyalties in Conflict: French Jewry and the Refugee Crisis, 1933–1935." *LBIY* 36 (1991).

———. "Prelude to Vichy: France and the Jewish Refugees in the Era of Appeasement." *Journal of Contemporary History* 20 (1985).

Carsten, F. L., *Faschismus in Österreich: Von Schönerer zu Hitler.* Munich, 1978.

Cecil, Lamar. *Albert Ballin: Business and Politics in Imperial Germany 1888–1918.* Princeton, N.J., 1967.

Chernow, Ron. *The Warburgs: The Twentieth-Century Odyssey of a Remarkable Jewish Family.* New York, 1993.

Chickering, Roger. *We Men Who Feel Most German: A Cultural Study of the Pan-German League, 1886–1914.* Boston, 1984.

Chronik der Stadt Stuttgart 1933–1945. Stuttgart, 1982.

Clare, George. *Last Waltz in Vienna: The Rise and Destruction of a Family, 1842–1942*. New York, 1981.

Cocks, Geoffrey. *Psychotherapy in the Third Reich: The Göring Institute*. New York, 1985.

Cohn, Norman. *The Pursuit of the Millennium: Revolutionary Messianism in Medieval and Reformation Europe and its Bearing on Modern Totalitarian Movements*. New York, 1961.

———. *Warrant for Genocide: The Myth of the Jewish World Conspiracy and the Protocols of the Elders of Zion*. London, 1967.

Cohn, Werner. "Bearers of a Common Fate? The 'Non-Aryan' Christian 'Fate-Comrades' of the Paulus-Bund, 1933–1939." *LBIY* 33 [London] 1988).

Conway, J. S. *The Nazi Persecution of the Churches 1933–1945*. London, 1968.

Cramer, Hans Donald. *Das Schicksal der Goslarer Juden 1933–1945*. Goslar, 1986.

Dahm, Volker. "Anfänge und Ideologie der Reichskulturkammer." *VfZ* 34, no. 1 (1986).

Dahms, Hans-Joachim. "Einleitung [Introduction]." In *Die Universität Göttingen unter dem Nationalsozialismus. Das verdrängte Kapitel ihrer 250 jährigen Geschichte*, edited by Heinrich Becker, Hans-Joachim Dahms, and Cornelia Wegeler. Munich, 1987.

Dawidowicz, Lucy. *The War Against the Jews 1933–1945*. New York, 1975.

Deak, Istvan. *Weimar Germany's Left-Wing Intellectuals: A Political History of the Weltbühne and Its Circle*. Berkeley, Calif., 1968.

Deichmann, Ute. *Biologen unter Hitler: Vertreibung, Karrieren, Forschung*. Frankfurt am Main, 1992.

Deutscher, Isaac. *The Non-Jewish Jew and Other Essays*. London, 1968.

Diamant, Adolf. *Gestapo Frankfurt am Main*. Frankfurt am Main, 1988.

Diner, Dan. "Constitutional Theory and State of Emergency in the Weimar Republic: The Case of Carl Schmitt." *Tel Aviver Jahrbuch für Deutsche Geschichte* 17 (1988).

———. "Grundbuch des Planeten: Zur Geopolitik Karl Haushofers." In *Weltordnungen: Über Geschichte und Wirkung von Recht und Macht*. Frankfurt am Main, 1993.

Dioudonnat, Pierre-Marie. *"Je suis partout 1930–1944": Les Maurrassiens devant la tentation fasciste*. Paris, 1973.

Dippel, John V. H. *Bound Upon a Wheel of Fire: Why So Many German Jews Made the Tragic Decision to Remain in Germany*. New York, 1996.

Dipper, Christoph. "Der deutsche Widerstand und die Juden." *Geschichte und Gesellschaft* 9 (1983).

Donn, Linda. *Freud and Jung: Years of Friendship, Years of Loss*. New York, 1988.

Drobisch, Klaus. "Die Judenreferate des Geheimen Staatspolizeiamtes und des Sicherheitsdienstes der SS 1933 bis 1939." *Jahrbuch für Antisemitismusforschung* 2 (1993).

Drobisch, Klaus, et al. *Juden unterm Hakenkreuz: Verfolgung und Ausrottung der deutschen Juden, 1933–1945*. Frankfurt am Main, 1973.

Duroselle, Jean-Baptiste. *La Décadence 1932–1939*. Paris 1979.

Dunker, Ulrich. *Der Reichsbund jüdischer Frontsoldaten, 1919–1938*. Düsseldorf, 1977.

Duwell, Kurt. "Jewish Cultural Centers in Nazi Germany. Expectations and Accomplishments." In *The Jewish Response to German Culture: From the Enlightenment to the Second World War*, edited by Yehuda Reinharz and Walter Schatzberg. Hanover, N.H., 1985.

Dwork, Deborah. *Children with a Star: Jewish Youth in Nazi Europe*. New Haven, Conn., 1991.

Edelheim-Mühsam, Margarete T. "Die Haltung der jüdischen Presse gegenüber der nationalsozialistischen Bedrohung." In *Deutsches Judentum: Aufstieg und Krise*, edited by Robert Weltsch. Stuttgart, 1963.

Esh, Shaul. "Eine neue literarische Quelle Hitlers? Eine methodologische Überlegung." *Geschichte in Wissenschaft und Unterricht* 15 (1964).

Ettinger, Elzbieta. *Hannah Arendt/Martin Heidegger*. New Haven, Conn., 1995.

Evans, Richard J. *In Hitler's Shadow: West German Historians and the Attempt to Escape from the Nazi Past*. New York, 1989.

Eyck, Erich. *Geschichte der Weimarer Republik*. 2 vols. Erlenbach, 1962.

Feingold, Henry L. *Bearing Witness: How America and its Jews Responded to the Holocaust*. Syracuse, N.Y., 1995.

Ferk, Gabriele. "Judenverfolgung in Norddeutschland." In *Norddeutschland im Nationalsozialismus*, edited by Frank Bajohr. Hamburg, 1993.

Fest, Joachim C. *Hitler*. New York, 1974.

Field, Geoffrey G. *Evangelist of Race: The Germanic Vision of Houston Stewart Chamberlain*. New York, 1981.

Fischer, Albert. *Hjalmar Schacht und Deutschlands "Judenfrage": Der "Wirtschaftsdiktator" und die Vertreibung der Juden aus der deutschen Wirtschaft*. Cologne, 1995.

Flade, Roland. *Die Würzburger Juden: Ihre Geschichte vom Mittelalter bis zur Gegenwart*. Würzburg, 1987.

Fleming, Gerald. *Hitler and the "Final Solution"* (Berkeley, Calif., 1984).

Franz-Willing, Georg. *Die Hitlerbewegung*. Vol. 1, *Der Ursprung 1919–1922*. Hamburg, 1962.

Freeden, Herbert. "Das Ende der jüdischen Presse in Nazideutschland." *Bulletin des Leo Baeck Instituts* 65 (1983).

Frei, Norbert. *Der Führerstaat: Nationalsozialistische Herrschaft 1933 bis 1945*. Munich, 1987.

———. *Nationalsozialistische Eroberung der Provinzpresse: Gleichschaltung, Selbstanpassung und Resistenz in Bayern*. Stuttgart, 1980.

Friedlander, Henry. *The Origins of Nazi Genocide: From Euthanasia to the Final Solution*. Chapel Hill, N.C., 1995.

Friedländer, Saul. "From Anti-Semitism to Extermination: A

Historiographical Study of Nazi Policies Toward the Jews." *Yad Vashem Studies* 16 (1984).

———. "The Demise of the German Mandarins: The German University and the Jews, 1933–1939." In *Von der Aufgabe der Freiheit: Politische Verantwortung und Bürgerliche Gesellschaft im 19. und 20. Jahrhundert,* edited by Christian Jansen et al. Berlin, 1995.

———. *L'Antisémitisme Nazi: Histoire d'une psychose collective.* Paris, 1971.

———. *History and Psychoanalysis: An Inquiry into the Possibilities and the Limits of Psychohistory.* New York, 1978.

———. *Pius XII und das Dritte Reich: Eine Dokumentation.* Reinbek/Hamburg, 1965.

———. "Political Transformations During the War and Their Effect on the Jewish Question." In *Hostages of Modernization: Studies on Modern Anti-Semitism 1870–1933/39,* edited by Herbert A. Strauss. 2 vols. Berlin, 1993.

———. "Some Thoughts on the Historicization of National Socialism." In *Reworking the Past: Hitler, the Holocaust and the Historians,* edited by Peter Baldwin. Boston, 1990.

Friedländer, Saul, ed. *Probing the Limits of Representation. Nazism and the "Final Solution."* Cambridge, Mass., 1992.

Frye, Bruce B. "The German Democratic Party and the 'Jewish Problem' in the Weimar Republic." *LBIY* 21 ([London] 1976).

Funkenstein, Amos. "Anti-Jewish Propaganda: Pagan, Christian and Modern." *Jerusalem Quarterly* 19 (1981).

———. "Changes in the Christian anti-Jewish Polemics in the Twelfth Century." In *Perceptions of Jewish History.* Berkeley, Calif., 1993.

Gay, Peter. *Freud: A Life for Our Time.* New York, 1988.

———. *Freud, Jews and Other Germans: Masters and Victims in Modernist Culture.* New York, 1978.

———. *Weimar Culture: The Outsider as Insider.* New York, 1968.

Gehler, Michael. "Murder on Command: The Anti-Jewish Pogrom in Innsbruck, 9–10 November, 1938." *LBIY* 38 (1993).

Geisel, Eike. "Premiere und Pogrom." In *Premiere und Pogrom: Der Jüdische Kulturbund 1933–1941,* edited by Eike Geisel and Heinrich M. Broder. Berlin, 1992.

———. "Ein Reich, ein Ghetto . . . ." In *Premiere und Pogrom: Der Jüdische Kulturbund 1933–1941,* edited by Eike Geisel and Heinrich M. Broder. Berlin, 1992.

Gelber, Yoav. "The Reactions of the Zionist Movement and the Yishuv to the Nazis' Rise to Power." *Yad Vashem Studies* 18 (1987).

———. "The Zionist Leadership's Response to the Nuremberg Laws." *Studies in the Holocaust Period* 6 (1988) (Hebrew).

Gellately, Robert. *The Gestapo and German Society: Enforcing Racial Policy 1933–1945.* Oxford, 1990.

———. "The Gestapo and German Society: Political Denunciations in the

Gestapo Case Files." *Journal of Modern History* 60, no. 4 (Dec. 1988).

Genschel, Helmut. *Die Verdrängung der Juden aus der Wirtschaft im Dritten Reich*. Göttingen, 1966.

Gerlach, Wolfgang. *Als die Zeugen schwiegen: Bekennende Kirche und die Juden*. Berlin, 1987.

Gilbert, Martin. "British Government Policy towards Jewish Refugees (November 1938–September 1939). *Yad Vashem Studies*, vol. 13, 1979.

Giles, Geoffrey J. "Professor und Partei: Der Hamburger Lehrkörper und der Nationalsozialismus." In *Hochschulalltag im Dritten Reich: Die Hamburger Universität 1933–1945*, edited by Eckart Krause, Ludwig Huber, and Holger Fischer. Berlin, 1991.

———. *Students and National Socialism in Germany*, Princeton, N.J., 1985.

Gilman, Sander L. *The Jew's Body*. New York, 1991.

Goldhagen, Erich. "Weltanschauung und Endlösung: Zum Antisemitismus der nationalsozialistischen Führungsschicht," *VfZ* 24 (1976).

Goldhagen, Daniel Jonah. *Hitler's Willing Executioners: Ordinary Germans and the Holocaust*. New York, 1996.

Gordon, Sarah. *Hitler, Germans, and the Jewish Question*. Princeton, N.J, 1984.

Gottlieb, Moshe R. *American Anti-Nazi Resistance, 1933–1941: An Historical Analysis*. New York, 1982.

Götz von Olenhusen, Albrecht. "Die 'nichtarischen' Studenten an den deutschen Hochschulen: Zur nationalsozialistischen Rassenpolitik 1933–1945." *VfZ* 14 (1966).

Graml, Hermann. *Anti-Semitism in the Third Reich*. Cambridge, Mass., 1992.

Gruchmann, Lothar. *Justiz im dritten Reich 1933–1940: Anpassung und Unterwerfung in der Ära Gürtner*. Munich, 1988.

———. "Blutschutzgesetz und Justiz: Zur Entstehung und Auswirkung des Nürnberger Gesetzes vom 15. September 1935." *VfZ* 3 (1983).

Gruner, Wolf. "'Lesen brauchen sie nicht zu können.' Die 'Denkschrift über die Behandlung der Juden in der Reichshauptstadt auf allen Gebieten des öffentlichen Lebens' von Mai 1938." *Jahrbuch für Antisemitismusforschung 4* (1995).

———. "Die Reichshauptstadt und die Verfolgung der Berliner Juden 1933–1945." In *Jüdische Geschichte in Berlin: Essays und Studien*, edited by Reinhard Rürup. Berlin, 1995.

Grunewald, Max. "The Beginning of the 'Reichsvertretung.'" *LBIY* 1 ([London] 1956).

Grüttner, Michael. *Studenten im Dritten Reich*. Paderborn, 1995.

Gutman, Robert W. *Richard Wagner: The Man, His Mind and His Music*. New York, 1968.

Gutteridge, Richard. "German Protestantism and the Jews in the Third Reich." In *Judaism and Christianity Under the Impact of National Socialism 1919–1945*, edited by Otto Dov Kulka and Paul R. Mendes-Flohr. Jerusalem, 1987.

————. *Open Thy Mouth for the Dumb! The German Evangelical Church and the Jews 1879–1950*. Oxford, 1976.

Hamburger, Ernest, and Peter Pulzer. "Jews as Voters in the Weimar Republic." *LBIY* 30 (1985).

Hamilton, Nigel. *The Brothers Mann: The Lives of Heinrich and Thomas Mann, 1871–1950 and 1875–1955*. London, 1978.

Hanke, Peter. *Zur Geschichte der Juden in München zwischen 1933 und 1945*. Munich, 1967.

Hass, Gerhart. "Zum Russlandbild der SS." In *Das Russlandbild im Dritten Reich*, edited by Hans-Erich Volkmann. Cologne, 1994.

Hayes, Peter. "Big Business and 'Aryanization' in Germany 1933–1939." *Jahrbuch für Antisemitismusforschung* 3 (1994).

————. *Industry and Ideology: IG Farben in the Nazi Era*. Cambridge, England, 1987.

Hayman, Ronald. *Thomas Mann: A Biography*. New York, 1995.

Heck, Alfons. *The Burden of Hitler's Legacy*. Frederick, Colo., 1988.

Heiber, Helmut. *Universität unterm Hakenkreuz*. Part 1, *Der Professor im Dritten Reich: Bilder aus der Akademischen Provinz*. Munich, 1991. Part 2, *Die Kapitulation der Hohen Schulen: Das Jahr 1933 und seine Themen*. Munich, 1992.

————. *Walter Frank und sein Reichsinstitut für Geschichte des neuen Deutschlands*. Stuttgart, 1966.

Heilbron, John L. *The Dilemmas of an Upright Man: Max Planck as Spokesman for German Science*. Berkeley, Calif., 1986.

Heim, Susanne. "Deutschland muss ihnen ein Land ohne Zukunft sein: die Zwangsemigration der Juden 1933 bis 1938." In *Arbeitsmigration und Flucht: Beiträge zur nationalsozialistischen Gesundheits- und Sozialpolitik*, edited by Eberhard Jungfer et al. Vol. 11. Berlin, 1993.

Helmreich, Ernst Christian. *The German Churches under Hitler: Background, Struggle and Epilogue*. Detroit, 1979.

Herbert, Ulrich. *Best. Biographische Studien über Radikalismus, Weltanschauung und Vernunft 1903–1989*. Bonn, 1996.

Herf, Jeffrey. *Reactionary Modernism: Technology, Culture and Politics in Weimar and the Third Reich*. Cambridge, England, 1984.

Hermand, Jost. "'Bürger zweier Welten?' Arnold Zweigs Einstellung zur deutschen Kultur." In *Juden als Träger bürgerlicher Kultur in Deutschland*, edited by Julius Schoeps. Bonn, 1988.

Heuer, Wolfgang. *Hannah Arendt: Mit Selbstzeugnissen und Bilddokumenten*. Reinbek/Hamburg, 1987.

Hilberg, Raul. *The Destruction of the European Jews*. Chicago, 1961.

————. *Die Vernichtung der europäischen Juden*, 3 vols. Frankfurt am Main, 1990.

————. *Perpetrators, Victims, Bystanders: The Jewish Catastrophe 1933–1945*. New York, 1992.

Hoelzel, Alfred. "Thomas Mann's Attitudes toward Jews and Judaism: An Investigation of Biography and Oeuvre." *Studies in Contemporary Jewry* 6 (1990).

Hoffmann, Hilmar. *"Und die Fahne führt uns in die Ewigkeit." Propaganda im NS-Film.* Frankfurt am Main, 1988.

Hofstadter, Richard. *The Paranoid Style in American Politics and Other Essays.* Chicago, 1979.

Höhne, Heinz. *The Order of the Death's Head: The Story of Hitler's SS.* New York, 1970.

————. *Die Zeit der Illusionen: Hitler und die Anfänge des Dritten Reiches 1933–1936.* Düsseldorf, 1991.

Höllen, Martin. "Episkopat und T4." In *Aktion T4 1939–1945:. Die "Euthanasie"-Zentrale in der Tiergartenstrasse 4,* edited by Götz Aly. Berlin, 1987.

Hollstein, Dorothea. *"Jud Süss" und die Deutschen: Antisemitische Vorurteile im nationalsozialistischen Spielfilm.* Frankfurt am Main, 1971.

Horwitz, Gordon J. *In the Shadow of Death: Living Outside the Gates of Mauthausen.* London, 1991.

Hoss, Christiane. "Die jüdischen Patienten in rheinischen Anstalten zur Zeit des Nationalsozialismus." In *Verlegt nach Unbekannt: Sterilisation und Euthanasie in Galkhausen 1933–1945,* edited by Mathias Leipert, Rudolf Styrnal, and Winfried Schwarzer. Cologne, 1987.

Hyman, Paula. *From Dreyfus to Vichy: The Remaking of French Jewry, 1906–1939.* New York, 1979.

Jäckel, Eberhard. *Hitler in History.* Hanover, N.H., 1984.

————. *Hitler's Worldview: A Blueprint for Power.* Cambridge, Mass., 1981.

Jacobsen, Hans-Adolf. *Karl Haushofer: Leben und Werke.* 2 Vols. Boppard am Rhein, 1979.

James, Harold. "Die Deutsche Bank und die Diktatur 1933–1945." In Lothar Gall et al., *Die Deutsche Bank 1870–1995.* Munich, 1995.

Jansen, Christian. *Professoren und Politik: Politisches Denken und Handeln der Heidelberger Hochschullehrer 1914–1935.* Göttingen, 1992.

Jansen, Hans. "Anti-Semitism in the Amiable Guise of Theological Philo-Semitism in Karl Barth's Israel Theology Before and After Auschwitz." In *Remembering for the Future: Jews and Christians During and After the Holocaust.* Vol. 1. Oxford, 1988.

Jarausch, Konrad H. "Jewish Lawyers in Germany, 1848–1938: The Disintegration of a Profession." *LBIY* 36 (1991).

Jelavich, Peter. *Munich and Theatrical Modernism: Politics, Playwriting and Performance 1890–1914.* Cambridge, Mass., 1985.

Joachimsthaler, Anton. *Korrektur einer Biographie: Adolf Hitler, 1908–1920.* Munich, 1989.

Jochmann, Werner. "Die Ausbreitung des Antisemitismus." In *Deutsches Judentum in Krieg und Revolution 1916–1923,* edited by Werner E. Mosse. Tübingen, 1971.

Jones, Larry E. *German Liberalism and the Dissolution of the Weimar Party System, 1918–1933.* Chapel Hill, N.C., 1988.

Jungk, Peter Stephan. *Franz Werfel: A Life in Prague, Vienna and Hollywood.* New York, 1990.

Kaplan, Marion. "Sisterhood Under Siege: Feminism and Antisemitism in Germany, 1904–38." In *When Biology Became Destiny: Women in Weimar and Nazi Germany*, edited by Renate Bridenthal, Atina Grossmann, and Marion Kaplan. New York, 1984.

Kater, Michael H. *Different Drummers: Jazz in the Culture of Nazi Germany.* New York, 1992.

———. "Everyday Anti-Semitism in Prewar Nazi Germany: The Popular Bases," *Yad Vashem Studies*, vol. 16, 1984.

———. *The Nazi Party: A Social Profile of Members and Leaders, 1919–1945*, Oxford, 1983.

———. *Studentenschaft und Rechtsradikalismus in Deutschland 1918–1933.* Hamburg, 1975.

Katz, Jacob. *Jews and Freemasons in Europe 1723–1939.* Cambridge, Mass., 1970.

———. *Out of the Ghetto: The Social Background of Jewish Emancipation 1770–1870.* New York, 1978.

———. *From Prejudice to Destruction: Anti-Semitism 1700–1933.* Cambridge, Mass., 1980.

Katz, Shlomo Z. "Public Opinion in Western Europe and the Evian Conference of July 1938." *Yad Vashem Studies*, vol. 9 (1973).

Katzburg, Nathaniel. *Hungary and the Jews: Policy and Legislation.* Ramat-Gan, Israel, 1981.

Keller, Stefan. *Grüningers Fall: Geschichten von Flucht und Hilfe.* Zurich, 1993.

Kershaw, Ian. *The "Hitler Myth": Image and Reality in the Third Reich.* Oxford, 1987.

———. *Hitler.* London, 1991.

———. *The Nazi Dictatorship: Problems and Perspectives of Interpretation*, London, 1993.

———. "The Persecution of the Jews and German Popular Opinion in the Third Reich." *LBIY* 26 (1981).

———. "'Working Toward the Führer': Reflections on the Nature of the Hitler Dictatorship." *Contemporary European History* 2, no. 2 (1993).

Klee, Ernst. *"Euthanasie" im NS-Staat: Die "Vernichtung lebensunwerten Lebens."* Frankfurt am Main, 1985.

Klee, Ernst. *"Die SA Jesu Christi": Die Kirche im Banne Hitlers.* Frankfurt am Main, 1989.

Knipping, Ulrich. *Die Geschichte der Juden in Dortmund während der Zeit des Dritten Reiches.* Dortmund, 1977.

Knütter, Hans-Helmuth. *Die Juden und die Deutsche Linke in der Weimarer Republik, 1918–1933.* Düsseldorf, 1971.

Koehl, Robert L. *The Black Corps: The Structure and Power Struggles of the Nazi SS.* Madison, Wis., 1983.

Koonz, Claudia. *Mothers in the Fatherland: Women, the Family and Nazi Politics.* New York, 1987.

Korzec, Pawel. *Juifs en Pologne: La Question juive pendant l'entre-deux guerres.* Paris, 1980.

Kranzler, David. "The Jewish Refugee Community of Shanghai, 1938–1945." *Wiener Library Bulletin* 36 (1972–73).

Krausnick, Helmut. "Judenverfolgung." In *Anatomie des SS-Staates*, edited by Hans Buchheim et al. 2 vols. Munich, 1967.

Krüger, Arndt. *Die Olympischen Spiele 1936 und die Weltmeinung.* Berlin, 1972.

Kudlien, Fridolf. *Ärzte im Nationalsozialismus.* Cologne, 1985.

Kulka, Otto Dov, "Public Opinion in Nazi Germany and the 'Jewish Question.'" *Jerusalem Quarterly* 25 (Fall 1982).

———. "Die Nürnberger Rassengesetze und die deutsche Bevölkerung im Lichte geheimer NS Lage- und Stimmungsberichte." *VfZ* 4 (1984).

Kwiet, Konrad. "Forced Labor of German Jews in Nazi Germany." *LBIY* 36 (1991).

Kwiet, Konrad, and Helmut Eschwege. *Selbstbehauptung und Widerstand: Deutsche Juden im Kampf um Existenz und Menschenwürde 1933–1945.* Hamburg, 1984.

Laak-Michael, Ursula. *Albrecht Haushofer und der Nationalsozialismus.* Stuttgart, 1974.

LaCapra, Dominick. *Representing the Holocaust: History, Theory, Trauma.* Ithaca, N.Y. 1994.

Lacouture, Jean. *Léon Blum.* Paris, 1977.

Lamberti, Marjorie. *Jewish Activism in Imperial Germany: The Struggle for Civil Equality.* New Haven, Conn., 1978.

Lang, Jochen von, ed. *Eichmann Interrrogated: Transcripts from the Archives of the Israeli Police.* New York, 1983.

Langer, Walter C. *The Mind of Adolf Hitler: The Secret Wartime Report.* New York, 1972.

Lauber, Heinz. *Judenpogrom "Reichskristallnacht": November 1938 in Grossdeutschland.* Gerlingen, 1981.

Laval, Michel. *Brasillach ou la trahison du clerc.* Paris, 1992.

Lemmons, Russel. *Goebbels and "Der Angriff."* Lexington, Ky., 1994.

Lenger, Friedrich. *Werner Sombart 1863–1941: Eine Biographie.* Munich, 1994.

Levi, Erik. *Music in the Third Reich.* New York, 1994.

———. "Music and National Socialism: The Politicisation of Criticism and Performance." In *The Nazification of Art: Art, Design, Music, Architecture and Film in the Third Reich*, edited by Brandon Taylor and Wilfried van der Will. Winchester, England, 1990.

Levy, Richard S., trans. and ed. Introduction to Binjamin W. Segel, *A Lie and a Libel: The History of the Protocols of the Elders of Zion.* Lincoln, Nebr., 1995.

Lewy, Guenter. *The Catholic Church and Nazi Germany.* New York, 1964.

Lill, Rudolf, and Michael Kissener, eds. *20 Juli 1944 in Baden und Württemberg.* Constance, 1994.

Lipstadt, Deborah E. *Beyond Belief: The American Press and the Coming of the Holocaust 1933–1945.* New York, 1986.

Loewenberg, Peter. "The Kristallnacht as a Public Degradation Ritual." *LBIY* 32 (1987).

Loewenstein, Kurt. "Die innere jüdische Reaktion auf die Krise der deutschen Demokratie." In *Entscheidungsjahr 1932: Zur Judenfrage in der Endphase der Weimarer Republik*, edited by Werner E. Mosse. Tübingen, 1965.

Lohalm, Uwe. *Völkischer Radikalismus: Die Geschichte des deutsch-völkischen Schutz- und Trutz-Bund 1919–1923.* Hamburg, 1970.

Longerich, Peter. *Hitlers Stellvertreter: Führung der Partei und Kontrolle des Staatsapparates durch den Stab Hess und die Partei-Kanzlei Bormann.* Munich, 1992.

Lookstein, Haskel. *Were We Our Brothers' Keepers? The Public Response of American Jews to the Holocaust, 1938–1944.* New York, 1985.

Lösener, Bernhard. "Als Rassereferent im Innenministerium." *VfZ* 3 (1961).

Lowenstein, Steven M. "The Struggle for Survival of Rural Jews in Germany 1933–1938: The Case of Bezirksamt Weissenburg, Mittelfranken." In *The Jews in Nazi Germany 1933–1943*, edited by Arnold Paucker. Tübingen, 1986.

Ludwig, Carl. *Die Flüchtlingspolitik der Schweiz in den Jahren 1933 bis 1955: Bericht an den Bundesrat zuhanden der Eidgenössischen Räte.* Bern, 1957.

Mahler, Raphael. *The Jews of Poland Between the Two World Wars.* Tel Aviv, 1968 (Hebrew).

Maier, Charles S. *The Unmasterable Past: History, Holocaust, and German National Identity.* Cambridge, Mass., 1988.

Mann, Golo. *Reminiscences and Reflections: A Youth in Germany.* New York, 1990.

Marcus, Joseph. *Social and Political History of the Jews in Poland 1919–1939.* Berlin, 1983.

Margalioth, Abraham. "The Problem of the Rescue of German Jewry during the Years 1933–1939; the Reasons for the Delay in their Emigration from the Third Reich." In *Rescue Attempts During the Holocaust*, edited by Yisrael Guttman and Efraim Zuroff. Jerusalem, 1977.

———. *Between Rescue and Annihilation: Studies in the History of German Jewry 1932–1938.* Jerusalem, 1990 (Hebrew).

———. "The Reaction of the Jewish Public in Germany to the Nuremberg Laws." *Yad Vashem Studies* 12 (1977).

Marrus, Michael R. "Vichy Before Vichy: Antisemitic Currents in France during the 1930s." *Wiener Library Bulletin* 33 (1980).

———. *The Holocaust in History.* Hanover, N.H., 1987.

———. "The Strange Story of Herschel Grynszpan." *American Scholar* 57, no. 1 (Winter 1987–88).

Marrus, Michael, R., and Robert O. Paxton. *Vichy France and the Jews.* New York, 1981.

Maser, Werner. *Adolf Hitler, Legende, Mythos, Wirklichkeit.* Munich, 1971.

Maurer, Trude. *Ostjuden in Deutschland 1918–1933.* Hamburg, 1986.

———. "Abschiebung und Attentat: Die Ausweisung der polnischen Juden und der Vorwand für die 'Kristallnacht.'" In *Der Judenpogrom 1938: von der "Reichskristallnacht" zum Völkermord,* edited by Walter Pehle. Frankfurt am Main, 1987.

———. "Die Juden in der Weimarer Republik." In *Zerbrochene Geschichte: Leben und Selbstverständnis der Juden in Deutschland,* edited by Dirk Blasius and Dan Diner. Frankfurt am Main, 1991.

Mayer, Arno J. *Why Did the Heavens Not Darken?: The "Final Solution" in History.* New York, 1988.

McKale, Donald M. "From Weimar to Nazism: Abteilung III of the German Foreign Office and the Support of Antisemitism, 1931–1935." *LBIY* 32 (1987).

Mehrtens, H., and S. Richter, eds. *Naturwissenschaft, Technik und NS-Ideologie.* Frankfurt am Main, 1980.

Meier, Heinrich. *Carl Schmitt, Leo Strauss und "Der Begriff des Politischen": Dialog unter Abwesenden.* Stuttgart, 1988.

Mendelsohn, Ezra. *The Jews of East Central Europe Between the World Wars.* Bloomington, Ind., 1983.

Mendes-Flohr, Paul R. "Ambivalent Dialogue: Jewish-Christian Theological Encounter in the Weimar Republic." In *Judaism and Christianity Under the Impact of National Socialism 1919–1945,* edited by Otto Dov Kulka and Paul R. Mendes-Flohr. Jerusalem, 1987.

Merkl, Peter. *Political Violence Under the Swastika: 581 Early Nazis.* Princeton, N.J., 1975.

Meyer, Michael A. *The Origins of the Modern Jew: Jewish Identity and European Culture in Germany, 1749–1824.* Detroit, 1967.

Michael, Robert "Theological Myth, German Anti-Semitism and the Holocaust: The Case of Martin Niemöller." *Holocaust and Genocide Studies* 2 (1987).

Michaelis, Meir. *Mussolini and the Jews: German-Italian Relations and the Jewish Question in Italy 1922–1945.* London, 1978.

Michel, Bernard. *Banques et banquiers en Autriche au début du XXe Siècle.* Paris, 1976.

Milton, Sybil. "Menschen zwischen Grenzen: Die Polenausweisung 1938." *Menora: Jahrbuch für deutsch-jüdische Geschichte 1990* (1990).

———. "Vorstufe zur Vernichtung: Die Zigeunerlager nach 1933." *VfZ* 43, no. 1 (1995).

Mommsen, Hans. *Beamtentum im Dritten Reich.* Stuttgart, 1966.

———. "Die Geschichte des Chemnitzer Kanzleigehilfen K.B." In *Die Reihen fast geschlossen: Beiträge zur Geschichte des Alltags unterm Nationalsozialismus*, edited by Detlev Peukert and Jürgen Reulecke. Wuppertal, 1981.

———. "Der Nationalsozialistische Staat und die Judenverfolgung vor 1938." *VfZ* 1 (1962).

———. "The Realization of the Unthinkable." In Hans Mommsen, *From Weimar to Auschwitz*. Princeton, N.J., 1991.

———. "Reflections on the Position of Hitler and Göring in the Third Reich." In *Reevaluating the Third Reich*, edited by Thomas Childers and Jane Caplan. New York, 1993.

Monier, Frédéric. "Les Obsessions d'Henri Béraud." *Vingtième Siècle: Revue d'Histoire* (Oct.-Dec. 1993).

Morse, Arthur. *While Six Million Died: A Chronicle of American Apathy*. New York, 1968.

Moser, Jonny. "Österreich." In *Dimension des Völkermords: Die Zahl der Jüdischen Opfer des Nationalsozialismus*, edited by Wolfgang Benz. Munich, 1991.

Mosse, George L. "Die Bildungsbürger verbrennen ihre eigenen Bücher. In *"Das war ein Vorspiel nur": Berliner Colloquium zur Literaturpolitik im "Dritten Reich,"* edited by Horst Denkler and Eberhard Lämmert. Berlin, 1985.

———. *The Crisis of German Ideology: Intellectual Origins of the Third Reich*. New York, 1964.

———. "Die Deutsche Rechte und die Juden." In *Entscheidungsjahr 1932: Zur Judenfrage in der Endphase der Weimarer Republik*, edited by Werner E. Mosse. Tübingen, 1965.

———. "German Socialists and the Jewish Question in the Weimar Republic." *LBIY 16* (1971).

———. "The Influence of the Völkisch Idea on German Jewry." In *Germans and Jews: The Right, the Left and the Search for a "Third Force" in Pre-Nazi Germany*. New York, 1970.

———. "Jewish Emancipation: Between *Bildung* and Respectability." In *The Jewish Response to German Culture: From the Enlightenment to the Second World War*, edited by Yehuda Reinharz and Walter Schatzberg. Hanover, N.H., 1985.

Mosse, Werner E. *Jews in the German Economy: The German-Jewish Economic Elite 1820–1935*. Oxford, 1987.

———. "Die Juden in Wirtschaft und Gesellschaft." In *Juden im Wilhelminischen Deutschland 1890–1914*, edited by Werner E. Mosse. Tübingen, 1976.

Müller, Ingo. *Hitler's Justice: The Courts of the Third Reich*. Cambridge, Mass., 1991.

Müller, Roland. *Stuttgart zur Zeit der Nationalsozialismus*. Stuttgart, 1988.

Müller-Hill, Benno. *Murderous Science: Elimination by Scientific Selection of Jews, Gypsies and Others, Germany 1933–1945.* Oxford, 1988.

Neliba, Günter. *Wilhelm Frick: Der Legalist des Unrechtsstaates: Eine politische Biographie.* Paderborn, 1992.

Nettl, J. Peter. *Rosa Luxemburg.* 2 vols. Oxford, 1966.

Nicholas, Lynn H. *The Rape of Europa: The Fate of Europe's Treasures in the Third Reich and the Second World War.* New York, 1994.

Nicosia, Francis R. "Ein nützlicher Feind: Zionismus im nationalsozialistischen Deutschland 1933–1939." *VfZ* 37, no. 3 (1989).

———. "Revisionist Zionism in Germany (II): Georg Kareski and the Staatszionistische Organisation, 1933–1938." *LBIY* 32 ([London] 1987).

———. *The Third Reich and the Palestine Question.* London, 1985.

Niederland, Doron. "The Emigration of Jewish Academics and Professionals from Germany in the First Years of Nazi Rule." *LBIY* 33 (1988).

Niewyk, Donald L. *The Jews in Weimar Germany,* Baton Rouge, La., 1980.

Nipperdey, Thomas. *Deutsche Geschichte 1866–1918.* Vol. 2, *Machtstaat vor der Demokratie.* Munich, 1992.

Noakes, Jeremy. "The Development of Nazi Policy Towards the German-Jewish 'Mischlinge' 1933–1945." *LBIY* 34 (1989).

———. "Nazism and Eugenics: The Background to the Nazi Sterilization Law of 14 July 1933." In *Ideas into Politics: Aspects of European History 1880–1950,* edited by R. J. Bullen, H. Pogge von Strandmann, and A. B. Polonsky. London, 1984.

———. "Wohin gehören die 'Judenmischlinge'? Die Entstehung der ersten Durchführungsverordnungen zu den Nürnberger Gesetzen." In *Das Unrechtsregime: Internationale Forschung über den Nationalsozialismus,* edited by Ursula Büttner, Werner Johe, and Angelika Voss. 2 vols. Hamburg, 1986.

Nolte, Ernst. "Eine frühe Quelle zu Hitlers Antisemitismus." *Historische Zeitschrift* 192 (1961).

Norden, Günther van. "Die Barmer Theologische Erklärung und die 'Judenfrage.'" In *Das Unrechtsregime: Internationale Forschung über den Nationalsozialismus,* edited by Ursula Büttner, Werner Johe, and Angelika Voss., 2 vols. Hamburg, 1986.

Obst, Dieter. *"Reichskristallnacht": Ursachen und Verlauf des antisemitischen Pogroms vom November 1938.* Frankfurt am Main, 1991.

———. "Die 'Reichskristallnacht' im Spiegel deutscher Nachkriegsprozessakten und als Gegenstand der Strafverfolgung." *Geschichte in Wissenschaft und Unterricht* 44, no. 4 (1993).

Oestreich, Carl. "Die letzten Stunden eines Gotteshauses." In *Von Juden in München,* edited by Hans Lamm. Munich, 1959.

Ofer, Dalia. *Escaping the Holocaust: Illegal Immigration to the Land of Israel 1939–1944.* New York, 1990.

Orlow, Dietrich. *The History of the Nazi Party: 1933–1945.* 2 vols. Pittsburgh, 1969–73.

Ott, Hugo. *Laubhüttenfest 1940: Warum Therese Loewy einsam sterben musste.* Freiburg, 1994.

———. *Martin Heidegger: Unterwegs zu seiner Biographie.* Frankfurt am Main, 1988.

Passelecq, Georges, and Bernard Suchecky. *L'Encyclique Cachée de Pie XI: Une occasion manquée de l'Eglise face à l'antisémitisme.* Paris, 1995.

Pätzold, Kurt. *Faschismus, Rassenwahn, Judenverfolgung: Eine Studie zur politischen Strategie und Taktik des faschistischen deutschen Imperialismus (1933–1945).* Berlin (East), 1975.

Pauley, Bruce F. *From Prejudice to Persecution: A History of Austrian Antisemitism.* Chapel Hill, N.C., 1992.

Pehle, Walter, ed. *Der Judenpogrom 1938: Von der "Reichskristallnacht" zum Völkermord.* Frankfurt am Main, 1988.

Perutz, M. F. "The Cabinet of Dr. Haber." *New York Review of Books,* June 20, 1996.

Peukert, Detlev J. K. *Inside Nazi Germany: Conformity, Opposition and Racism in Everyday Life.* New Haven, Conn., 1987.

———. "The Genesis of the 'Final Solution' from the Spirit of Science." In *Reevaluating the Third Reich,* edited by Thomas Childers and Jane Caplan. New York, 1993.

Phelps, Reginald H. Before Hitler Came: Thule Society and Germanenorden." *Journal of Modern History* 35 (1963).

Plewnia, Margarete. *Auf dem Weg zu Hitler: Der 'völkische' Publizist Dietrich Eckart.* Bremen, 1970.

Poliakov, Léon. *Histoire de l'Antisémitisme.* Vol. 4, *L'Europe Suicidaire 1870–1933.* Paris, 1977.

Pommerin, Reiner. *Sterilierung der "Rheinlandbastarde": Das Schicksal einer farbigen deutschen Minderheit 1918–1937.* Düsseldorf, 1979.

Prieberg, Fred K. *Musik im NS-Staat.* Frankfurt am Main, 1982.

Proctor, Robert N. *Racial Hygiene: Medicine under the Nazis.* Cambridge, Mass., 1988.

Pulzer, Peter. *The Rise of Political Anti-Semitism in Germany and Austria.* Cambridge, Mass., 1988.

Rathkolb, Oliver. *Führertreu und gottbegnadet: Künstlereliten im Dritten Reich.* Vienna, 1991.

Reichmann, Eva G. "Diskussionen über die Judenfrage 1930–1932." In Mosse, *Entscheidungsjahr 1932: Zur Judenfrage in der Endphase der Weimarer Republik,* edited by Werner E. Mosse. Tübingen, 1965.

Reuth, Ralf Georg. *Goebbels,* Munich, 1990.

Richarz, Monika, ed. *Jüdisches Leben in Deutschland: Selbstzeugnisse zur Sozialgeschichte 1918–1945.* Stuttgart, 1982.

Ritchie, J. M., *German Literature Under National Socialism.* London, 1983.

Röhm, Eberhard, and Jörg Thierfelder. *Juden-Christen-Deutsche,* 3 vols. Vol. 1, *1933–1935.* Stuttgart, 1990.

Röhl, John C. G. "Das Beste wäre Gas!" *Die Zeit*, Nov. 25, 1994.

Rose, Paul Lawrence. *Revolutionary Antisemitism in Germany from Kant to Wagner.* Princeton, 1990.

Rose, Paul Lawrence. *Wagner, Race and Revolution*, London, 1992.

Rosenkranz, Herbert. "Austrian Jewry: Between Forced Emigration and Deportation." In *Patterns of Jewish Leadership in Nazi Europe 1933–1945*, edited by Yisrael Guttman and Cynthia J. Haft. Jerusalem, 1979.

Rosenstock, Werner. "Exodus 1933–1939: A Survey of Jewish Emigration from Germany." *LBIY* 1 ([London] 1956).

Rürup, Reinhard. *Emanzipation und Antisemitismus: Studien zur "Judenfrage" der bürgerlichen Gesellschaft.* Göttingen, 1975.

Rüthers, Bernd. *Carl Schmitt im Dritten Reich: Wissenschaft als Zeitgeist-Bestärkung?* Munich, 1990.

Sabrow, Martin. *Der Rathenaumord: Rekonstruktion einer Verschwörung gegen die Republik von Weimar.* Munich, 1994.

Safranski, Rüdiger. *Ein Meister aus Deutschland: Heidegger und seine Zeit.* Munich, 1994.

Safrian, Hans. *Die Eichmann-Männer.* Vienna, 1993.

Sauder, Gerhard, ed. *Die Bücherverbrennung.* Munich, 1983.

Sauer, Paul. "Otto Hirsch (1885–1941), Director of the 'Reichsvertretung.' " *LBIY* 32 (1987).

Schäfer, Hans Dieter. "Die nichtfaschistische Literatur der "jungen Generation" im nationalsozialistischen Deutschland." In *Die deutsche Literatur im Dritten Reich*, edited by Horst Denkler and Karl Prumm. Stuttgart, 1976.

Schleunes, Karl A. *The Twisted Road to Auschwitz: Nazi Policy Toward German Jews 1933–1939.* Urbana, Ill., 1970.

Schmuhl, Hans-Walter. *Rassenhygiene, Nationalsozialismus, Euthanasie.* Göttingen, 1987.

———. "Reformpsychiatrie und Massenmord." In *Nationalsozialismus und Modernisierung*, edited by Michael Prinz and Rainer Zitelmann. Darmstadt, 1991.

Schoenberner, Gerhard. *Der gelbe Stern: Die Judenverfolgung in Europa 1933–1945.* Frankfurt am Main, 1982.

Scholder, Klaus. *Die Kirchen und das Dritte Reich.* Vol. 1, *Vorgeschichte und Zeit der Illusionen 1918–1934.* Frankfurt am Main, 1977.

———. "Judaism and Christianity in the Ideology and Politics of National Socialism." In *Judaism and Christianity Under the Impact of National Socialism 1919–1945*, edited by Otto Dov Kulka and Paul R. Mendes-Flohr. Jerusalem, 1987.

Schonauer, Franz. "Zu Hans Dieter Schäfer: Bücherverbrennung, staatsfreie Sphäre und Scheinkultur." In Horst Denkler und Eberhard Lämmert (eds.), *"Das war ein Vorspiel nur": Berliner Colloquium zur Literaturpolitik im Dritten Reich*, edited by Horst Denkler and Eberhard Lämmert. Berlin, 1985.

Schönwälder, Karen. *Historiker und Politik: Geschichtswissenschaft im Nationalsozialismus.* Frankfurt am Main, 1992.

Schor, Ralph. *L'Antisémitisme en France pendant les années trente.* Bruxelles, 1992.

Schorcht, Claudia. *Philosophie an den Bayerischen Universitäten 1933–1945.* Erlangen, 1990.

Schorske, Carl E. *Fin-de-Siècle Vienna: Politics and Culture.*New York, 1980.

Schottländer, Rudolf. "Antisemitische Hochschulpolitik: Zur Lage an der Technischen Hochschule Berlin 1933/34." In *Wissenschaft und Gesellschaft: Beiträge zur Geschichte der Technischen Universität Berlin 1879–1979,* edited by Reinhard Rürup. 2 vols. Berlin, 1979.

Schuker, Stephen A. "Origins of the 'Jewish Problem' in the Later Third Republic." In *The Jews in Modern France,* edited by Frances Malino and Bernard Wasserstein. Hanover, N.H., 1985.

Schüler, Winfried. *Der Bayreuther Kreis von seiner Entstehung bis zum Ausgang der Wilhelminischen Ära.* Münster, 1971.

Schulin, Ernst. *Walter Rathenau: Repräsentant, Kritiker und Opfer seiner Zeit.* Göttingen, 1979.

Schwabe, Klaus. "Der Weg in die Opposition: Der Historiker Gerhard Ritter und der Freiburger Kreis." In *Die Freiburger Universität in der Zeit des Nationalsozialismus,* edited by Eckhard John et al. Freiburg, 1991.

Schwabe, Klaus, Rolf Reichardt, and Reinhard Hauf, eds. *Gerhard Ritter: Ein politischer Historiker in seinen Briefen.* Boppard am Rhein, 1984.

Segev, Tom. *The Seventh Million: The Israelis and the Holocaust.* New York, 1993.

Seidler, Edward. "Die medizinische Fakultät zwischen 1926 und 1948." In *Die Freiburger Universität in der Zeit des Nationalsozialismus,* edited by Eckhard John et al.Freiburg, 1991.

Sereny, Gitta. *Albert Speer: His Battle with Truth.* New York, 1995.

———. *Into That Darkness: From Mercy Killing to Mass Murder.* New York, 1974.

Sheehan, Thomas. "Heidegger and the Nazis." *New York Review of Books,* June 16, 1988.

Shell, Susan. "Taking Evil Seriously: Schmitt's Concept of the Political and Strauss's True Politics." In *Leo Strauss: Political Philosopher and Jewish Thinker,* edited by Kenneth L. Deutsch and Walter Nigorski. Lanham, Md., 1994.

Shirakawa, Sam H. *The Devil's Music Master. The Controversial Life and Career of Wilhelm Furtwängler.* New York, 1992.

Sösemann, Bernd. "Liberaler Journalismus in der politischen Kultur der Weimarer Republik." In *Juden als Träger bürgerlicher Kultur in Deutschland,* edited by Julius H. Schoeps. Bonn, 1989.

Solmssen, Arthur R. G. *A Princess in Berlin.* Harmondsworth, England, 1980.

Sorkin, David. *The Transformation of German Jewry, 1780–1840.* New York, 1987.

Soucy, Robert. *French Fascism: The Second Wave, 1933–1939*. New Haven, Conn., 1995.

Speer, Albert. *Inside the Third Reich*. London, 1970.

Spotts, Frederic. *Bayreuth: A History of the Wagner Festival*. New Haven, Conn., 1994.

Steel, Ronald. *Walter Lippmann and the American Century*. Boston, 1980.

Steinberg, Michael P. *The Meaning of the Salzburg Festival: Austria as Theater and Ideology, 1890–1938*. Ithaca, N.Y. 1990.

Steinert, Marlis. *Hitlers Krieg und die Deutschen: Stimmung und Haltung der deutschen Bevölkerung im Zweiten Weltkrieg*. Düsseldorf, 1970.

Steinweis, Alan E. *Art, Ideology and Economics in Nazi Germany: The Reich Chamber of Culture and the Regulation of the Culture Professions in Nazi Germany*. Chapel Hill, N.C., 1988.

Steinweis, Alan E. "Hans Hinkel and German Jewry, 1933–1941." *LBIY* 38 (1993).

Stern, Fritz. *Gold and Iron: Bismarck, Bleichröder and the Building of the German Empire*. New York, 1977.

————. *The Politics of Cultural Despair*. Berkeley, Calif., 1961.

————. *Dreams and Delusions: The Drama of German History*. New York, 1987.

Stern, J. P. *Hitler, the Fuehrer and the People*. Glasgow, 1975.

Sternhell, Zeev. *Ni droite, ni gauche: L'idéologie fasciste en France*. Paris, 1983.

Stolzenberg, Dietrich. *Fritz Haber: Chemiker, Nobelpreisträger, Deutscher, Jude*. Weinheim, 1994.

Strauss, Herbert A. "Jewish Emigration from Germany: Nazi Policies and Jewish Responses (I)." *LBIY* 25 ([London] 1980).

Suchy, Barbara. "The Verein zur Abwehr des Antisemitismus (II): From the First World War to its Dissolution in 1933." *LBIY* 30 (1985).

Tal, Uriel. *Christians and Jews in Germany: Religion, Politics and Ideology in the Second Reich, 1870–1914*. Ithaca, N.Y., 1975.

————. "Law and Theology: On the Status of German Jewry at the Outset of the Third Reich." *Political Theology and the Third Reich*. Tel Aviv, 1989 (Hebrew).

————. "On Structures of Political Theology and Myth in Germany Prior to the Holocaust." In *The Holocaust as Historical Experience*, edited by Yehuda Bauer and Nathan Rotenstreich. New York, 1981.

Thalmann, Rita. "Du Cercle Sohlberg au Comité France-Allemagne: une évolution ambigüe de la coopération franco-allemande." In Hans Manfred Bock et al., eds. *Entre Locarno et Vichy: les relations culturelles franco-allemandes dans les années 1930*. Paris, 1993.

Theweleit, Klaus. *Male Fantasies*. 2 vols. Minneapolis, 1987–89.

Toekes, Rudolph L. *Béla Kun and the Hungarian Soviet Republic*. New York, 1967.

Toury, Jacob. *Die politischen Orientierungen der Juden in Deutschland: Von Jena bis Weimar*. Tübingen, 1966.

————."Ein Auftakt zur Endlösung: Judenaustreibungen über nichtslawische Grenzen, 1933 bis 1939." In *Das Unrechtsregime: Internationale Forschung über den Nationalsozialismus*, edited by Ursula Büttner, Werner Johe, and Angelika Voss. 2 vols. Hamburg, 1986.

Tuchel, Johannes, and Reinhold Schattenfroh. *Zentrale des Terrors: Prinz-Albrecht-Strasse 8: Hauptquartier der Gestapo.* Berlin, 1987.

Vago, Béla. *The Shadow of the Swastika: The Rise of Fascism and Anti-Semitism in the Danube Basin, 1936–1939.* London, 1975.

Volkov, Shulamith. "Die Verbürgerlichung der Juden in Deutschland als Paradigma." In *Jüdisches Leben und Antisemitismus im 19. und 20. Jahrhundert.* Munich, 1990.

Vondung, Klaus. "Der literarische Nationalsozialismus." In *Die deutsche Literatur im dritten Reich*, edited by Horst Denkler and Karl Prumm. Stuttgart, 1976.

————. *Magie und Manipulation: Ideologischer Kult und politische Religion des Nationalsozialismus.* Göttingen, 1971.

Waite, Robert G. L. *The Psychopathic God: A Biography of Adolf Hitler.* New York, 1976.

Wasserstein, Bernard. *Britain and the Jews of Europe 1939–1945.* Oxford, 1988.

Weber, Eugen. *Action Française: Royalism and Reaction in Twentieth-Century France.* Stanford, Calif., 1962.

Weckbecker, Arno. *Die Judenverfolgung in Heidelberg 1933–1945.* Heidelberg, 1985.

Weinberg, David H. *A Community on Trial: The Jews of Paris in the 1930s.* Chicago, 1977.

Weindling, Paul. *Health, Race and German Politics between National Unification and Nazism, 1870–1945.* Cambridge, England, 1989.

————."Mustergau Thüringen: Rassenhygiene zwischen Ideologie und Machtpolitik," in *Medizin und Gesundheitspolitik in der NS-Zeit*, edited by Norbert Frei. Munich, 1991.

Weiner, Marc A. *Richard Wagner and the Anti-Semitic Imagination.* Lincoln, Nebr., 1995.

Weltsch, Robert. "A Goebbels Speech and a Goebbels Letter." *LBIY* 10 (1965).

————. "Vorbemerkung zur zweiten Ausgabe (1959)," In Siegmund Kaznelson, *Juden im Deutschen Kulturbereich: Ein Sammelwerk*. Berlin, 1962.

Wiesemann, Falk. "Juden auf dem Lande: Die wirtschaftliche Ausgrenzung der jüdischen Viehhändler in Bayern." In *Die Reihen fast geschlossen: Beiträge zur Geschichte des Alltags unterm Nationalsozialismus*, edited by Detlev Peukert and Jürgen Reulecke. Wuppertal, 1981.

Wildt, Michael. *Die Judenpolitik des SD 1935 bis 1938.* Munich, 1995.

Winkler, Heinrich-August. *Weimar 1918–1933: Die Geschichte der ersten deutschen Demokratie.* Munich, 1993.

Winter, Jay. *Sites of Memory, Sites of Mourning: The Great War in European Cultural History.* Cambridge, England, 1995.

Wippermann, Wolfgang. *Das Leben in Frankfurt zur NS-Zeit.* Vol. 1, *Die nationalsozialistische Judenverfolgung.* Frankfurt am Main, 1986.

Wistrich, Robert S. *The Jews of Vienna in the Age of Franz Josef.* Oxford, 1989.

Wyman, David S. *Paper Walls: America and the Refugee Crisis 1938–1941.* New York, 1985.

Yahil, Leni. "Madagascar—Phantom of a Solution for the Jewish Question." In *Jews and Non-Jews in Eastern Europe,* edited by Bela Vago and George L. Mosse. New York, 1974.

———. *The Holocaust: The Fate of European Jewry,* New York, 1990.

Yuval, Israel J. "Vengeance and Damnation, Blood and Defamation: From Jewish Martyrdom to Blood Libel Accusations." *Zion* 58, no. 1 (1993) (Hebrew).

Zapf, Lilli. *Die Tübinger Juden: Eine Dokumentation.* Tübingen, 1974.

Zechlin, Egmont. *Die Deutsche Politik und die Juden im Ersten Weltkrieg.* Göttingen, 1969.

Zelinsky, Hartmut. *Richard Wagner: Ein deutsches Thema 1876–1976.* Vienna, 1983.

Zimmerman, Moshe. "Die aussichtslose Republik—Zukunftsperspektiven der deutschen Juden vor 1933." In *Menora: Jahrbuch für deutsch-jüdische Geschichte 1990.* Munich, 1990.

Zimmermann, Michael. *Verfolgt, vertrieben, vernichtet: Die nationalsozialistische Vernichtungspoltik gegen Sinti und Roma.* Essen, 1989.

Zitelmann, Rainer. *Hitler: Selbstverständnis eines Revolutionärs.* Stuttgart, 1990.

Zuccotti, Susan. *The Italians and the Holocaust: Persecution, Rescue and Survival.* New York, 1987.

Ph.D. Dissertations

Combs, William L. *The Voice of the SS: A History of the SS Journal "Das Schwarze Korps."* 2 vols. Ann Arbor, Mich., 1985.

Engelman, Ralph Max. *Dietrich Eckart and the Genesis of Nazism.* Ann Arbor, Mich., 1971.

Maron, Ephraim. "The Press Policy of the Third Reich on the Jewish Question and Its Reflection in the Nazi Press." Tel Aviv University, 1992 (Hebrew).

Ne'eman Arad, Gulie. "American Jewish Leadership and the Nazi Menace." Tel Aviv University, 1994.

Pierson, R. L. *German Jewish Identity in the Weimar Republic.* Ann Arbor, Mich., 1972.

Index